DATE DUE FOR RETURN

ROBOTICS

Designing the Mechanisms for Automated Machinery

Second Edition

ROBOTICS

Designing the Mechanisms for Automated Machinery

Second Edition

Ben-Zion Sandler

The Hy Greenhill Chair in Creative Machine and Product Design
Ben-Gurion University of the Negev, Beersheva, Israel

ACADEMIC PRESS

San Diego   London   Boston
New York   Sydney   Tokyo   Toronto

A Solomon Press Book

Copyright © 1999 by Academic Press

Copyright © 1991 by Prentice-Hall, Inc.

ACADEMIC PRESS
525 B. Suite 1900, San Diego, California 92101-4495, USA
http://www.apnet.com

Academic Press
24-28 Oval Road, London NW1 7DX, UK
http://www.hbuk.co.uk/ap/

Book designed by Sidney Solomon and Raymond Solomon

Library of Congress Cataloging-in-Publication Data
Sandler, B. Z., 1932–
    Robotics : designing the mechanisms for automated machinery /
Ben-Zion Sandler. — 2nd ed.
        p.   cm.
    Includes bibliographical references and index.
    ISBN 0-12-618520-4
    1. Automatic machinery—Design and construction.   I. Title.
TJ213.S1157   1999
670.428 72—dc21   98-45839
                                                                            CIP

Contents

Preface to the Second Edition

This book provides information on the stages of machinery design for automated manufacturing and offers a step-by-step process for making it optimal. This is illustrated by numerous examples of technical concepts taken from different manufacturing domains. The author, being a university teacher, sees that teaching curricula and textbooks most often do not provide the answers to the questions: How are things built? How do they work? How does one best approach the design process for a specific machine? Most textbooks emphasize computation theories and techniques and deal less with the physical objects that the theories describe.

During recent years, some new techniques have been developed and put into widespread use. The book thus covers such modern concepts as rapid modeling; automated assembly; parallel-driven robots; and mechatronic systems for reducing dynamic errors of a mechanical link by continuous, close-to-optimal, control of its oscillation parameters by electronic means. The author understands that writing and publishing procedure can involve a time lag between the contents of the book and the real, rapidly developing world. The revised edition of the book is based on an evaluation of both current principles and newly developed concepts.

Some experiments carried out in the laboratory and described here also serve as illustrations for the relevant topics; for instance:

- Automotive mechanical assembly of a product by a manipulator (robot),
- Systems for reducing vibrations,
- Parallel-driven robots.

In this edition, greater use is made of calculation examples. Calculations performed mostly with the help of the MATHEMATICA program have a number of advantages: they are time-saving, are especially useful in solving nonlinear equations, and are capable of providing a graphic display of processes. Problems and solutions are integrated into the text so as to provide a better understanding of the contents by quantitatively illustrating the solutions and procedures. This also helps in solving other problems of a similar nature; it improves and shortens some mathematical deductions; and it contributes greatly to an understanding of the subject. For instance, one can find here:

- Solutions of essentially nonlinear equations describing the behavior of a piston in pneumatic systems;
- Equations describing the behavior of a body on a vibrating tray, widely used in, for example, vibrofeeding devices, which can be effectively solved by this computation tool (substituting boring traditional calculations);
- Description of the behavior of a slider on its guides (a common structure in machinery) when dry friction exists in this pair, resulting in limited accuracy in the slider's displacement;
- Equations (and an example of a solution) describing the free oscillations of a robot's arm when reaching the destination point. This is important for accuracy and productivity estimations;
- Solutions of nonlinear equations describing the behavior of an electric drive equipped with an asynchronous motor, etc.

The second edition is now more informative, more reliable, and more universal.

I wish to express my deep gratitude and appreciation to my colleagues at the Mechanical Engineering Department of the Ben-Gurion University of the Negev for their spiritual support and cooperation in creating this book; to the Paul Ivanier, Pearlstone Center for Aeronautical Engineering Studies, Department of Mechanical Engineering, Center for Robotics and Production Management Research; to Inez Murenik for editorial work on the manuscript; to Eve Brant for help in production and proofreading; to Sidney Solomon and Raymond Solomon for sponsoring the book and for their skill in the production/design processes and project management. Finally, I thank my wonderful wife and family whose warmth, understanding and humor helped me throughout the preparation of this book.

Ben-Zion Sandler
December, 1998

1

Introduction: Brief Historical Review and Main Definitions

1.1 What Robots Are

The word "robot" is of Slavic origin; for instance, in Russian, the word работа (rabota) means labor or work. Its present meaning was introduced by the Czechoslovakian dramatist Karel Čapek (1890–1938) in the early twentieth century. In a play entitled *R.U.R.* (Rosum's Universal Robots), Čapek created automated substitutes for human workers, having a human outlook and capable of "human" feelings. Historically, in fact, the concept "robot" appeared much later than the actual systems that are entitled to answer to that name.

Our problem is that there is as yet no clear, efficient, and universally accepted definition of robots. If you ask ten people what the word "robot" means, nine will most likely reply that it means an automatic humanoid creature (something like that shown in Figure 1.1), or they will describe a device that may be more accurately defined as a manipulator or an automatic arm (Figure 1.2). *Encyclopaedia Britannica* [1] gives the following definition: "A robot device is an instrumented mechanism used in science or industry to take the place of a human being. It may or may not physically resemble a human or perform its tasks in a human way, and the line separating robot devices from merely automated machinery is not always easy to define. In general, the more sophisticated and individualized the machine, the more likely it is to be classed as a robot device."

Other definitions have been proposed in "A Glossary of Terms for Robotics," prepared for the Air Force Materials Laboratory, Wright-Patterson AFB, by the (U.S.) National Bureau of Standards [2]. Some of these definitions are cited below.

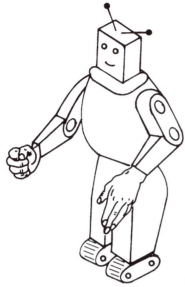

FIGURE 1.1   Android-type robot.

"Robot—A mechanical device which can be programmed to perform some task of manipulation or locomotion under automatic control." [*Note:* The meaning of the words "can be programmed" is not clarified. Programs can differ in their nature, and we will discuss this aspect later in greater detail.]

"Industrial robot— A programmable, multi-function manipulator designed to move material, parts, tools, or specialized devices through variable programmed motions for the performance of a variety of tasks."

"Pick and place robot—A simple robot, often with only two or three degrees of freedom, which transfers items from place to place by means of point-to-point moves. Little or no trajectory control is available. Often referred to as a 'bangbang' robot."

"Manipulator—A mechanism, usually consisting of a series of segments, jointed or sliding relative to one another, for the purpose of grasping and moving objects usually in several degrees of freedom. It may be remotely controlled by a computer or by a human." [*Note:* The words "remotely controlled. . . by a human" indicate that this device is not automatic.]

"Intelligent robot—A robot which can be programmed to make performance choices contingent on sensory inputs."

"Fixed-stop robot—A robot with stop point control but no trajectory control. That is, each of its axes has a fixed limit at each end of its stroke and cannot stop except at one or the other of these limits. Such a robot with N degrees of freedom can therefore

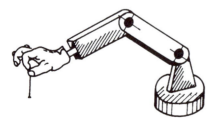

FIGURE 1.2   Manipulator or automatic arm.

stop at no more than *2N* locations (where location includes position and orientation). Some controllers do offer the capability of program selection of one of several mechanical stops to be used. Often very good repeatability can be obtained with a fixed-stop robot."

"*Android*—A robot which resembles a human in physical appearance."

"*Sensory-controlled robot*—A robot whose program sequence can be modified as a function of information sensed from its environment. Robot can be servoed or nonservoed. (See Intelligent robot.)"

"*Open-loop robot*—A robot which incorporates no feedback, i.e., no means of comparing actual output to command input of position or rate."

"*Mobile robot*—A robot mounted on a movable platform."

"*Limited-degree-of-freedom robot*—A robot able to position and orient its end effector in fewer than six degrees of freedom."

We will not discuss here the problem of the possibility (or impossibility) of actually creating a robot with a "human soul." The subject of our discussion will be limited mainly to industrial robots, including those which belong to the family of bangbang robots. The application of these robots in the modern world must meet the requirements of industry, including functional and manufacturing demands and economic interests. Obviously, esthetics and environmental considerations are also involved. The mechanical component of the design of robotic systems constitutes the main focus of our consideration.

Historically, the development of robot systems and devices may be considered as the merging of two distinct lines of creativity: 1) early automation and watchmaking, and 2) innovations and refinements in industrial machinery. A brief description of some of these devices will be useful for illustrating these two lines. As long ago as 400–350 B.C. Archytas of Tarentum, a Greek, built a wooden model of a pigeon actuated by a steam jet. In about the first century A.D., Hero of Alexandria designed a number of devices actuated by water, falling weights, and steam. In about 500 A.D. in Gaza the Byzantines erected a huge water-operated clock in which the figure of Hercules struck the hour in an automatic manner. Roaring lions and singing birds were employed to impress foreigners by the Byzantine emperor Theophilus (829–842). Roger Bacon (1220–1292) created a talking head, and at approximately the same time Albertus Magnus (1200–1280) created an iron man. These two manmade creatures may be classified as "androids." A "magic fountain" was designed in 1242 by a Parisian goldsmith, Guillaume Boucher. The German astronomer and mathematician Johann Muller (1436–1476) built a flying iron eagle. In the Fifteenth century, a truly portable source of mechanical power was invented and applied—the coiled tempered-steel spring. This energy source stimulated the creation of a number of sophisticated mechanical automatons. In 1738, Jacques de Vancanson (1709–1782) built a "flute player" capable of playing a dozen songs. During the eighteenth century, another group of gifted men, Jacquet-Droz, his son Pierre, his grandson Henri-Louis, and Jean-Frederic Leshot, created several androids which wrote, drew, or played musical instruments. The list of automatically actuated animals, men, birds, and so forth is never-ending, and we do not need to discuss it in detail, but two important conclusions do emerge:

1. This line of technical creativity was intended for entertainment purposes, and nothing productive was supposed to be achieved by these devices.

2. A large body of technical skills and experience, and many innovations, were accumulated by the craftsmen engaged in the production of such automatons. This amalgamation of knowledge, skills, and experience found application in the second line of development mentioned above—development of, and the drive for perfection in, industry.

We have reason to consider the clepsydra (a type of water clock) as the earliest representative of robotic devices. Supposedly invented in 250 B.C., it was able to recycle itself automatically. The centrifugal-speed governor for steam engines invented in 1788 by James Watt, together with the system of automatically controlled valves, made the steam engine the first automatic device capable of keeping an almost constant rotating speed of the fly wheel regardless of changes in the load. Analogously, the internal combustion engines invented in the nineteenth century serve as an example of another automatically recycling device realizing repeatedly the suction, compression, and ignition of the fuel mixture. The Industrial Revolution stimulated the creation of a number of automatically operated machines first in the textile industry and later in machine tools and other industrial operations. The most brilliant invention of this type was Jacquard's loom, which had a punched-paper-tape-controlled system for flexible fabric-pattern production. We will return to this example a number of times, but it is worth mentioning here that this machine, which was introduced into industry as long ago as 1801, was based on an idea which is applicable to almost every definition of a robot, i.e., the machine is programmable and is intended for the execution of a variety of fabric patterns.

In 1797, Henry Mandslay designed and built a screw-cutting lathe. The main feature of this machine was that it had a lead screw for driving the carriage on which the cutter was fastened and which was geared to the spindle of the lathe. Actually, this is a kind of template or contour machining. By changing the gear ratio practically any thread pitch could be obtained, i.e., the lead screw controlled a changeable program. Obviously, this is the precursor of the tracer techniques used widely in lathes and milling machines. The later tools are to some extent robotic systems. The further refinement of this machine tool led to the creation of automatic lathes of a purely mechanical nature for the mass production of parts such as bolts, screws, nuts, and washers. These machines were, and still are, mechanically programmed, and after two to three hours the currently produced pattern can be exchanged for another. Many such machines were first produced between the years 1920 and 1930.

In the 50s, after World War II, numerically controlled (NC) machine tools such as lathes and milling machines were first introduced into industry. These machines were, and still are, more flexible from the point of view of program changeability. At this level of refinement, the relative positioning between the tool and the blank had to be made by point-to-point programming of the displacements. When computerized numerically controlled (CNC) machines replaced NC machines, the programming became more sophisticated—the trajectories were then computed by the computer of the machine. At this level of refinement the operator had to define both the kind of the trajectory (say, a straight line or an arc) and the actual parameters of the trajectory (say, the coordinates of the points connecting the straight line or the center coordinates and the radius of the arc, etc.). Other improvements were made in parallel, e.g., continuous measurement of the processed parts to fix the moment at which a tool needed sharpening, replacing, or tuning; computation of the optimal working conditions such as cutting speeds, feeds, and depths; and changing tools to cater to the processing sequence.

We have described the development of the lathe as representative of the world of automatically operated industrial machines. Similarly, we could have chosen the development of textile machinery or, perhaps the most outstanding example of all, of printing. Techniques for the printing of books and newspapers had their origin in Europe (we do not know their history in China) in the fifteenth century when Johannes Gutenberg invented the first printing press. In the beginning the typesetting process was purely manual, being based on the use of movable type. This method remained essentially unchanged until the twentieth century. The problem of mechanizing typesetting was first tackled by Ottmar Mergenthaler, an American inventor who "cast thin slugs of a molten fast-cooling alloy from brass matrices of characters activated by a typewriter-like keyboard; each slug represented a column line of type." This machine was known as a linotype machine (patented in 1884). In 1885, a short time later, another American, Tolbert Lawton, created the monotype printing press in which type is cast in individual letters. Further development led to the creation of machines operated by electronic means, which resulted in higher productivity, since one machine could process the material of a number of compositors. Indeed, the computerized printing systems available today have completely changed the face of traditional typography.

In Koren's book *Robotics for Engineers,* [3] we find some additional definitions of robots. For instance, an industrial robot is defined as "a programmable mechanical manipulator, capable of moving along several directions, equipped at its end with a work device called the end effector (or tool) and capable of performing factory work ordinarily done by human beings. The term robot is used for a manipulator that has a built-in control system and is capable of stand-alone operation." Another definition of a robot—taken from the Robotics International Division of the Society of Manufacturing Engineers—is also given in that book, i.e., a robot is "a reprogrammable multifunctional manipulator designed to move materials, parts, tools, or specialized devices through variable programmed motions for the performance of a variety of tasks."

We read in Koren's book that it is essential to include in the definition of a robot key words such as "motion along several directions," "end effector," and "factory work." Otherwise "washing machines, automatic tool changers, or manufacturing machines for mass production might be defined as robots as well, and this is not our intention." The question we must now pose is: What is wrong with defining a washing machine, a tool changer, or an automatically acting manufacturing machine as a robot? Are they not machines? Would it be right to say that washing machines do not belong to the family of robots when they act according to the concepts accredited to modern devices of this sort? And would it be justified to relate the concept shown in Figure 1.3 to the robot family? We will return to this example later when we discuss the concept of an automatic or a robotic system for the realization of a particular industrial task.

We are, in fact, surrounded by objects produced by machines, many of which completely fit the above-cited definitions of robots of higher or lower levels of sophistication. For example:

- Cans for beer or preserved foodstuffs
- Ball bearings and ballpoint pens

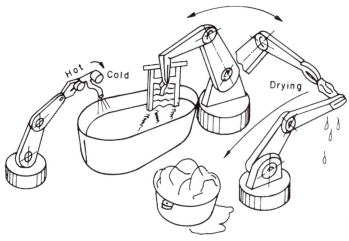

FIGURE 1.3 A washing process executed by manipulators.

- Screws, nuts, washers, nails, and rivets
- Socks and shoes
- Electronic chips, resistors, capacitors, and circuit plates
- Candies and ice cream

The list can be extended through batteries and photographic films to many, many other products that are fully or partially produced by some automatically acting machines. The question arises how to determine on a more specific basis whether a particular machine is a robot and, if so, what kind of robot it is. For this purpose, we need to take into consideration some general criteria without which no system can exist. To make the consideration clear we must classify automatic machines in terms of their intellectual level. This classification will help us to place any concept of automation in its correct place in relation to other concepts. An understanding of this classification will help us to make sense of our discussion.

1.2 Definition of Levels or Kinds of Robots

Every tool or instrument that is used by people can be described in a general form, as is shown in Figure 1.4. Here, an energy source, a control unit, and the tool itself are connected in some way. The three components need not be similar in nature or in level of complexity. In this section, when examining any system in terms of this scheme, we will decide whether it belongs to the robot family, and if so, then to which branch

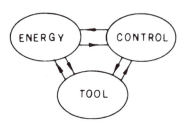

FIGURE 1.4 Energy-control-tool relations.

of the family. It is easy to see that this scheme can describe any tool: a hammer, a spade, an aircraft, a computer, a missile, a lunar vehicle, or a razor. Each of these examples has an energy source, a means of control, and the tools for carrying out the required functions. At this stage we should remember that there is no limit to the number of elements in any system; i.e., a system can consist of a number of similar or different energy sources, like or unlike means of control for different parameters, and, of course, similar or different tools. The specific details of this kind of scheme determine whether a given system can be defined as a robot or not. Let us now look at Figure 1.5 (examples I to X) which shows the various possibilities schematically.

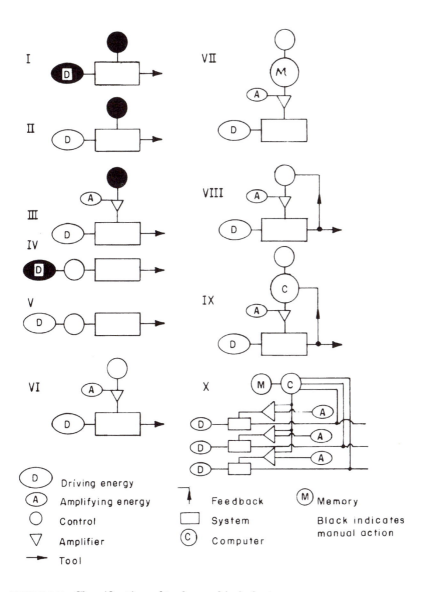

FIGURE 1.5   Classification of tools used in industry.

1. The energy source is a person, and his or her hands are the means of control; for example, a hammer, a shovel, a spade, a knife, or a sculptor's chisel. Indeed, when a person manipulates a hammer, the trajectory of this tool, the power of its impact, and the pace of action are controlled by the operator. In this case, the feedback or the sensors which inform the operator about the real location of the hammer, its speed, and its accumulated energy are the muscles of the arm, the hand, the shoulder, and the eyes. Obviously, this is also true for a spade or a chisel.

2. The energy source is a motor, but the means of control are still in human hands; for example, a simple lathe, a motor-powered drill, a dentist's drill (would anybody really be prepared to entrust the operation of such a tool to some automatic controller?), a motor-driven sewing machine, an electric or mechanically driven razor. To some extent, this group of machines also includes machines driven by muscle power of another person (or animal) or even driven by the legs of the same person.

3. The energy source is a motor and the means of control are manual, but are artificially amplified; for example, prostheses controlled by muscle electricity, or the power steering of a car fit this case to a certain extent.

4. The energy source is a person but the control function occurs (in series) via the system; for example, a manually driven meat chopper, or a manual typewriter. Here, some explanation is required. Rotating the handle of the meat chopper, for example, the operator provides the device with the power needed for transporting the meat to the cutter, chopping it, and squeezing it through the device's openings. The speed of feeding or meat transporting is coordinated with the chopping pace by the pitch of the snake and the dimensions and form of the cutter. Analogously, when the key of the typewriter is pressed, a sequence of events follows: the carbon ribbon is lifted, the hammer with the letter is accelerated towards the paper, and the carriage holding the paper jumps for one step. This sequence is built into the kinematic chain of the device.

5. The energy source is a motor, and the control is carried out in series by the kinematics of the system; for example, an automatic lathe, an automatic loom, an automatic bottle-labelling machine, and filling and weighing machines. This family of devices belongs to the "bang-bang" type of robots. Such systems may be relatively flexible. For instance, an automatic lathe can be converted from the production of one product to the manufacture of another by changing the camshaft. Figure 1.6 shows examples of different parts produced by the same lathe. Figure 1.7 presents examples of items produced by this type of automatic machines, i.e., a) a paper clip, b) a safety pin, c) a cartridge, d) roller bearings, e) a toothed chain, and f) a roller chain.

6. The energy source is a motor, and the control is achieved automatically according to a rigid program and is amplified; for example, an automatic system controlled by master controllers, i.e., electric, pneumatic, or hydraulic relays. Such systems are flexible in a limited domain.

7. The same as in (6), but the controller is flexible or programmable; for example, automatic tracking systems. An illustration of such a system is given in Figure 1.8. The shape of a wooden propeller vane is tracked by a tracer (or feeler), and the displacements of the tracer as it maintains gentle contact with the outline of the wooden part are amplified and transformed via the control into displacements of the metal cutter. Other examples are Jacquard's programmable loom and numerically controlled (NC) machines.

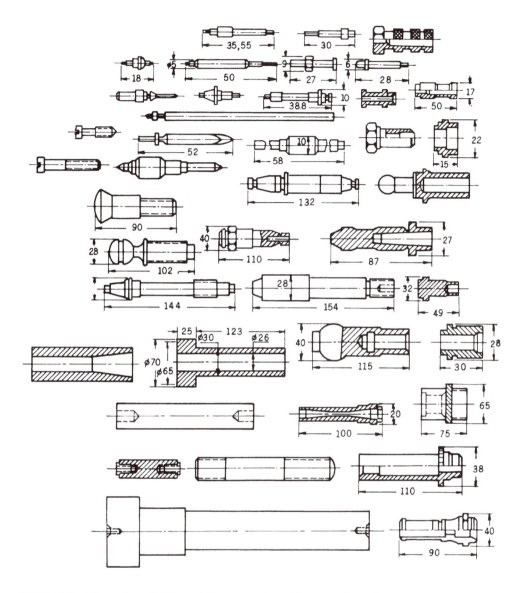

FIGURE 1.6.   Examples of different items produced by an automatic lathe (case 5 in Figure 1.5).

8. The same as in (4) and (7), with the addition of feedbacks, i.e., sorting, blocking, and measuring and tuning systems. Here we will give two examples. The first is an automatic grinding machine with automatic tuning of the grinding wheel which requires continuous measurement of the processed dimension (say, the diameter) and of the displacement of the wheel. In addition, the wheel can be sharpened and the thickness of the removed layer of the wheel can be taken into account. The second example is the blocking of a loom when a thread of the warp or of the weft (or of both) tears.

9. The same as in (8), with the addition of a computer and/or a memory; for example, automatic machines able to compute working conditions such as cutting regimes, or

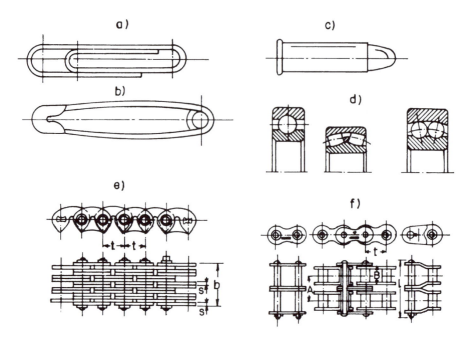

FIGURE 1.7. Examples of different items manufactured by the same automation level (case 5 in Figure 1.5). a) Paper clip; b) Safety pin; c) Cartridge; d) Roller bearings; e) Toothed chain; f) Roller chain.

the moving trajectories of grippers, or cutters. To this group of machines also belong those systems which are "teachable." For instance, a painting head can be moved and controlled manually for the first time; this movement will then be "remembered" (or even recalculated and improved); and thereafter the painting will be carried out completely automatically, sometimes faster than during the teaching process.

10. This level is different from (9) in that it is based on communication between machines and processes executing control orders to bring a complete system into har-

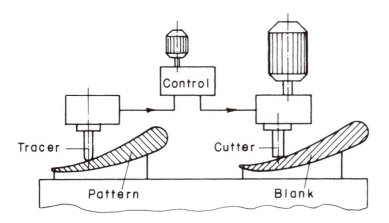

FIGURE 1.8. Layout of a tracing system (case 6 in Figure 1.5).

monious action. This case is shown schematically in Figure 1.5. As an example we can take an automatic line for producing pistons for internal combustion engines.

We must emphasize here that there are no rigid borders between one case and another. For example, a machine can as a whole belong to group (5), but for some specific task it may be provided with a feedback, say, signalling the lack of blanks followed by stopping of the action to avoid idle work. Another example is a car which is manually controlled but has an automatically acting engine. The solution to the argument about the definition of a robot probably lies somewhere between case (5) and case (7) in the above-given classification. Thus, it would be more useful to employ the terminology "automatically acting manufacturing machines (AAMM) and systems" instead of the foggy concept of robot. The means which provide the action of such a system at almost every level of complexity can be of purely mechanical, electromechanical, electronic, pneumatic, hydraulic or of mixed nature.

Irrespective of the level or kind of AAMM—numerically controlled or a computerized flexible manufacturing system (FMS)—its working part is mechanical. In other words, regardless of the control "intelligence" the device carries out a mechanical action. For example, the crochet hooks of a knitting machine execute a specific movement to produce socks; X-Y tables realize a mechanical motion corresponding to a program to position a circuit base so that electronic items can be assembled on it; and the cutter of a milling machine runs along a defined trajectory to manufacture a machine part. Cutters, grippers, burners, punches, and electrodes are tools and as such their operation is the realization of mechanical motion. (Even if the tool is a light beam, its source must be moved relative to the processed part.)

Being adherents of mechanics, we deem it appropriate at this stage to make a short digression into the glory of mechanics. In our times, it is customary to sing hymns of praise to electronics, to computer techniques, and to programming. Sometimes, we tend to forget that, regardless of the ingenuity of the invented electronics or created programs, or of the elegance of the computation languages, or of the convenience of the display on the terminal screen, all these elements are closely intertwined with mechanics. This connection reveals itself at least in two aspects. The first is that the production of electronic chips, plates, and contacts, i.e., the so-called hardware, is carried out by highly automated mechanical means (of course, in combination with other technologies) from mechanical materials. The second aspect is connected with the purely mechanical problems occurring in the parts and elements making up the computer. For instance, the thermal stresses caused by heat generation in the electronic elements cause purely mechanical problems in circuit design; the contacts which connect the separate blocks and plates into a unit suffer from mechanical wear and contact pressure, and information storage systems which are often purely mechanical (diskette and tape drives, and diskette-changing manipulators) are subject to a number of dynamic, kinematic, and accuracy problems. Another example is that of pushbuttons which are a source of bouncing problems between the contacts, which, in turn, lead to the appearance of false signals, thus lowering the quality of the apparatus. Thus, this brief and far-from-complete list of mechanical problems that may appear in the "brains" of advanced robots illustrates the importance of the mechanical aspects of robot design.

The AAMM designer will always have to solve the following mechanical problems:

- The nature of the optimal conceptual solution for achieving a particular goal;
- The type of tools or organs to be created for handling the subject under processing;
- The means of establishing the mechanical displacements, trajectories, and movements of the tools;
- The ways of providing the required rate of motion;
- The means of ensuring the required accuracy or, in other words, how not to exceed the allowed deviation in the motion of tools or other elements.

1.3   Manipulators

Let us return here to the definition of a manipulator, as given in Section 1.1. A manipulator may be defined as "a mechanism, usually consisting of a series of segments, jointed or sliding relative to one another, for the purpose of grasping and moving objects usually in several degrees of freedom. It may be remotely controlled by a computer or by a human" [2]. It follows from this definition that a manipulator may belong to systems of type 1 or 4, as described in Section 1.2, and are therefore not on a level of complexity usually accepted for robots. We must therefore distinguish between manually activated and automatically activated manipulators.

Manually activated manipulators were created to enable man to work under harmful conditions such as in radioactive, extremely hot or cold, or poisonous environments, under vacuum, or at high pressures. The development of nuclear science and its applications led to a proliferation in the creation of devices of this sort. One of the first such manipulators was designed by Goertz at the Argonne National Laboratory in the U.S.A. Such devices consist of two "arms," a control arm and a serving arm. The connection between the arms provides the serving arm with the means of duplicating, at a distance, the action of the control arm, and these devices are sometimes called teleoperators. (Such a device is a manually, remotely controlled manipulator.) This setup is shown schematically in Figure 1.9, in which the partition protects the operator sitting on the manual side of the device from the harmful environment of the working zone. The serving arm in the working zone duplicates the manual movements of the operator using the gripper on his side of the wall. The window allows the operator to follow the processes in the working zone. This manipulator has seven degrees of freedom, namely, rotation around the X-X axis, rotation around the joints A, translational motion along the Y-Y axis, rotation around the Y-Y axis, rotation around the joints B, rotation around the Z-Z axis, and opening and closing of the grippers. The kinematics of such a device is cumbersome and is usually based on a combination of pulleys and cables (or ropes).

In Figure 1.10 we show one way of transmitting the motion for only three (out of the total of seven) degrees of freedom. The rotation relative to the X-X axis is achieved by the cylindrical pipe 1 which is placed in an immovable drum mounted in the partition. The length of the pipe determines the distance between the operator and the servo-actuator. The inside of the pipe serves as a means of communication for exploiting the other degrees of freedom. The rotation around the joints A-A is effected by a connecting rod 2 which creates a four-bar linkage, thus providing parallel movement of the arms. The movement along this Y-Y axis is realized by a system of pulleys and cable 3, so that by pulling the body 4, say, downwards, we cause movement of the body

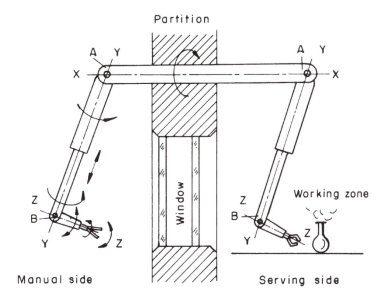

FIGURE 1.9.   Manually actuated manipulator/teleoperator.

5 in the same direction. This is a result of the fastening of the bodies 4 and 5 to the corresponding branches of the cable 3. By adding more pulleys and cables, we can realize additional degrees of freedom. Obviously, other kinematic means can be used for this purpose, including electric, hydraulic, or pneumatic means. Some of these means will be discussed later.

The mimicking action of the actuator arm must be as accurate as possible both for the displacements and for the forces the actuator develops. The device must mimic the movement of a human arm and palm for actions such as pouring liquids into special vessels, keeping the vessels upright, and putting them in definite places. Both

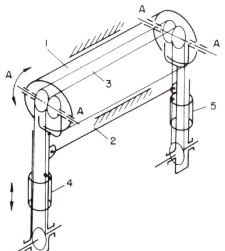

FIGURE 1.10.   Kinematic example of a three-degrees-of-freedom teleoperator (see Figure 1.9).

in principle and in reality the teleoperator is able to perform many other manipulations. Obviously, the number of degrees of freedom attributable to a manipulator is considerably less than the 27 degrees of freedom of the human arm. The operator of such a device thus has to be specially skilled at working with it. At present, engineers are nowhere near creating a manipulator with 27 degrees of freedom, which would be able to replace, at least in kinematic terms, the human arm. An additional problem is that a human arm, unlike a manipulator, is sensitive to the pressure developed, and the temperature and the surface properties of the object it is gripping. To compensate for the limited possibilities of the teleoperators, the workplace and the objects to be manipulated have to be simplified and organized in a special way.

The distance between the control and serving arms can range from one meter to tens of meters, and the maximum weight the manipulator can handle is 7–8 kg; i.e., the maximum weight the average person is able to manipulate for a defined period of time. The friction forces and torques can reach 1–4 kg and 10–20 kg cm, respectively, i.e., values which reduce the sensitivity of the device. Mechanical transmissions are the simplest way of arranging the connections between the control and serving arms. When the distance between the arms is large, the deformations become significant; for example, for a distance of 1.5–2 m, a force of 8 kg causes a linear deformation of 50–60 mm and angular deflections of 3–8°.

An additional problem occurs as a result of the mass of the mechanical "arms." To compensate for these weights, balance masses are used (in Figure 1.10 they are fastened to the opposite branches of the cables where the bodies 4 and 5 are fastened). This, in turn, increases both the forces of inertia developed when the system is in action and the effort the operator has to apply to reach the required operating speed. Thus, an ideal device which would be able to mimic, at any distance, the exact movement of the operator's arm is still a dream.

Let us now make a brief survey of automatically acting manipulators. The primary criterion used to distinguish between different types of manipulators is the coordinate system corresponding to the different kinds of degrees of freedom. The simplest way of discussing this subject is to look at schematic representations of some of the possible cases. Figure 1.11, for example, illustrates the so-called spherical system. It is easy to imagine a sphere with a maximal radius of $r_1 + r_2$ which is the domain in which, in

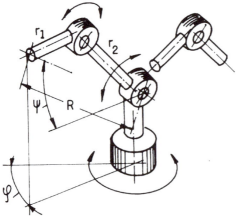

FIGURE 1.11. Layout of a spherical manipulator.

principle, any point inside the sphere can be reached by a gripper fixed to the end of an arm. In reality, there are certain restrictions imposed by the real dimensions of the links and the restraints of the joints which result in a dead zone in the middle of the sphere. Sometimes the angle of rotation ϕ is also restricted (possibly because, for instance, of the twisting of pipes or cables providing energy and a means of control to the links).

In Figure 1.12 we show a cylindrical manipulator. This kind of manipulator is also called a serpentine. When the links are straightened so that the arm reaches its maximal length $r_1 + r_2$, we can imagine a cylinder drawn by the manipulator for variables φ and Z. This cylindrical volume delineates a space in which the manipulator can touch every point. In reality, here as in the previous case, a dead zone appears in the neighborhood of the vertical axis for the same reasons mentioned above. The angle of rotation ~ may also be restricted for analogous reasons.

In Figure 1.13 a Cartesian-type manipulator is shown. A parallelepipedon based on the maximal possible displacements along the X, Y, and Z axes can be imaged. Here no rotational movements exist. Every point of the space inside the parallelepipedon is reached by corresponding combinations of coordinates.

Combinations of different coordinate systems are often used in the design of manipulators. In Figure 1.14 we see a combination of rotational and translational movement to provide variable value R. Part 1 can rotate around its longitudinal axis, creating an additional degree of freedom. Figure 1.15 gives another example of a combination of coordinate systems—this time a Cartesian and cylindrical manipulator. There are obviously other possible combinations, and we will discuss some of them later on.

Let us now look at the concept of the "fracture" of a degree of freedom. For instance, an indexing mechanism which rotates through a definite angle before stopping and carrying out a point-to-point rotation can be defined as a half-a-degree-of-freedom device. Such a device is shown in Figure 1.16.

The manipulators described above are driven by electric or other kinds of motors: thus, they do not depend on human power, and the drive is able to overcome useful, harmful and inertial resistance to develop the required speed of action. There are, however, problems with devices of this nature which do not arise with the manually

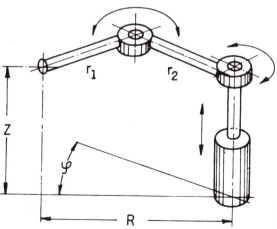

FIGURE 1.12 Layout of a cylindrical manipulator.

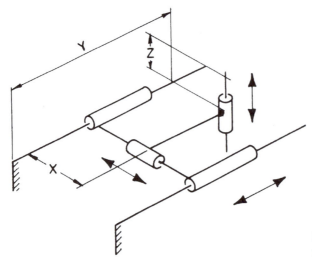

FIGURE 1.13   Layout of a Cartesian manipulator.

activated devices described previously; namely, the control of the movements must be organized artificially, and what would be a natural action for a manually operated device becomes a complicated technical problem in a nonmanual manipulator. Thus, the sequence, speed, and directions of the movements of the links must be found. For example, it takes a person one or two seconds to light a match, but it takes a manipulator about 30 seconds to carry out the same action.

   Another difficulty is that the mechanical manipulator is usually not able to feel resistance while handling different objects. This is one of the most important problems being tackled in the development of modern robotics. Sensitive elements able to gauge the forces, texture, or response of objects to be handled have still to be created for industrial purposes.

   Certain success has been achieved in the development of manipulators acting in concert with means for artificial vision. The task of the artificial-vision manipulator is

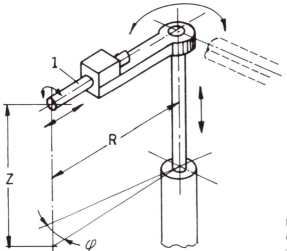

FIGURE 1.14   Layout of a combined cylindrical and linear coordinate manipulator.

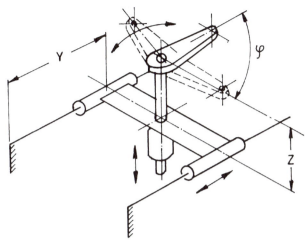

FIGURE 1.15 Layout of a combined Cartesian and cylindrical coordinate manipulator.

to recognize the location, shape, and orientation of the object to be handled. The manipulator and its gripper (or any other tool) are controlled in a way corresponding to the information obtained from the "vision means." The simplest example illustrating the means of action of such a manipulator is a situation in which a number of cubes of different sizes and colors, which are placed at random in a plane (inside an area which can be reached by the manipulator), must be collected and put in a definite order in a certain place by the manipulator. The "work" of the manipulator would be much easier and faster if the details (i.e., the cubes) were organized by some other system so that at a certain moment a particular detail or part being processed would be at a defined place (known to the manipulator) in a certain position (oriented). This brings us to the concepts of automatic feeding and orientation of parts.

We still have another point to discuss in relation to manipulators—that of the number of degrees of freedom of these devices. Why do we usually deal with a maximum of six, seven, or, rarely, eight degrees of freedom? (Remember the human arm has 27 degrees of freedom.) The reasons for this restriction lie in the fact that as the number of degrees of freedom increases:

- The design of the transmission becomes more complicated;
- The number of backlashes to be included in the kinematics and structure of the system increases;

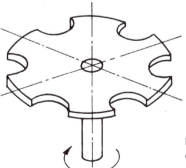

FIGURE 1.16 Indexing device: an example of a fraction-of-a-degree-of-freedom mechanism.

- The mechanical system becomes more flexible or less stiff, and
- The dynamic behavior, the accuracy, the repeatability, and the speed of response of the system deteriorate.

The answer to that question leads us to the next question: to obtain, say, eight degrees of freedom, should one use two manipulators with four degrees of freedom or even four manipulators with two degrees of freedom? The facts show that in certain situations it is indeed preferable to use a team of low-degrees-of-freedom manipulators rather than one sophisticated multi-degrees-of-freedom system.

Let us take an example to help us understand this statement. Figure 1.17 presents a schematic diagram of an automatic machine for assembling electronic circuits. A magazine (hopper) 1 contains a number of printed circuits (there are sufficient circuits in the magazine to provide a reasonable nonstop action time of the machine). A manipulator 2 with 2 degrees of freedom (provided with a vacuum suction cup) takes each circuit from the hopper (with an upwards/downwards movement) to the X-Y table 3 (to the left). (On the X-Y table there are two pins 4 for fixing the circuit by means of corresponding openings 5.) After the assembling is finished, the manipulator takes the completed plate to the pile 12 where the completed circuits are stacked. The X-Y table is also a two-degrees-of-freedom manipulator whose purpose is to place the circuit under the corresponding assembling devices. Both manipulators are of the Cartesian type. The barrel 6 is a hopper in which the electronic items are stored. The items are

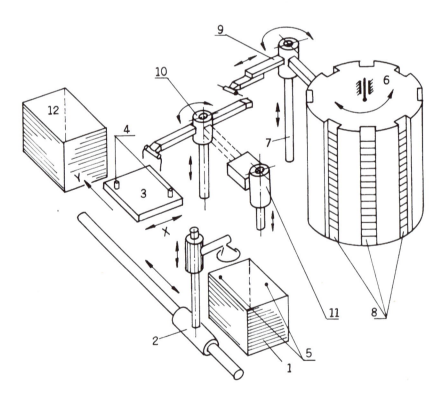

FIGURE 1.17 Layout of electronic circuit assembling machine composed of a number of low-degrees-of-freedom manipulators.

placed in the barrel in such a way that specific items are arranged in each sector 8 of the barrel. This barrel has only one degree of freedom—it rotates such that it brings the corresponding sector 8 to the "pick-up" manipulator 7. The manipulator 7 has three degrees of freedom: it rotates, it moves vertically, and it has a gripper which moves radially along the lever 9. The barrel and the manipulator act in the following sequence. The barrel 6 rotates to the required position, and then the gripper of the device 7 picks up the item. For this purpose the lever 9 faces the corresponding sector of barrel 6 and moves vertically until the gripper finds itself opposite the required item (obviously the level of the items in the magazines changes constantly as production proceeds, and the memory of the machine keeps track of the levels of the items in each magazine of the barrel 6). After the item has been "caught," the radially moving gripper removes it from the barrel. The lever rotates through 90°, and then stops at some definite height which allows it to pass the item to the manipulator 10. The manipulator 10 has two degrees of freedom, i.e., it rotates, and it has three possible positions at which it stops: at the first it obtains the item from the lever 9, at the second the leads of the electronic item are prepared for assembling by the device 1 1, and at the third the actual assembling takes place. For this latter operation, the lever of the manipulator 10 drops so that the leads hit the corresponding openings in the *X-Y* table which is already in position. (The devices 6, 7 and 10 are of the cylindrical kind.) The device 11 cuts the leads to the proper length, and then bends them downwards to create the required pitch. This action is shown schematically in Figure 1.18 a and b, respectively.

After the items have been installed on the circuit plate, the leads underneath the plate must be bent to fasten the items to the plate and to prepare them for soldering (which will provide proper reliable electric contact between the printed circuit and the electronic items). Figure 1.19 presents a possible solution for bending of the leads. Since there are different types of electronic items and the leads are different lengths and

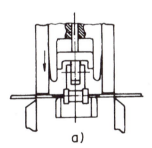

a)

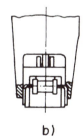

b)   **FIGURE 1.18**   Tool for cutting leads.

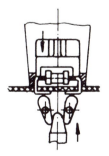

FIGURE 1.19   Tool for bending leads.

location, it is necessary to have an automatically controlled computerized system to carry out the complete operation.

This example (which does not pretend to be either the sole or the best concept of circuit assembly) permits us to derive some very significant conclusions:

1. Single manipulators working in concert facilitate the performance of simultaneous operations, thus saving time;
2. Simple manipulators are faster because their masses are smaller (no complicated transmission or drives are necessary) and their stiffness is greater (fewer backlashes, smaller dimensions);
3. There are groups of devices which carry out universal tasks, e.g., hoppers, magazines, feeders, carriers, conveyors, and grippers;
4. There are devices or specialized manipulators which carry out some specific tasks, such as assembling, bending, and cutting.

1.4   Structure of Automatic Industrial Systems

It is possible to describe a generalized layout of an automatic machine almost regardless of the level of control to which it belongs. We will thus show that the building-block approach is an effective means of design of automatic machine tools. The following building blocks for devices may be used in the layout of automatic machines:

- Feeding and loading of parts (or materials) blocks,
- Functional blocks,
- Inspection (or checking) blocks,
- Discharge blocks,
- Transporting (or removing) blocks.

Blocks that are responsible for the feeding and loading of materials in the form of rods, wires, strips, powders, and liquids, or of parts such as bolts, washers, nuts, and special parts must also be able to handle processes such as orientation, measuring, and weighing.

Functional devices are intended for processing, namely, assembling, cutting, plastic deformation, welding, soldering, pressing in, and gluing.

Inspection or checking blocks ensure that the part being processed is the correct one and that the part is in the right position. These devices also check tools for readiness, wear, etc. The necessity for such devices for purposes of safety and efficiency is obvious.

A discharge device is obviously used for releasing an item from a position and preparing the position for a new manufacturing cycle.

Transporting devices provide for the displacement of semi-finished items and parts during the manufacturing process. These devices are responsible for ensuring that the parts are available in a certain sequence and that each part is in the correct place under the relevant tool, device, or arrangement at a certain time.

There are different approaches to combining these building blocks in the design of an automatic machine. Let us consider some of the more widely used combinations. In Figure 1.20 we show a circular composition. Here, the feeding 1, functional 2, inspection 3, and discharge 4 devices are located around the transporting block 5.

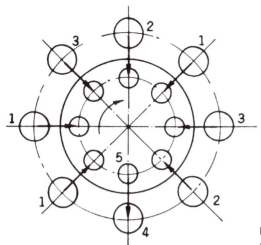

FIGURE 1.20 Circular configuration for an automatic tool.

Obviously, in many cases there can be several feeding, functional, and inspection blocks in one machine. The transporting block, which in this case is a rotating table driven by an indexing mechanism, moves in an interruptive manner. Its movement can be described by the speed-change form shown in Figure 1.21. Here, it can be clearly seen that the table moves periodically and each period consists of two components of time, t_1—the duration of movement, and t_2—the duration of the pause. Obviously, both the functional devices and the loading and discharging blocks can act only when the rotating table is not in motion and the parts (or semifinished items) are in position under the tools that handle them. Thus, the ratio

$$t_2 / t_1 = \eta \qquad [1.1]$$

becomes very important since it describes the efficiency of the transporting block. The higher the ratio, the smaller are the time losses for nonproductive transportation. In the periodically acting systems some time is required for the idle and auxiliary strokes that the tools have to execute. For instance, a drill has to approach a part before the actual drilling operation, and it has to move away from the part after the drilling has been accomplished. These two actions may be described as auxiliary actions because no positive processing is carried out during their duration. On the other hand, no positive processing can be carried out without these two actions. The time the device spends on these two strokes can, however, be reduced by decreasing the approaching

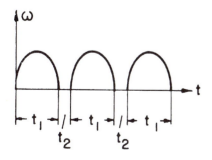

FIGURE 1.21 Speed-versus-time diagram for an indexing mechanism.

and withdrawal distances and by increasing the speeds of approach and withdrawal. Similarly, the transportation of a part from a drilling position to, say, a threading position is an idle stroke. In principle, neither the threading process nor the drilling operation requires this transportation component, which appears only as a result of the chosen design concept.

Let us denote the idle and auxiliary time losses τ, then

$$t_2 = T + \tau,$$

[1.2]

where T is the pure processing time.

From Equation (1.1) we obtain:

$$\frac{T + \tau}{t_1} = \eta \quad \text{or} \quad T = \eta t_1 - \tau.$$

[1.3]

We can now introduce the concept of a processing efficiency coefficient η_1 in the form

$$\eta_1 = T / (t_1 + \tau) = \frac{\eta t_1 - \tau}{t_1 + \tau}.$$

[1.4]

A modification of the composition discussed above may also be used. In this modification the blocks 1, 2, 3, and 4 are partly or completely placed inside the rotating table, as shown in Figure 1.22. This modification is more convenient because it facilitates free approach of the items to the tools and to all the devices, while the devices do not obscure the working zones. However, the drives, the kinematics, and the maintenance of this type of composition are more complicated. Another possibility is to build the transporting device 5 in a linear shape as a sort of a conveyer, as is shown in Figure 1.23. In this configuration the devices 1, 2, 3, and 4 are located on the same side of the conveyer (although there is no reason that they should not be located on both

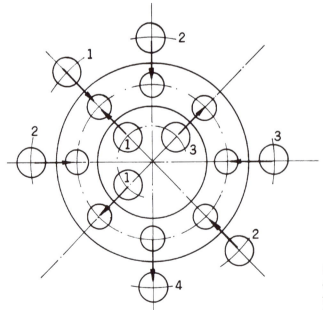

FIGURE 1.22 Circular configuration for an automatic tool with partial internal location of blocks.

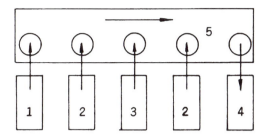

FIGURE 1.23   A linear configuration for an automatic tool.

sides of the block 5), thus facilitating maintenance from the other side. Obviously, Expression (1.1) is also valid in this case.

The ideal situation occurs when $t_1 = 0$, i.e., where there are no time losses during the process and the process is nonperiodic or continuous. Figure 1.24 presents an example of such a machine—the rotary printing machine (for newspaper printing). Here, 1 and 2 are the paper feeding blocks which incorporate a number of guiding rollers; 3 and 4 are the printing blocks which include the printing ink feeders and distributors for both sides of the paper 2 and the impression cylinders 4 and 5 is the discharge block, which receives the completed product—folded and cut newspapers.

The productivity of such a system is measured in speed units, say $V = 5$ m/sec of printed paper in a rotary printing machine, or 10 m/sec of wire for a drawing bench, or 15 m/sec for rolled stock produced on a mill. If the length of the paper needed for one newspaper (or any other piece-like product) is 1, then the productivity P is:

$$P = V / 1 \text{ [piece per time unit]}. \tag{1.5}$$

Obviously, the time T_c needed per produced unit is:

$$T_c = 1 / P. \tag{1.6}$$

This is the actual time spent for production, while the systems working periodically require, per product unit, a time interval T_p.

$$T_P = t_1 + t_2 = (t_1 + \tau)(1 + \eta_1) = t_1(1 + \eta), \tag{1.7}$$

which includes nonproductive items t_1 and τ.

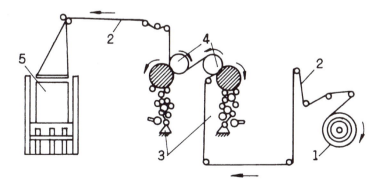

FIGURE 1.24   Rotary printing machine as an example of a continuous (nonperiodic) system.

The comparison of Expressions (1.6) and (1.7) proves that, for equal concepts $(T_c \approx T)$, the continuous process is about $(1 + \eta)/\eta$ times more effective. This fact makes the continuous approach very attractive, and a great deal of effort has been spent in introducing this approach for the manufacturing and production of piece-type objects.

Sometimes it is even possible to design a continuously acting machine for complicated manufacturing processes. The main idea underlying this type of automative machine tool is represented in Figure 1.25. The machine consists of a number of rotors 1, each of which is responsible for a single manufacturing operation. The design of each rotor depends on the specific operation, and its diameter and number of positions or radius depend on the time that specific operation requires; i.e., if the operation n takes T_n seconds, the radius r of rotor number n can be calculated from the following expression:

$$T_n = l_n / V = r_n \phi_n / V, \qquad [1.8]$$

where V = the peripheral velocity of the rotors,

l_n = the length of the arc where the product is handled for the n-th rotor,

r_n = the radius of the n-th rotor, and

ϕ_n = the angle of the arc of the n-th rotor where the product is handled.

In addition, there are rotors 2 which provide for transmission of the product from one operation to another. The machine is also filled with a feeding device 3, where the blanks are introduced into the processing and with a discharging device 4, where the finished (or semi-finished) product is extruded from the machine. Thus, the main feature of such a continuously acting system is that the manufacturing operations take place during continuous transportation of the product. Therefore, there are no time losses for pure transportation.

Let us now look at an example of this type of processing. Figure 1.26, which shows the layout of a continuous tablet production process, can serve as an example of a device for continuous manufacturing of noncontinuous products. Figure 1.26 presents a cross section through one of the rotors. The rotor consists of two systems of plungers, an upper system 1 and a lower series 2. The plungers fit cylinders 3 which are made into a rotating body 4 (this body is, in fact, the rotor). The device operates in the fol-

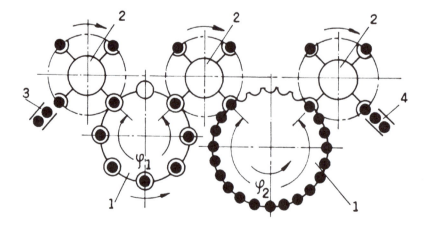

FIGURE 1.25   Layout of a rotary machine for periodic manufacturing processes.

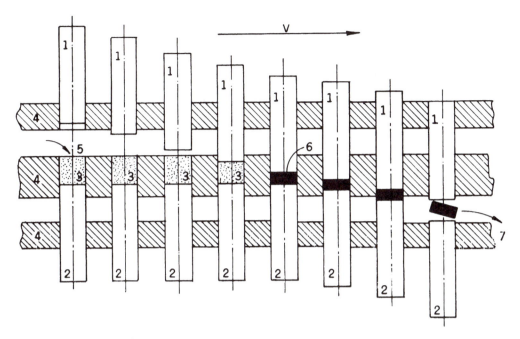

FIGURE 1.26   Layout of a continuous tablet manufacturing process.

lowing way. At some point, one of the cylindrical openings 5 is filled with the required amount of powder for the production of one tablet. (This process is carried out by means of the movement of the rotor.) Then, the upper plungers begin to descend while the lower plungers create the bottom of the pressing die. When the pressing of the tablet 6 is finished, both plungers continue their downward movement to push the finished product out of the die in position 7. All these movements of plungers take place while the rotor is in motion.

We can also imagine an intermediate case. This case is illustrated by the example of a drilling operation shown schematically in Figure 1.27. The rotor 1, which rotates with a speed V, is provided with pockets 2 in which the blanks 3 are automatically

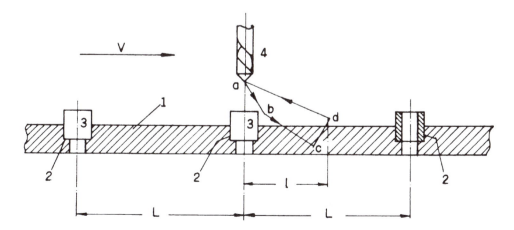

FIGURE 1.27   Layout of a pseudo-continuous drilling process.

placed. The drilling head 4 (which can be considered as a two-degrees-of-freedom manipulator) carries out a complex movement. The horizontal component of this movement is equal to the rotor's speed V on the section a-b-c-d. The vertical component is made up as follows: fast approach of the drill to the blank on the section a-b (the drill's auxiliary stroke), drilling speed on the section b-c (processing stroke), and high-speed extraction of the drill on the section c-d (second auxiliary stroke). As soon as the second auxiliary stroke has been completed, the opening in the blank has been processed, and the drill must return to the initial point a to meet the next blank and begin the processing style. The time of one cycle is:

$$T = L / V,$$ [1.9]

where L is the linear distance between the pockets. The time t that the drilling head follows the rotor can be calculated from the obvious expression:

$$t = l / V,$$ [1.10]

where l is the distance through which the drilling head follows the rotor. The time τ remaining for the drilling head to return is:

$$\tau = T - t.$$ [1.11]

Thus, the returning speed (the horizontal component) of the drilling head V_1 is:

$$V_1 = \frac{l}{\tau}.$$ [1.12]

This case combines the two main approaches in automatic machining. Irrespective of the nature of the conceptual design of the automatic machine, it consists (as was stated above) of feeding, transporting, inspecting, tooling, and discharging blocks. The design of these blocks and some relevant calculations will be the subject of our discussion in the following chapters.

1.5   Nonindustrial Representatives of the Robot Family

In this section we will discuss in brief some robot systems that almost do not belong to the family of industrial robots, namely,

1. Mobile robots
2. Exoskeletons
3. Walking machines
4. Prostheses

Each of these kinds of special robot can be classified in terms of certain criteria. We will thus describe the main features of these machines and the principal parameters and concepts characterizing them.

1.   Mobile robots

These devices have a wide range of applications which are often not of an industrial nature. We can classify the types of mobile robotic machines described here in terms of their means of propulsion, i.e., wheels or crawler tracks.

A mobile robot may be controlled in one of the following ways:

- Remotely controlled by wires, cable, or radio;
- Automatically (autonomously) controlled or programmed; or
- Guided by rails, or inductive or optic means.

Mobile robots find application in the following situations:

- In harmful or hostile environments, such as under water, in a vacuum, in a radioactive location, or in space; or
- Handling explosive, poisonous, biologically dangerous, or other suspect objects.

Let us consider here, in general, the industrial applications of mobile robots. In some factories mobile robots are used as a means of transportation of raw materials, intermediate and finished products, tools, and other objects. One of the problems arising here is that of navigation. One side of the problem is the technical and algorithmic solution to the creation of an automatic control system. (This solution will be described in greater detail in Chapter 9.1.) The other side of the problem is the choice of the strategy for designing a pathway for such a vehicle. As an aid in clarifying this specific problem, let us look at the layout given in Figure 1.28. In our illustration an automatic waiter must serve nine tables in a cafeteria.

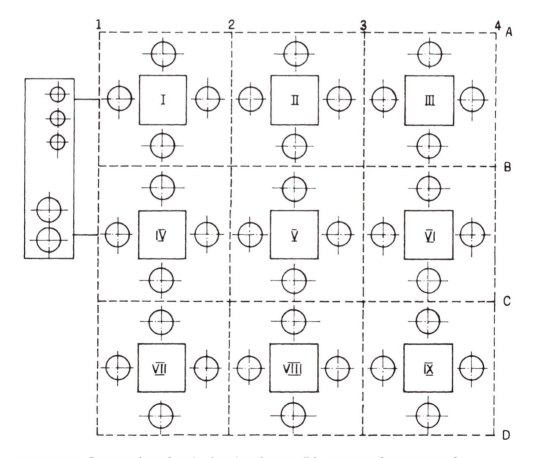

FIGURE 1.28   Layout of a cafeteria showing the possible courses of movement of an automatic waiter.

Analyzing Figure 1.28, we can offer the "waiter" several possible courses. For example, from the bar or counter it can move through the following points:

1. A1-A2-A3-A4-B4-B3-B2-B1-C1-C2-C3-C4-D4-D3-D2-D1, or,
2. A1-A2-A3-A4-B4-C4-D4-D3-D2-D1-C1-B1-B2-B3-C3-C2-C1, etc.

The criteria we must satisfy here are:

- The minimal service time (which includes the distribution of meals, the collection of dishes, the distribution of bills, and the collection of money);
- The optimal number of dishes on the trays;
- The minimal disturbance to the customers.

If the mobile robot is propelled on tracks, then a separate drive is required for navigation of the tracks. If it is propelled by wheels, a steering wheel is required (3-wheel designs are conventional), or each wheel is equipped with an independent drive of a special kind (see Figure 9.55).

In the case of the above-described cafeteria, control could be effected, for instance, by colored strips on the floor combined with a system capable of counting the number of times the robot crosses each strip. The memory of the "waiter's" computer is programmed with the action it has to perform after each crossing point. The commands would be of the kind:

- Go ahead;
- Stop;
- Turn left or right; or
- Turn around.

If such a vehicle were combined with a manipulator, we would obtain a very flexible device which would be able to handle, for example, a suspicious object such as a bomb.

2. *Exoskeletons*

Let us imagine a person working in a "hostile environment," such as under water or in space. His safety suit must withstand high pressures (external or internal). Obviously, the joints of such a suit, its weight, and its resistance to the environment will hamper the movement of the person. Thus, special means must be provided to compensate for these harmful forces. These means can comprise an external energy source linked to a type of amplifier which permits the person enclosed in the safety suit to act almost normally as a result of the fact that the real forces developed by the device are significantly larger than those developed by the working person. We can then "extrapolate" the situation of the hostile environment to normal circumstances to provide a person with a means of protection from the environment or with a means of handling heavy objects which are far beyond the limits of a normal person. Whatever the specific application of use of the device, it must: 1) free the skeleton of the person from physical overloads; 2) amplify the person's muscular efforts; and 3) provide feedback to enable the user to gauge the reaction of the object being manipulated.

The first requirement described above indicates that the device has an auxiliary function to the human skeleton, hence the name, exoskeleton. One possible design comprises a double-layered structure. The internal layer, which includes the control mechanism, makes direct contact with the operator. The external cover follows the movements of the internal layer and is responsible for amplifying the forces. This kind of system was first used in about 1960 in the Cornell Aeronautical Laboratory, U.S.A. For instance, the American exoskeleton known as Hardiman enables its operator to lift weights up to 450 kg. It has about 30 degrees of freedom (arms, legs, and body) and permits the operator to move at about 1.5 km/hour. The power system is hydraulic. Figure 1.29 shows the basic structure of an exoskeleton. It consists of a frame 1 to which the links to moving parts of the body are connected: the thighs 2, shins 3, feet 4, shoulders 5, elbows 6, and hands 7. Hydrocylinders 8 are used to drive the links. The power supply is provided by the compressor station 9 fastened to the back of the exoskeleton. The control of the cylinders shown in the figure, and those which are not shown (such as the rotation of the elbow around its longitudinal axis) is carried out by the person enveloped in the exoskeleton. By moving his limbs which are connected to corresponding links of the mechanical device, the person activates a system of amplifiers which in turn actuates the corresponding cylinders. The principle of the operation of the hydraulic amplifier will be explained in Chapter 4. Means of exploiting the biocurrents of human muscles for this purpose are now being investigated.

3. *Walking machines*

The wheel was invented about 6,000 years ago. This invention, coupled to an animal as a source of driving power, increased the possibility of load displacement about ten times. However, this invention created the problem of providing roads. To circumvent this complication (since roads cannot cover every inch of countryside) caterpillar tracks were invented. (This solution reduces the pressure under the vehicle by about eight times.) Thereafter efforts were devoted to creating a walking machine able to simulate the propelling technique of animals in such a way that the machine could move over rough terrain. The idea of creating a walking vehicle is not new. We will take as an example the walking mechanism synthesized by the famous mathematician Chebyshev (1821–1894). Figure 1.30 presents the kinematic layout of this mechanism, while the photographs in Figure 1.31 show its realization produced in the laboratory of the Department of Mechanical Engineering of the Ben-Gurion University of the Negev. This mechanism fulfills the main requirement of a properly designed walking device; i.e., in practice, the height of the mass center of the platform 1 (see Figure 1.30) does not change relative to the soil. This ingenious mechanism, however, is not able to change direction or move along a broken surface. (It is an excellent exercise for the reader to find a means of overcoming these two obstacles.)

The link proportions shown in Figure 1.30 are obligatory for this walking machine. The walking technique is more effective than wheel- or track-based propulsion, not only because obstacles on the surface can easily be overcome (for instance, legs climbing stairs), but also because the nature of the contact between the leg and the surface is different from that between a wheel or tracks and a road. As can be seen from Figure 1.32, the rolling wheel is continuously climbing out of the pit it digs in front of itself. This process entails, in turn, the appearance of a resistance torque T as a result of the force F acting on the lever l. On the other hand, any type of walking mechanism is a periodically acting system. At this stage we should remember that dynamic loads increase in direct proportion to the square of the speed. In addition the design of such a walking leg is much more complicated than that of a rotating wheel. Thus, we cannot

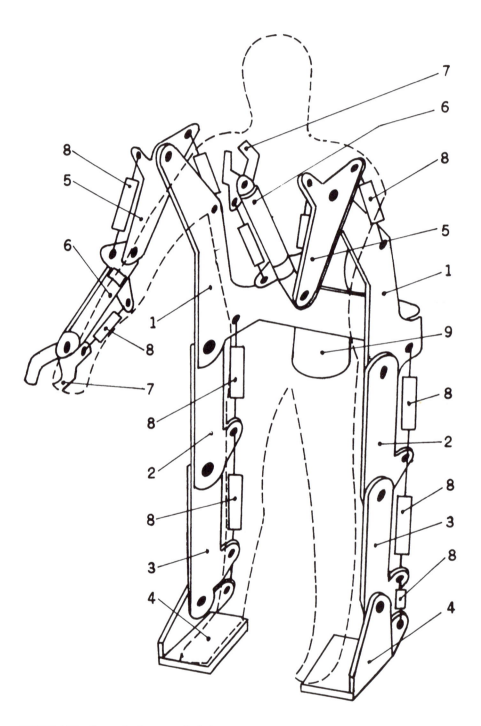

FIGURE 1.29   Layout of an exoskeleton.

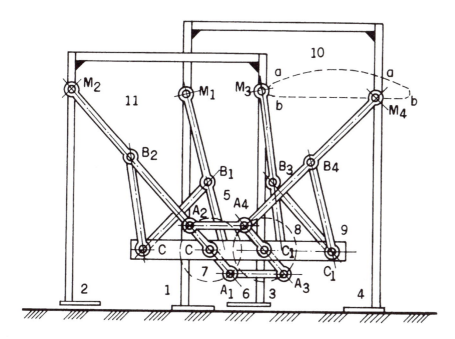

$$A_1 B_1 = B_1 C = B_1 M_1 = A_2 B_2 = B_2 C = B_2 M_2 = A_3 B_3 =$$
$$= B_3 C_1 = B_3 M_3 = A_4 B_4 = B_4 C_1 = B_4 M_4 = 1 ;$$
$$A_1 C' = A_2 C' = A_3 C_1' = A_4 C_1' = 0.355 ;$$
$$C C' = C_1 C_1' = 0.785 ; \quad A_2 A_4 = A_1 A_3 = C' C_1' = 0.634 .$$

FIGURE 1.30   Chebyshev's walking mechanism.

expect the same speed of motion from a walking vehicle as from a wheeled vehicle. This is the price we pay for the maneuverability and mobility of a leg. Human beings, animals, and insects are the prototypes for research in this field. Such research also includes work on the principles of vibrating and jumping.

Various attempts have been made to solve the problem of artificial automatic walking. These attempts include one-, two-, three-, four-, six- and eight-legged vehicles. Some principles of two-legged walking will be discussed later in Chapter 9. When such machines are designed, the following problems must be solved:

1. Kinematics of the legs;
2. Control system providing the required sequence of leg movements;
3. Control of mechanical stability, especially for movement along a broken surface or an inclined plane.

One of the possible solutions is similar to that for an exoskeleton; i.e., the driver moves his limbs and the vehicle repeats these movements. Such a vehicle becomes more cumbersome as the number of legs is increased. A four-legged lorry built by General Electric, which weighs 1,500 kg, develops a speed of about 10 km/hour and

FIGURE 1.31   General view of a Chebyshev walking mechanism.

can handle a load of about 500 kg. Six legs give very stable movement, because in this case three support points can be provided at any given moment. On the other hand, a six-legged vehicle (like an insect) can become very complex, since it facilitates about 120 styles of walking by changing and combining the sequence of operation of the legs. This type of robot came into its own when Space-General Corporation created a family of "Lunar Walkers" consisting of six, and later of eight, legs.

4.   Prostheses

As a natural continuation of the above-discussed exoskeletons and walking machines, we come to the idea of an automatically controlled device replacing a

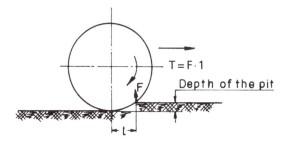

FIGURE 1.32   Layout of the means of propulsion by a rolling wheel.

damaged or lost human limb. For these specific cases, it is of crucial importance to shorten the time for transforming a command into an action. The time taken for the chain of operations—from the disabled person's desire to carry out an action to the action itself—must be minimal. In addition, the means and the system itself must be compact and light—it must not create additional difficulties for the owner. The most effective way of achieving this aim is to exploit the control commands given via the bioelectricity of the relevant muscles, to amplify them, to analyze them, and to transform them into the desired actions of a mechanical device—a prosthesis. Figure 1.33 shows the layout of such a device for replacing a disabled hand. Here 1 are electrodes fastened to the skin of the elbow, 2 is an amplifier, 3 is a rectifier, 4 an integrator, 5 an artificial hand, and 6 a servomechanism. The servomechanism is shown in greater detail in Figure 1.34. In this figure, 1 and 2 are the same elements as 1 and 2 in Figure 1.33, while 3 is electrohydraulic valves, 4 is an electromotor driving hydraulic pumps, 5 are pumps and 6 is a cylinder with a piston. The piston rod of 6 actuates the artificial hand 7. The valves 3 are provided with control needles 8 which open or close the

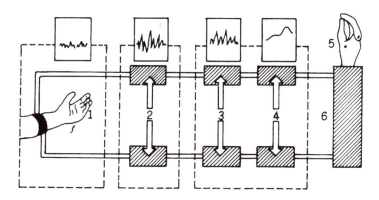

FIGURE 1.33 Layout of the control of a hand prosthesis.

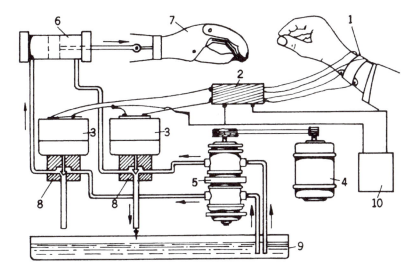

FIGURE 1.34 Layout of the power supply for a hand prosthesis.

oil (or other liquid) passages in proportion to the electrosignal on the inputs of the valves 3. The oil is collected in a reservoir 9. The power needed to feed the system is supplied by batteries 10.

As far as is known, the first idea for this type of artificial hand was patented in 1957 in the USSR by N. Kobrinsky, M. Breydo, B. Gurfinkel, A. Sysin, and J. Jacobson. Later, such hands were created in Yugoslavia, Canada and the U.S.A.

1.6   Relationship between the Level of Robot "Intelligence" and the Product

The question may arise as to what level of robot will be needed for each particular area of manufacturing or processing. There is a feeling that, in the immediate (or not-too-distant) future, all goods will be produced by "intelligent' robots, and no man-power will be involved in manufacturing processes. In such a scenario, a human being would command and control the processes by means of computers and pushbuttons. He would tell the computerized machine to produce, say, strawberry ice cream, and the machine would be instructed by its electronic (or other?) "brain" to mix the correct amounts of the right ingredients, to heat or cool them to the right temperatures, to treat the mixture at the right pressures, and so on. On another day, the system might receive an order to prepare potato chips, and again the computer would find, in its memory, the right recipes, would choose the optimal one, would tell the right machine to do the right job, and so on.

In our opinion this is a childish (although attractive) approach to the future. There will always be industrial needs for robots of low intellectual level and perhaps also for manually powered or hand-controlled machines. The reason for this is not solely indi-vidual preference for some mechanical devices. (For instance, there are many people who do not like cars with an automatic transmission but prefer those with a manual gear box.) There are also some objective economic reasons for thinking that "manual" machines will not be phased out. For example, people will always wear socks or some-thing like socks. Whatever the material from which they are made, socks will always be needed in amounts of tens of millions or more. Similarly, people will always need writing tools—be it goose feathers, fountain pens, ballpoint pens, or electronic pens—and hundreds of millions of these tools will be needed. Screws, nuts, washers, and nails are manufactured each year in millions and millions of each shape and size. Food con-tainers—bags, cans, and bottles—are another example of mass production. Ironically, even electronic chips used for making robot "brains" are often produced in such large quantities that flexibility is not required during manufacturing.

All the products mentioned above are characterized by being in demand in large amounts and over relatively long periods of production. The question is thus whether it is cheaper and more effective to use specialized, relatively inflexible (or completely inflexible) machines for such purposes rather than sophisticated flexible robots. The answer, in our opinion, is "yes." The problem is to define whether it is appropriate for a particular industrial need to design and build a high-level advanced robot or a "bang-bang" type of robot.

Let us denote:

τ = the lifetime of the automatic equipment or its concept, i.e., the time the machine or its concept is useful;

T = the lifetime of the product produced by the automatic equipment, or the time the product keeps being sold on the market, or the market demand for the product is maintained;

T_1 = the time needed to design the equipment (to work out its concept), to build it, and to implement it in industry.

For rough estimates we can derive obvious conclusions:

1. For $\tau \leq T$, bang-bang-type robotics is justified because it is cheaper, can be implemented faster and, what is most important, no flexibility is needed: even if the potential for flexibility does exist, it will not be utilized because the product keeps going on the market longer than the machine is able to produce.
2. For $\tau > T$, the use of flexible manufacturing systems is justified. The machine lives longer than the specific product manufactured. Thus the same machine would be able to produce something else. It is expensive but its flexibility facilitates return on the expenses.
3. For bang-bang robots, but not for flexible manufacturing systems, the condition $T > T_1$ must be met.

Thus, for example, for the production of paper clips, the design engineer must decide what level of complexity of machine is required, and the answer will most likely be:

- An open loop,
- Pure mechanical control,
- Rigid layout equipment.

In this case, the designer must ensure that $T > \tau$ (so that as many copies of the machine as are needed can be made), and that $T > T_1$. If on the other hand, the product is any kind of molding die, the answer would instead be:

- A computerized, closed-loop, automatically controlled milling machine, if possible with an automatic tool exchange, dimensional control, and feedbacks, etc.

In the latter case, the condition $T \leq \tau$ is ensured due to flexibility of the equipment and the possibility of fitting it easily and cheaply to any (or many) kinds of die. The condition

$$T < T_1$$

is ensured because the specific dies are needed for relatively short time periods while the machine is built for practically indefinite time T interval.

Figure 1.35 presents a graphic interpretation of the inequalities used. The area limited by bisectors I and II contains the cases for which bang-bang or point-to-point robots can (and, in the author's opinion, should) be used. For the other cases, effective manufacturing requires higher-level robots, i.e., robots with programmable and computerized systems.

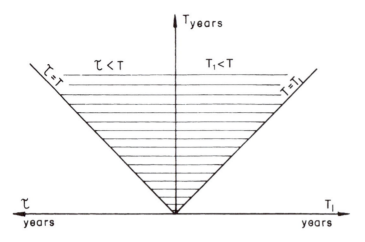

FIGURE 1.35   Diagram for estimation of the level at which a machine tool should be designed.

References

1. "Robots," *Encyclopaedia Britannica,* vol. 15, William Benton Publisher, p. 910, 1973.
2. "A Glossary of Terms for Robotics," prepared for Air Force Materials Laboratory, Wright-Patterson Air Force Base, Ohio. National Bureau of Standards, Washington, D.C., 1980.
3. Koren, Y., *Robotics for Engineers.* McGraw-Hill, New York, 1985.

2

Concepts and Layouts

2.1   Processing Layout

In designing any automatic manufacturing system, the first step is to determine the general, basic concept of the manufacturing process. This concept includes the general nature, shape, and type of the tools taking part in the manufacturing process, their action or operation sequence, and the conditions under which manufacturing must take place.

This concept must be properly determined until the object is comprehensively analyzed. Selecting any optimal manufacturing or production process sometimes involves interfering with the design of the object being produced and maybe even modifying it; thus, one may need to reconsider the material of a given part (plastic instead of metal, say), or its precision, shape, and so on. In other words, the manufacturing process dictates the design of the product. Let us now consider a number of examples that illustrate this important point. (Recall that the same object or part can usually be produced by different manufacturing processes, tools, and techniques.)

At this stage in the design of our automatic machine, we need to visualize the concept underlying the manufacturing process. The best way to do it is to express this concept graphically. Our sketch must show:

- Every processing and auxiliary operation;
- The tools or elements which carry out these operations;
- The special conditions or requirements which must hold during processing;
- Basic calculations—speeds of displacement and rotation, linear and angular displacements, forces and torques, dimensions, etc.

While the layout sketch need not be to scale, it must clearly explain our concept of the process. Some examples follow. The first of these is a discontinuous process where the operations appear stepwise one after the other. The second example, by contrast, is a continuous process for producing spiral springs.

Example 1

Consider the manufacturing of a chain (Figure 2.1). This chain is made of brass wire 1.5 mm in cross section, other dimensions being clear from the figure. The manufacturing process must begin with a feeding operation. We propose to take a coil of wire and pull out wire from this coil. The next operation is straightening of the wire. In Figure 2.2 position a) shows the feeding device, here consisting of two rollers 1 driven so as to provide the required length of wire. As it is being pulled, the wire passes through a device 2 which imparts multiple plastic bends to the wire in two perpendicular planes by a system of specially shaped pins. Such "torture" of the wire causes it to "forget" its previous curved form and, assuming the device is properly tuned, provides straight wire after the feeder. A length of wire corresponding to that needed for one link is measured out in our example, 85 mm. The next operation is cutting the measured section by cutter 3. While fed and cut, the wire is passed through a slot created on one side by a shaped support 4 and on the other side by two parts of a horseshoe-like tool 5. The shape of support 4 matches the inner contour of a single link. At the next operation the pair of tools 5 moves to the right, bending the wire section around support 4. Towards the end of the bending process tool 5 pinches the "horns" of the link inwards, completing the shape of the link. Pins 6 fetch the tools 5 to perform this movement. The latter operations are illustrated in Figure 2.2b) and c) . Next a punch 7 drops down and bends the ends of the link. An opening 8 guides the punch and provides a place for the link (see Figure 2.2d)). The rest of the link must be held under restriction so as to produce bending. In our case restriction is effected by the slot in which the link is located during this procedure. Now a pusher 9 is actuated, causing the link to move to the right till it falls down opening 8 (the punch is now retracted and does not move again). At this point the tools 5 return to their initial positions, preparing the working space for the following link.

The previously produced links are arranged on a guide 12 as shown in Figure, 2.2e). Thus, the new link, as it falls, inserts its "horns" through the previously fallen link. In turn, the semi-ready chain is promoted by a specially shaped wheel 10 (Figure 2.2f)). The teeth of this wheel pull the chain, the wheel rotates, and the "horns" of the links are brought in contact with a sprocket 11 which bends the "horns," thereby complet-

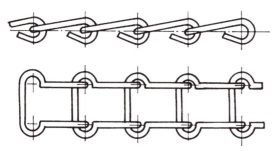

FIGURE 2.1 Chain produced by an automatic process.

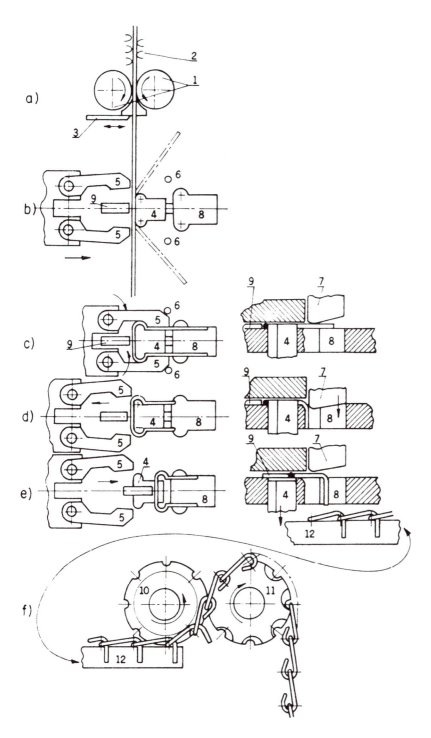

FIGURE 2.2   Processing layout of an automatic machine for manufacture of the chain shown in Figure 2.1.

ing the process of production of this chain, The completed chain is received by the sprocket 11 rotating at an appropriate speed.

Example 2

Consider an automatic machine for the mass production of springs made of steel wire of, say, 0.2–0.3 mm diameter. The dimensions of the springs must be adjustable within certain ranges; see examples of springs, Figure 2.3. The pitch of a particular spring is also adjustable; see Figure 2.3b). Figure 2.3c) shows a spring with a variable diameter. The concept underlying the manufacturing process is as follows (the layout is shown schematically in Figure 2.4). Feeding and straightening of the wire are carried out by a process analogous to Example 1 (production of a chain). Rollers 1 pull the wire 2 through a straightener and push the wire into guides 4. Next the wire meets a number of tools 5, 6, 7, and 8. The distance between the outlet of the guides 4 and tool 5 is short enough to avoid buckling of the wire. Tool 5 is provided with a slot on its tip which bends the wire, creating a certain curvature and inclination of the wind. Tools 6, 7, and 8 serve to define the winds more exactly. By adjusting the angles φ of the tools and the positions along the $Z–Z$ and $X–X$ axes, one obtains springs of different diameters D and pitches t. If these parameters are tuned and then left unaltered throughout the winding process, we obtain a spring of constant dimensions. If at least one of the parameters (say the positioning along the $X–X$ axis) is modified during winding, we obtain a spring of variable pitch. By changing the position along the $Z–Z$ axis during the process, we can obviously produce a spring of variable diameter. When the required number of winds is made a command is given to a cutter 9 which then moves down and cuts the wire against a support 10. From this moment a new cycle begins and a new spring of identical parameters is produced. The previous spring falls freely into a receive box.

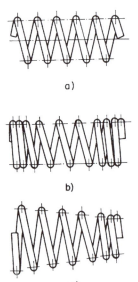

a)

b)

c)

FIGURE 2.3   Several types of springs produced by an automatic process.

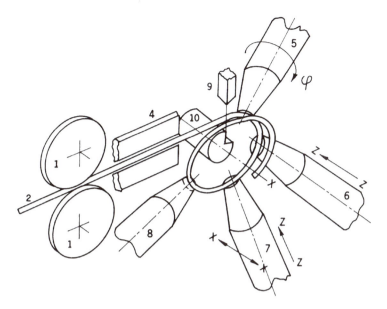

FIGURE 2.4 Layout of a spring manufacturing machine.

Example 3

Aneroid barometers are often used for pressure measurements and as altimeters. Figure 2.5a) shows a possible design consisting of two corrugated membranes. To make the description more specific we will define the diameter of the device D to be 60 mm (such barometers are useful, for instance, in meteorological probing of the upper layers of the atmosphere). The two membranes are sealed hermetically connected along the perimeter. The sealing or connecting techniques can vary: soldering, welding, or gluing. A vacuum is created in the inner volume of the aneroid barometer. This can be achieved after assembly of the membranes by making an opening somewhere in one of the membranes and then pumping out the air. Another way is to assemble the membranes inside a vacuum chamber, which involves automatic soldering or welding of the membranes. This example will illustrate several important steps typical of the conceptual design stage. To return to the aneroid barometer, it consists of two membranes which may be beryllium bronze or alloy-treated steel (the alloys usually consist of chromium, nickel, and titanium). For simplicity we will henceforth refer to the latter as steel membranes. The designer of the barometer is weighing the pros and cons for these two materials. With regard to bronze the advantages are: it is more easily stamped than steel at the stage of membrane production; and it is easily soldered by quick solder, usually based on tin alloys. Its disadvantage is that the elasticity modulus of the material depends to a significant extent on temperature involving the use of a special thermocompensating device connected to the aneroid barometer. Let us look at graphs showing the characteristics of an aneroid barometer (Figure 2.5b)). When the pressure p outside the device changes, the interval h between the membranes changes too. This dependence must stay linear within a certain range of pressure changes. What happens in the case of bronze is that the straight line changes its location and inclination when the temperature of the aneroid barometer changes.

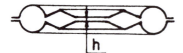

FIGURE 2.5a) Aneroid barometer sensor.

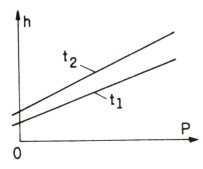

FIGURE 2.5b) Pressure versus displacement characteristics of an aneroid barometer.

The advantage of steel membranes is that their elasticity modulus is practically independent of temperature changes and thus no thermocompensators are needed. However, steel is not suitable for soldering, and the only way to join steel membranes is by welding. Among possible welding techniques, it would seem that the most effective one is so-called seam resistance welding. This process requires much more sophisticated equipment than soldering does.

There are differences between the two materials with regard to stamping as well. The stamping properties of steel alloys are poorer than those of bronze. However, this difficulty can be overcome by using specially designed stamps. Nowadays welding of membranes can be carried out by laser beam, in which case it is a matter of indifference what material is selected. (Because resistance welding depends on the specific electrical resistance of the materials being treated, bronze obviously has a lower resistance than steel. The lower the specific resistance, the worse the conditions for resistance welding because of the smaller amounts of heat produced at the contact point. Therefore resistance welding is not suitable here.) In this case we have to set the cost of these two welding approaches against their productivity.

Example 4

The rapid development of electronics in the 1950s spurred the invention of printed circuits. Further automation of electronic equipment production led to the development of systems for automatic assembling of electronics items on printed boards. The last step in this process is soldering of the leads (the contacts of the electronic items) to the corresponding places on the printed circuit. This too must be done automatically. There are several possible methods of doing this:

1. Immersion of the lower side of the board containing the above items into a molten tin pool; this, of course, means protecting the surface of the molten tin from oxidation, for instance by carrying out the process in an atmosphere of inert gas (shown schematically in Figure 2.6).
2. Setting up a wave of molten tin of a certain height; the printed board is then moved tangentially so as to touch the wave. This is shown in Figure 2.7. The

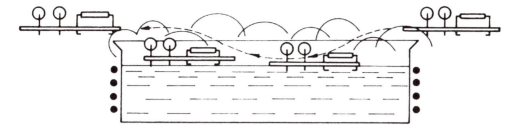

FIGURE 2.6   Soldering of printed circuits—immersion method.

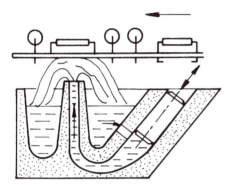

FIGURE 2.7   Molten tin wave created by a pump.

wave is produced by means of a special pump made, for instance, of ceramic. This method, which uses molten tin from the deep layers of the pool, obviously provides a clean stream of tin.

3. The wave is produced by magneto-hydrodynamic means, with the device shown schematically in Figure 2.8. It consists of a specially shaped pool where the tin is located. The bottom of the pool has two hollow protuberances. The poles of

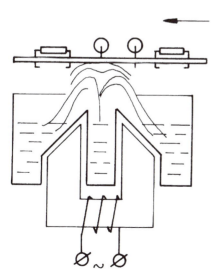

FIGURE 2.8   Molten tin wave created by magneto-hydrodynamics.

a correspondingly shaped electromagnet are inserted into these cavities from below. The electric coil of the electromagnet is fed by an alternating current of say 50 or 60 Hz. The tin creates a secondary wind in which the current reaches very high values (recall that the ratio of the currents in the two coils is inversely proportional to the number of winds in every coil). This current heats the tin and brings it to the melting point. On the other hand, due to the electromagnetic forces which appear as a result of the interaction between the alternating current and the magnetic field, molten tin is thrown out of the gap between the protuberances, creating a steady wave of fresh (clean) tin.

Example 5

The Sendzimir process is used to galvanize a steel strip. The strip is unwound from a coil 1 as shown in Figure 2.9. The first position 2 is for removing the oil and grease adhering to the strip; this is achieved by intensive heating leading to oxidation of the dirt. The second position 3 is for annealing the strip, the oxides being reduced by ammonia. In the third position 4 the strip is cooled to 500°C and afterwards immersed in a melted zinc bath 5. The molten zinc is maintained at a temperature of 450°C. On leaving the bath the strip can either be coiled 7 or cut by a special cutter 6. (How Things Work 2: The Universal Encyclopedia of Machines. Granada Publishing Ltd., 1975.)

Analyzing the foregoing examples, we arrive at some very important conclusions:

1. The design of the product influences the manufacturing process;
2. The same product or operation can be obtained by different manufacturing processes.

A manufacturing process can usually be broken down into a number of operations, each requiring a specific tool and time interval; this is clearly illustrated by our first example. The first operation is usually feeding. Sometimes it is the feeding of a continuous material like the wire in Examples 1 and 2, or the metal strip in Example 5; sometimes it is feeding separate parts, like the membranes in Example 3. Next come functional operations like cutting the wire, bending, soldering, welding, etc. The last operation usually consists of extracting the ready product or part from the machine. The total time needed for one cycle to be completed is at least the sum of operational times. To this sum we have to add non-operational time intervals, which are needed:

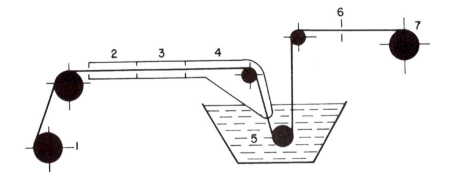

FIGURE 2.9 Processing layout of steel strip galvanization.

1. To shift the part from one operational position to the next;
2. To avoid interference between one operation and the next.

Obviously, these time intervals dictate the productivity of the machine. [The reader is again referred to Formulas (1.1–1.7), which deal with these concepts and with the time components influencing system productivity.]

2.2   How Does One Find the Concept of an Automatic Manufacturing Process?

There are several approaches to this problem. *The first approach* is to try to *copy the manual process*, that is, to imitate the manipulations of the human hands and arms by mechanical means. Weaving is a good illustration; another example is the forging hammer. As the blacksmith manipulated his hammer, so the first mechanical hammers (driven by horse or waterwheel) made practically the same monotonous up and down motions. Direct copying is, in general, less successful than roundabout solutions. For example, the forge hammer can be fastened to the piston rod of a hydraulic cylinder to effect reciprocating vertical (or, if needed, horizontal) motion. Combination of steam power with hydraulics permits us to achieve a force of the order of thousands of tons. By contrast, a relatively small force can be developed in pneumatic or electromagnetic hammers for fine smithery. The explanation lies in the fact that human hands work in concert with other senses such as touch, sight, and feel. In addition, the number of degrees of freedom possessed by the human hand develops a flexibility that has never been achieved by any mechanism. Thus the only way to succeed is to modify the conception of the manual process. Sewing is another well-known example; instead of the single thread and the needle with a hole at the tail used in manual sewing, two threads and a needle with a hole at the tip are used in mechanical and automatic sewing.

Consider, for instance, the problem of producing aneroid barometers by a soldering technique (Example 3). The manual process begins by cleaning the membrane's flange surface by soldering flux. This is usually done with a wad moistened with the flux. The next step is to tin-plate the flanges of the membranes, which is done with the usual soldering iron. The membrane is held in one hand while the other hand manipulates the wad or soldering iron. Both hands can gauge pressure and relative movements, while the eyes control all procedures; none of this is true in a mechanized process, and so the approach must be altered. Figure 2.10 shows a possible layout for a four-stage tin-plating process for beryllium–bronze membranes: In the first position a membrane is taken from a magazine 1 containing 500–800 membranes (remember, the thickness of a membrane is about 0.2 mm) by means of a vacuum suction cup 2. The four suction cups are in permanent rotation. In the second position the membrane is pressed against rollers 3 which free rotate, the lower part of the rollers being immersed in the flux liquid. Thus, the flange of the membrane gets cleaned. The third position is the soldering one. A soldering iron 4 made of copper rotates in a pool of melted tin covered by a special flux which protects the tin from oxidation. The soldering iron is electrically heated from the inside and thus the flange is tin-plated by rotation of both iron and membrane. The fourth and last position is meant for extraction of the tin-plated membrane, which falls freely when the vacuum is disconnected

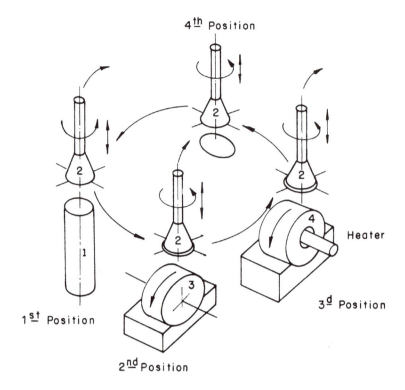

4th Position

Heater

3^d Position

1st Position

2nd Position

FIGURE 2.10   Possible layout for tinplating of bronze membranes.

from the cup 2. To take up a new membrane and to clean and solder the flanges, the vacuum suction cups carry out a reciprocating vertical motion at the appropriate time.

Here we come close to the *second approach* to the question framed in the title of this section: *copy existing concepts in the same industry or in adjacent industrial fields.* In the example we just considered, for instance, the idea of a rotating soldering iron was borrowed from an automatic machine for soldering longitudinal seams in cans for canned food. The printing technique originally invented for books was transferred to the textile industry for printing on fabric (or perhaps the other way around), and, more recently, the same technique was transferred to electronics, when printed circuits were introduced instead of wiring. However, it is interesting to note that automatization of the writing process for small amounts of copies involved basic alteration of the printing concept, and that the technical history of the typewriter bears only a broad resemblance to printing. This is a powerful approach to concept search. Take the following example: anyone needing to produce a metal part or to alter its shape will first consider cutting. To analyze this possibility, the designer or technician has at his disposal a lot of information about the cutters, about the work conditions as a function of the material under consideration, about dimensions, etc. Moreover, there are different ways to generate relative movement between the cutter and the blank. The designer need only compare the possibilities, compute the approximate costs and productivities and derive the conclusion; the menu of concepts is spread out before him. For instance, when toothed wheels must be produced automatically in large amounts, the menu of manufacturing methods consists of the following possibilities:

1. *Cutting*, first of the body of the wheel and next of the teeth. In turn, there are at least four ways to cut the teeth: by gear cutter, gear hob, gear shaping, or pinion-shaped gear cutter.
2. *Casting*, either of the body (the teeth being cut) or of the whole wheel. In turn, this can be (depending upon the material): loam or sand mold casting, metal (chill) mold casting, machine molding, lost-wax process or investment casting, die casting, or precision casting.
3. *Manufacturing by powder metallurgy.*
4. In some special cases—*sheet metal stamping.*
5. *Knurling* the teeth on a blank made of corresponding alloy.

Of course, the choice of an appropriate processing method depends on design requirements and restrictions (and vice versa). Thus the choice of a material will be dictated by strength, durability, accuracy, and other special conditions; obviously, if surface and accuracy call for grinding and superfinishing, operations which meet these requirements will have to be selected. However, as a rule the restrictions imposed by design do not detract from the point made above.

Finally, the *third approach* to the search for manufacturing concepts involves *scientific or engineering research*. This situation arises when:

- A completely new product is under consideration and no prototypes can be found in any technical field or industry;
- The existing prototypes of the processing techniques are too slow, too expensive, yield unsatisfactory quality, or require inaccessible materials or techniques.

To illustrate the first case, consider the following examples. Laser-beam machining is based on melting or vaporizing the cut material along the seam with an intense beam of light from a laser. Such beams can also be used to produce small-diameter holes. Direction and control of the beam is relatively easily accomplished by a combination of optics and computer techniques. This technology has only recently been introduced into industry and has opened new economic and industrial vistas. For instance, thanks to the absence of mechanical contact between tool and blank, the deformation is minimal, thus improving the accuracy of the product. The holes can be produced under any inclination (which is difficult to achieve by drilling). The absence of inertia and the electrical nature of the tool makes the process susceptible to electronic control and therefore very swift and accurate.

Electrical-discharge machining introduced an effective method of processing holes and cavities of almost any shape in almost any sort of metal regardless of hardness. This is of great importance in the manufacture of dies and molds. Another advantage of this method is that it is practically the only way to produce a hole with an arched axis. This technique involves direction of high-frequency electrical spark discharges from a metal or graphite tool. The tool serves as one electrode while the blank is the other electrode, both electrodes being immersed in dielectric liquid. A special mechanism maintains a spark gap of about 0.015 to 0.5 mm. Spark discharges melt or vaporize small parts of the blank under processing. This technology makes it possible to carry out processing which earlier required special complications in design, as illustrated by Figure 2.11. To get an opening of the form shown here using the old technology, the part had to be made in two sections (Figure 2.11b). The tool produces the opening by a rotating movement around its center, thus providing the needed curvature.

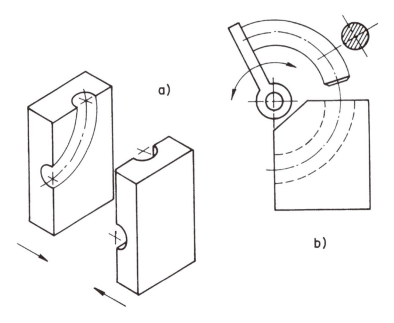

FIGURE 2.11 Layouts of process for producing curved openings.

Another unique application of laser-beam machining is in steel surface hardening. Here hardening is effected by intensive point heating and intensive cooling— quenching when the beam is removed from the point. The procedure leads to the creation of a very thin, glass-hard layer of material. No other known method (including high-frequency hardening) can achieve the same effect.

By the use of both laser-beam and electrical-discharge techniques one can produce openings of very small diameters in various materials regardless of hardness. The openings can be made at any inclination to the surface of the blank, especially in the case of laser-beam machining. To achieve such openings by drilling would be highly complicated—if at all possible.

Considering the second approach to the search for manufacturing concepts, we can formulate some kind of general principle. This principle states that the development of technical concepts (and perhaps that of other concepts as well) goes through a stage of accelerated improvement and perfection, followed by a stage of slow, expensive, and tedious stagnation. This is illustrated graphically in Figure 2.12. Here the horizontal axis gives the time (in years, let us say), while the vertical axis gives a criterion of effectiveness of the concept under consideration. When the near-exponential curve becomes almost horizontal (i.e., further investment of effort in concept improvement becomes unproductive), a new concept must be found. This is typical of situations calling for the second approach.

Other methods used consciously or unconsciously to discover new concepts include the rehabilitation of an old, "forgotten" technique. In 1902 a U.S. patent was issued to W. E. Heal for a process for manufacturing sheet and plate glass. The process consists of pouring the molten glass from the melting tank into a receiver containing another melted material of greater density than glass. In 1905 a similar patent was issued to

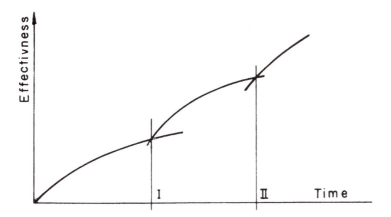

FIGURE 2.12   Concept effectiveness versus time.

H. K. Hitchcock for the same purpose, describing a continuous liquid bed for a plate glass formed on molten metal. The problem with these solutions is that for high-quality flatness and transparency, expensive processes must be used to grind and polish the raw glass sheet in order to correct defects due to the rollers which support the sheet during cooling. Obviously, the size of the rollers and the intervals between them will influence the quality of the sheet: the smaller the rollers and the smaller the intervals between them, the smaller the deflection of the sheet. From the idea of reducing the size of the support rollers was derived the concept of molecular supports, where one liquid, glass, supported by another liquid, molten metal (say tin; Figure 2.13 shows the possible development of the idea). For about 50 years this brilliant approach was almost forgotten because of its being technically premature. It was implemented after 1959 at Pilkington's British company, only after a lot of technical improvements had been made in the classical process of flat glass sheet production, such as continuous flow of molten glass from the furnace.

Another way of revolutionizing a concept is called inversion. For instance, an automatic lathe must have a feeding device ensuring minimal time expenditure on chuck reload after completion of a piece of work. This usually involves a rod magazine that feeds a rod into the chuck automatically and which, on completion of the piece, replaces it automatically by the next rod. The breakthrough solution here (especially when the processing time is small and comparable with the feeding time per manufactured unit) is to use a coil of material instead of separate rods. Such a coil of continuously wound rod or bar makes it possible to load the machine with much more material and to save time otherwise lost in rod changing and magazine refilling.

FIGURE 2.13   Development of the idea of liquid glass supported by another liquid.

However, this conceptual solution does not permit rotation of the spindle, it being too cumbersome to rotate a coil of, say, one meter diameter. We have to invert the process. The solution is to rotate the cutters, drills, etc., around the rod, which is kept immobile in the chuck.

Exercises

Explain the concept underlying the following automatized manufacturing processes:

1. Sewing machine.
2. Meat-chopping machine.
3. Automatic record-player.
4. Slot-machines of different kinds.
5. Automatic labeller (for bottles, say).
6. Machine for filling matchboxes.
7. Machine for wrapping 10 boxes of matches in a parcel.
8. Machine for producing nuts of, say, 5-mm thread diameter.
9. Machine for assembling a ball bearing.
10. Machine for sorting the balls of a ball bearing into, say, 5 groups corresponding to their size tolerances.

2.3   How to Determine the Productivity of a Manufacturing Process

We have thus determined the concept underlying our process, and the next step is to estimate the main parameters characterizing the productivity or efficiency of our concept. The way to do this is to construct the so-called sequence or timing diagram. The first example considered in Section 2.1 will illustrate the procedure. The diagram is given in Figure 2.14. Here each horizontal line corresponds to a specific mechanism. The first line describes the behavior of the wire-feeding mechanism, the second line that of the wire-cutting mechanism, and the third that of the mechanism for horizontal bending of the link. It is convenient to consider this bending as a two-stage procedure. The first stage (3a) is creation of the horseshoe-like shape, which involves only a rightward movement of the tools 5 (this procedure is called *swaging*). The fourth line corresponds to the action of the vertical bending mechanism, and the fifth to the support 4, which must be countersunk at a certain moment to make way for some other tool. The sixth line describes the action of the mechanism 9 whereby the link is pushed towards the opening (where it falls to the lower level and causes the assembly of the links into a chain), and the seventh line corresponds to the last operation, closing the links. The vertical axis of this diagram usually represents some kinematic value: speed, displacement or (less frequently) acceleration. The scale of these values can be different for each mechanism. The horizontal axis represents the angular values ψ (because of the periodical nature of the process) or time t, which is related to the angles ψ through the velocity ω of the distribution shaft as follows:

$$\psi = \omega t.$$

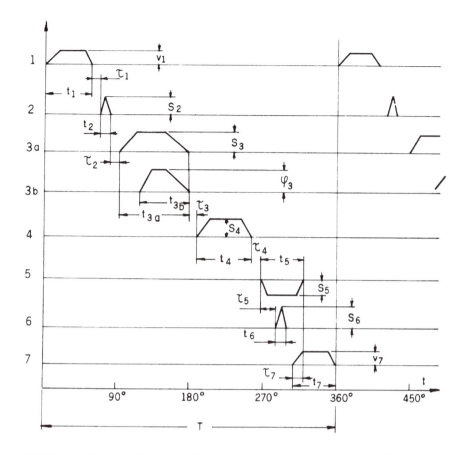

FIGURE 2.14 Timing diagram of the chain manufacturing machine (Figure 2.1).

For the example under discussion, the diagram in Figure 2.14 can be described as follows: The feeding mechanism, which consists of two rollers (see Figure 2.2), is actuated for about 75° of the revolution of the distribution shaft. During this action the speed of the feeding rollers grows from 0 to some nominal value (the acceleration takes about 15° of the period); this nominal value is kept constant for about 45° and afterwards, during 15° of the period, it decreases to 0 and the rollers remain immobile for the rest of the period. The cutter carries out the fast cutting movement during approximately 5° to 7° of the period, and after about the same angle it returns to its initial position. (Note that in the first case we are referring to speeds and in the second case to linear displacements.) The third mechanism, as explained, can be regarded as acting in two stages: the first (line 3a) consists of pure linear movement of the tools, which together with acceleration and deceleration takes about 120° of the period, and the second stage (Line 3b) of a combination of linear and angular movements of the tools. (The diagram describes linear displacement in the first stage and angular displacement in the second stage.) Now comes the turn of the vertical bending by punch 7, which is effected by vertical displacement of the punch and takes about 50° of the period. Note that in every bending process we envisage a time interval where the tool rests; this is done to provide stress relief in the bent material of the link. This process takes about 10° of the period. The fifth line shows the movement of the support 4 which must lie beneath the surface so as not to interfere with the pusher 9 as it shifts the semi-ready link towards and through opening 8; timewise, the sixth line corresponds to all the movements mentioned in this connection. Lastly, the seventh line gives the action of wheels 10 and 11. Here there are two alternatives:

1. The wheel rotates at constant speed. Thus, during the period T it pulls the chain over the length of one link.
2. The wheel provides interrupted motion; after the corresponding time interval the wheel reaches the speed V required to move one link, then rests for the remainder of the period (solid line in Figure 2.14). This takes up 55° of the period.

We have defined the duration of each operation in angular units, the whole cycle or period obviously taking 360°. To transform the angles into time units we have to define the time taken up by the total period T. To design a highly productive machine we desire T to decrease. On the other hand, certain restrictions limit the minimal value of T. These restrictions are of various kinds. One of the most important sources of restrictions is the kinematics and dynamics of the drives, whether purely mechanical, pneumatic, hydraulic, or electric. Another class of restrictions applies to purely physical (or chemical) events. For instance, in the example above (fifth line), we mentioned that the operation includes the time needed for the semiready link (Figure 2.2) to fall through opening (8) and connect up with the previously produced links of the chain. This time t^* does not depend on engineering techniques; it is in practice a function only of the distance h through which the link falls. Thus,

$$t^* = \sqrt{\frac{2h}{g}}, \qquad [2.1]$$

where $g = 9.8 \text{ m/sec}^2$.

For another illustration let us take Example 3, the welded aneroid. We saw that seam-resistance welding was the most appropriate technique here. It involves producing a line of welded points such that each point partly covers the next. Thus, if the diameter d of one point is 0.25 mm and the overlap η (which provides the safety factor) is, say, 0.3, and since, as noted, the diameter D of the membranes equals 60 mm, the length L of welded seam is

$$L = \pi D \cong 190 \text{ mm}, \qquad [2.2]$$

and the number of welding points N is

$$N = (L/d) \cdot (1 + \eta) = (190/0.25)(1 + 0.3) \cong 1000. \qquad [2.3]$$

The generator of the electric pulses, correspondingly shaped and amplified, is usually controlled by the alternate current of the industrial network. The frequency f of the welding pulses is about 50 Hz and therefore the time t^* needed to get N pulses can be calculated from

$$t^* = N/f \cong 1000/50 \cong 20 \text{ sec.}$$

In practice the seam overlap should be such that $t^* = 25$ sec. Thus, at least this is the minimum time needed to produce one aneroid. These two illustrations are simple enough to be easily solved by direct analytical approaches, although to do so an engineer would clearly need to know the general laws of physics and related disciplines.

However, in some cases the necessary information must be obtained experimentally. Take bronze aneroids assembled by soldering (Example 3): the heating and cooling times required to ensure proper coating with tin in automatic plating of the membranes cannot be calculated analytically, with the necessary accuracy. The only way to get reliable results is to carry out experiments under conditions closely approximating the machine under design.

Similarly, in the tin-plating of printed circuits after the electronic items have been mounted on the bases (Example 4), the time of exposure of the circuits to the tin wave as well as the cooling time of the circuit have to be determined experimentally.

The timing diagram indicates where time can be saved to increase productivity. In Figure 2.14 the auxiliary time intervals denoted by τ serve to prevent collisions between tools or between kinematic elements and when precisely defined can be reduced. Reducing the auxiliary times can be a very effective way of raising productivity. Another time-saving device is to reduce operation time intervals. For instance, the rate of wire input can be increased, thereby reducing the time t_1; however, this entails increasing the driving power of the feeding rollers. Similarly, we can shorten the strokes of the punch 7, tools 5, etc., reducing t_3 and t_4 correspondingly. However, this makes it necessary to apply additional driving power, which entails higher accelerations and decelerations and, in turn, heavy dynamic loads on parts in executing the desired movements. Thus, the timing diagram brings us to the next step in the design procedure, namely, designing the kinematics of the automatic machine, or the kinematic layout. The layouts for cyclic and continuous manufacturing processes are different. The advantages of continuous processes were discussed in Section 1.4 of Chapter 1. Example 2 illustrates a continuous process. Note, however, that certain processes involve a mix of concepts. For instance, Example 2 (manufacturing of cylindrical springs) illustrates the combination of cyclic mechanisms (wire cutting tools; Figure 2.4) and the continuous process of feeding the wire and bending it into a spring. The pitch-controlling mechanism involved in the production of spiral springs of variable pitch is also cyclic. It must be mentioned here that at the stage of manufacturing-process design we are concerned neither with the means which carry out the displacements, nor with speeds. forces, sequences, and durations; we just define what these parameters should be and estimate their values. It is at the kinematics stage that we deal with how to achieve our desired objective.

There are two ways to draw the timing diagram: one we have already discussed in the above example of chain manufacture and we can call it the *linear approach;* it is the one given in Figure 2.14. The other is to use the *circular approach.*

As one might expect from its name, this diagram is circular in shape. It is convenient to use inasmuch as it graphically illustrates the breakdown of the time period into specific operations, auxiliary actions, etc.; see the diagram for a washing machine in Figure 2.15. However, the disadvantages of this kind of timing diagram are as follows:

- It is difficult to render displacements, speed changes, temperature changes. etc. In fact, such diagrams are generally used to show on-off actions.
- Because of the different diameters of the circles that make up the diagram, the arcs corresponding to equal angles are of different length. This psychologically disturbing feature interferes with evaluation of the diagram.

After the duration of the sequence of operations has been in some way determined theoretically or experimentally, we may still conclude that the production output is

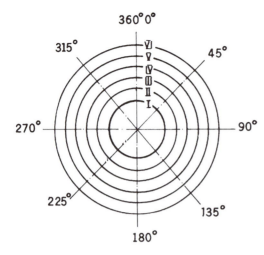

FIGURE 2.15 Circular timing diagram.
Washing machine: 1) Rinsing; 2) Heating;
3) Water inlet; 4) Water outlet; 5) Drying;
6) Washing powder insertion.

unsatisfactory. A possible approach which is often used is to carry out the different operations simultaneously. The general case for this principle is illustrated by the diagram in Figure 2.16. Let us suppose that the manufacture of some product by a certain machine takes the time T and consists of three steps A, B, and C; by running three such machines the product can be manufactured within an average time t where $t = T/3$. When the three machines are combined into a single machine, the advantage will be even greater because one machine (even a complicated one) is cheaper, takes less space, etc. Figure 2.16a) shows the case where the three operations A, B, and C are

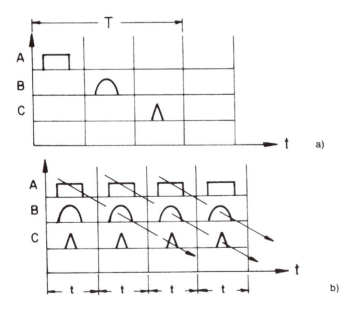

FIGURE 2.16. Timing diagrams. a) Operations run in series
in a machine; b) Simultaneous execution of a number of
operations in a single machine.

run in series one after the other (this is the usual way to think about a manufacturing process). Figure 2.16b) illustrates the simultaneous running of the operations to economize the production time. The example shown in Figure 2.10 conforms to this approach.

Exercises

Show the timing diagrams for the following automatic processes:

1. Four-stroke internal combustion engine. Use the circular approach.
2. Washing machine (any washing regime). Use the diagram given in Figure 2.15.
3. Sewing machine. Use the linear approach.
4. Spring-producing machine (Figure 2.4) for the spring shown in Figure 2.3c).
5. Automatic record player.

2.4   The Kinematic Layout

After the processing layout and timing diagram are finished comes the turn of the kinematic layout. At this stage the designer has to choose the means by which to effect the required movements of tools as defined in the processing layout. A variety of mechanical concepts are at the designer's disposal for this purpose. These may be known and established concepts; on the other hand, sometimes new concepts must be found. Any mechanism chosen for carrying out a specific movement needs a drive. We will now consider and compare those most commonly used, beginning with mechanical drives. (See Table 2.1.)

TABLE 2.1   Mechanical Systems

Advantages	*Disadvantages*
1. Relative clarity of the mechanical layout (compared with an electric circuit, for instance).	1. For spatially extended constructions, cumbersomeness of the kinematic solution.
2. Absence of the need for a specific power supply.	2. Difficulties in creating relatively rapid movements.
3. Possibility of implementing different kinds of movement or different displacement laws.	3. Difficulties in generating very large forces.
4. Relative ease of achieving accurate displacements.	4. The necessity for special protective devices to avoid breakage of expensive links.
5. Rigidity of the mechanical links.	
6. High accuracy of ratios in movement transmission.	

Some words of explanation: what we mean by clarity when it comes to mechanical systems may be demonstrated by the following extreme example. When a mechanism is rotated slowly, almost everyone is able to understand how it works, its logic; and when broken or out of order it is relatively easy to locate the broken or worn part simply by looking at it. By contrast, in electronic systems, extensive measurement and special knowledge are needed to pinpoint defects in, say, a mounted plate. This is why we generally replace the suspect plate by a new one in electronic systems, instead of replacing a part or even repairing it as in mechanical systems. A purely mechanical system is usually driven at a single point, the input, and no additional energy supply is needed along the kinematic chain; in some cases, however, a multidrive system is more effective. The layout will then include a multiple of electromotors or hydrocylinders. (Pneumatic and hydraulic systems, for instance, require air or liquid supply to every cylinder or valve along the system.)

We have at our disposal a wide range of known and examined solutions for achieving various movements. Moreover, when electric, hydraulic, or pneumatic drives are used to effect complex motions, they are generally combined with mechanical devices. This is because the latter make it possible to achieve accurate displacement thanks to the rigidity of mechanical parts.

Thanks to the use of high pressure, the transmission of large forces to considerable distances in hydraulic systems can be realized in small volumes where a purely mechanical solution would entail the use of massive parts, massive supports, massive joints, etc. The fact that liquid consumption is easily controlled ensures fine control of the piston speed, while the nature of liquid flow ensures smoothness of piston displacement. (See Table 2.2.) Thanks to the flexibility of the pipes almost any spatial and remote location of the cylinders can be arranged with ease using pipes made of flexible materials, including alteration of the location and turning of the elements when the machine is in use. On the other hand, once designed, a mechanical system is difficult to modify (at the least any modification would require special devices). Rigorously coordinated displacements between remotely located elements are problematic. Spatially oriented displacements of different elements require specially costly means. Hydraulic systems can use relatively cheap safety valves to prevent breakdown of elements due to acci-

---

TABLE 2.2 Hydraulic Systems

Advantages	*Disadvantages*
1. Possibility of generating very large forces.	1. Difficulties resulting from the use of high pressures.
2. Possibility of carrying out slow, smooth movements.	2. Mechanical supports or complicated control layout required for accurate displacements.
3. Relative simplicity of spatial location of moving elements.	3. Leakages can influence the pressure inside the system.
4. Possibility of changing velocities of displacements in a smooth manner.	4. Variation of the working liquid's viscosity due to temperature changes.
5. The fact that it is not explosive (pressure drops sharply when fluid leaks out).	

TABLE 2.3   Pneumatic Systems

Advantages	*Disadvantages*
1. Relative ease with which complicated spatial location of moving elements can be achieved (e.g., pipes can be bent into any shape).	1. Difficulties in effecting displacements subject to specific laws of motion.
2. Relative ease of execution of rapid movements (dependent on the thermodynamics of gases).	2 Need for mechanical supports to ensure accurate displacement.
3. Relative ease of generation of large forces (which are the product of the pressure and the area of the piston or diaphragm).	3. Dependence of operation on pressure in the piping.
	4. Need for special auxiliary equipment.
	5. Need for means to avoid leakage.
	6. Danger of explosion.

dental overload, whereas mechanical devices for the same purpose are much more cumbersome.

Advantages and disadvantages of pneumatic systems are summarized in Table 2.3. Two points are particularly worth emphasizing:

1. Rapid, even, long-distance displacement is easily achieved, thanks to the thermodynamics of compressed air;
2. For this very reason special measures must be taken to prevent explosions (in comparison with hydraulics).

The advantages of electrical and electronic systems far outweigh their disadvantages. (See Table 2.4.) The combination of electrical drive (servomotors of various types, servomagnets, servovalves) with electronic control at varying levels of intelligence (including computerized systems) makes them very attractive when flexibility is necessary. It is, of course, possible to combine all the drives described above in a single system so as to exploit the advantages of each. However, it is recommended that no more drive types be used than are justified. To illustrate this point, consider the kinematic layout of an automatic machine for producing springs. Obviously, a number of alternatives can be offered. We begin with the layout of a purely mechanical system driven by an electromotor (Figure 2.17). The motor 1 transmits motion by means of a belt drive 2 to a worm speed reducer consisting of a worm 3 and wheel 4. The latter drives the shaft 5 on which the wire-pulling wheel 6 is fastened. The other wire-pulling wheel 7 is also driven (to provide reliable friction) by a pair of gears 8. The shaft 5 serves

TABLE 2.4   Electrical Systems

Advantages	*Disadvantages*
1. Spatial locations of working elements easily achieved.	1. Problems of reliability.
2. High rate of automation easily obtained.	2. Need for relatively well-educated maintenance personnel.

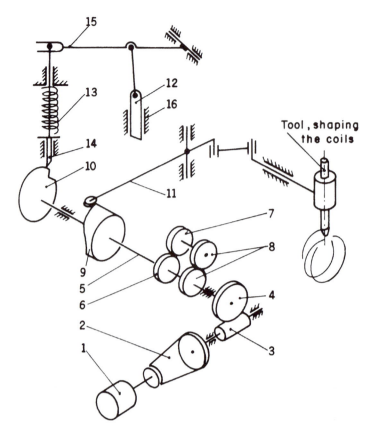

FIGURE 2.17 Kinematic layout of automatic spring
manufacturing machine (nonflexible case).

as the main motion-distribution shaft (MMDS). Cams 9 and 10 are fastened onto it so
as to create a certain phase angle between them. Cam 9 moves the coil-producing tool
by means of a lever system 11 so as to impart the right pitch to the spring. The other
cam 10 controls the wire cutter 12. The layout provides the following wire-cutting
process. During a processing period T cam 10 compresses a spring 13 on the rod of
the follower 14; when the follower 14 reaches the highest point on the profile it jumps
down from the step. At this moment spring 13 actuates the levers 15 and the cutter 12,
which slides along guides 16.

Note that the layout need not be kept to scale; the main point, when designing the
kinematic layout, is to include every element or link of the transmission and mechanism.
At this stage, too, the ratios, speeds, displacements, and sometimes accelerations must
be defined. The layout should also show every support and guide. Thus, the ratios of the
belt drive 2 (see Figure 2.17) and worm-speed reducer (3 and 4) must be specified in the
layout. For instance, if the initial speed of the motor 1 is about 1,500 RPM and the cycle
duration $T = 1.2$ sec, the belt drive and the reducer together must provide the ratio:

$$i \cong \frac{1500 \cdot 1.2}{60} = 30. \tag{2.4}$$

This ratio can be apportioned between the belt drive and reducer in, say, the following way:

$$i = i_1 \cdot i_2 = 1.25 \cdot 24 = 30,$$ [2.5]

where the ratio of the belt drive $i_1 = 1.25$ and that of the worm reducer $i_2 = 24$.

The gears 8 obviously deliver a 1:1 ratio, the wire-pulling rollers 6 and 7 being of identical diameter. Another important point we have to mention here is that the kinematic solution discussed above is not flexible. For instance, to add more coils to the end product we must increase the diameter of wire-feeding rollers 6 and 7, causing more wire to be introduced into the machine and producing more coils per spring. To change the pitches along the spring, cam 9 must be replaced by a corresponding cam. Note, however, that substitution of these elements of the kinematics entails relocating corresponding shafts and their bearings, in addition to relocating the guides of the wire 4 (see Figure 2.4). Briefly stated, the proposed concept restricts the flexibility of the machine. The only parameters which are easy to modify are the diameter and constant pitch of the spring, thanks to the design of the supports (5, 6, and 7 in Figure 2.4).

This difficulty can be overcome by adopting a different concept of the kinematic layout (assuming a higher degree of flexibility is needed). One possible solution is represented in Figure 2.18. Here the systems of the automatic machine are kept separate, the feeding mechanism having its own drive while the cutter and bending tools are moved independently. In more detail it can be explained in the following way. The motor 1 drives the feeding rollers 3 through a worm-gear reducer 2: as in the previous case the rollers are engaged by a pair of cylindrical gear wheels 4. An electromagnet 5 drives the cutter 7 along guides 8 with the help of a lever 6: the return of the cutter to its initial position is accomplished by a spring 9. The tools 10 shaping the spring (one or several) are fastened in corresponding guides 11. These guides can be moved along axis X (the tools are fixed to the guides by means of bolts 12). An independent motor 13 is used to carry out this movement. This motor drives link 14 which consists of a nut restricted in its axial movement and therefore able to realize pure rotation only. The thread of this nut is engaged with a lead screw 15. The latter, in turn, is restricted in its angular (rotational) motion by means of a key 16. Thus this screw realizes a pure axial motion, driving also the tool 10. The designer decides how many tools are to be driven independently. Analyzing this new kinematic layout and comparing it with the previous one, we arrive at a significant conclusion. The second layout permits easy modification of the duration of action of the motors, thereby delivering any (reasonable) length of spring, coil number, or pitch. For this purpose the control of the two motors and electromagnet must be correspondingly tuned. Obviously, instead of electric motors, hydraulic or pneumatic drives could be installed; the control unit would then have to be designed to fit the nature of these drives.

Here we must return to Chapter 1, Section 1.2, and analyze the examples given there in terms of the diagram given in Figure 1.5. The purely mechanical kinematic layout (Figure 2.17) clearly belongs to level 5. Indeed, the energy source is a motor and control is carried out in series by the kinematics (transmission, cams, and levers) of the system. Considering the case given in Figure 2.18 we see that the machine consists of at least three systems of level 6 (Figure 1.5). Two motors (1 and 13) and an electromagnet 5 impart the driving energy. In addition the motor 1 drives the program carrier, which consists of a conical speed variator comprising a cone 17 and friction-

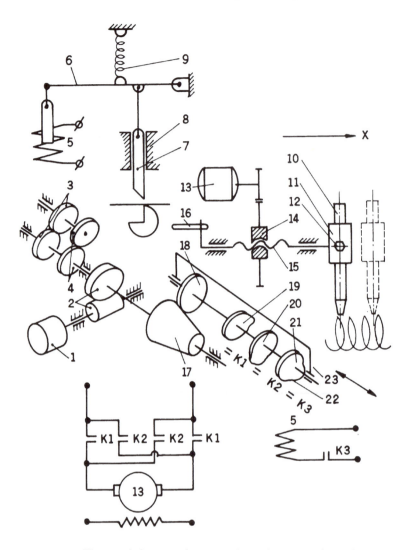

FIGURE 2.18 Kinematic layout of automatic spring manufacturing machine (programmable case).

ally driven disc 18. The latter rotates cams 19, 20, and 21. Each cam is provided with lobes 22. The relative positioning of these lobes can be changed in accordance with the requirements dictated by the parameters of the spring under manufacture. The task of the lobes is to actuate contacts K1, K2, and K3 (the contacts are indicated schematically under the corresponding cams). The disc 18 can be moved along axis Y, thereby altering the ratio between the cone and the disc and consequently the time needed for the cams to carry out one revolution. A stiff frame 23 is used to move the disc. (Of course, the design must provide for friction between the variator links at every relative position between the cone and the disc.) The time T of one revolution of a cam determines the time of a period of the machine. The larger the value of T, the longer the section of wire fed in during the period. The longer the wire section, the longer will be the spring (more coils) or the larger its diameter. During the revolution the cams actuate contacts K1, K2, and K3, thereby controlling the motor 13 and magnet 5.

As indicated by the electric layout in Figure 2.18 when contacts K1 are closed (directly or by means of a relay), motor 13 is brought into clockwise rotation, moving the tool 10 correspondingly, say, rightward. When contacts K2 are closed, motor 13 is reversed (shunt-excitated DC electromotors change their rotation direction by changing the voltage polarity on the rotor brushes). In turn, the electromagnet 5, which cuts the wire and thereby completes the production of the spring, is actuated by cam 21 due to its contacts K3. If the drives are pneumatic or hydraulic the control layouts will obviously include valves and pipes.

At this stage, the designer has completed the conceptual stage of the design and can pass over to pure design. No strict dividing line exists between one stage and the next (we saw that even in the earliest stages of creating the manufacturing layout we sometimes had to resort to engineering calculations), and no purely conceptual design stage exists. Nonetheless, the shift in emphasis is clear-cut enough to justify our drawing this distinction. The next step is to calculate and draw, regardless of whether this is done manually, by computer, or both. The next chapter is devoted to the selection of drives and corresponding calculations of the dynamics.

2.5 Rapid Prototyping

New production concepts of a different nature have recently been introduced into manufacturing processes. Among these concepts, some are modifications of already existing ideas, but others are completely revolutionary. As examples of the former group, we may cite computerized numerically controlled (CNC) cutting of a variety of materials, from wood to ceramics, with a laser beam and a water-plus-abrasive jet.

With regard to the latter group, we may describe the process of rapid modeling or three-dimensional processing of parts. This concept is rich in content and industrial potential, and it is therefore worthwhile discussing it in brief. It is based on a principle that has been possible to formulate largely as a consequence of the power of the computer.

The productivity of the first group of manufacturing processes mentioned above is vastly improved by the application of computers, although, at least in principle, these processes may be carried out in a manual mode. For the second group, it is *impossible* to execute the processes without a computer.

Modern manufacturing relies on a large number of molded parts made of plastics and metals. These parts sometimes have very complicated shapes and ornate surfaces. Such shapes cannot be processed on conventional machines, which makes any attempt to produce a single part of this kind very time and money consuming. For the same reason, the use of a mold to produce individual patterns, which may require changes after they are examined, is even more expensive (this is the case in which nonconventional tools are used and the process is expensive and time and labor consuming). In recent years, a new concept for providing the solution to this problem has been proposed. It is known as rapid prototyping, stereolithography, quick prototype tooling, or rapid modeling, and is described in the book *Solid Freeform Manufacturing*, by H. D. Kochan (Technical University Dresden, Germany, Elsevier Scientific Publishers).

To explain the idea underlying this manufacturing process, we use the model shown in Figure 2.19a. The model represents a helical wheel provided with specially formed

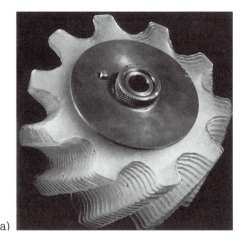

a)   b)

FIGURE 2.19 a) Illustration of a rapidly modeled subject. Pay attention to the clearly visible layers of the material comprising the wheel. Each layer is displaced by a certain angle, thus creating the image of a helical gear (here, for purposes of illustration, the thickness of the layers is exaggerated). b) Examples of patterns made by this technique before the final design (production of Conceptland Ltd., Ra'anana, Israel).

teeth, consisting of plane layers that are angularly shifted relative to one another. In other words, a three-dimensional model with a complicated shape is composed of a number of thin, planar, and simply shaped layers.

There are a number of different techniques that exploit this idea for the computer-aided processing of spatially cumbersome parts. We will describe here, in brief, the essence of the concept.

The memory of the computer is loaded with geometric information about the part to be processed so that the configuration of each thin (say, 0.3–0.5 mm) slice of the part can be numerically defined.

A possible layout for a process—based on this concept for creating lamellar bodies—for an intricate three-dimensional shape is shown in Figure 2.20. This layout consists of a vessel 1 filled with a special liquid 2, which polymerizes to a solid under ultraviolet irradiation. The surface of the liquid covers a plate 3, the vertical location of which

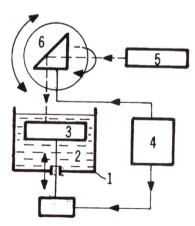

FIGURE 2.20 Layout of the rapid modeling process.
1) Vessel; 2) Polymerizing liquid; 3) Plate;
4) Computer; 5) Laser; 6) Rotating mirror.

is controlled by the system's computer 4. The ultraviolet beam generated by means of laser 5 is focussed with the aid of a mirror 6, which is also controlled by a computer, so that the beam moves on the surface of the liquid according to a given program. As a result of this operation, a thin plane layer is created with a predetermined shape. In the next step, the plate 3 moves down for a distance corresponding to the thickness of one layer, and the procedure is repeated. At this point in the process, the trajectory of the beam may be changed according to the configuration of the new layer. Thus, the body grows, layer by layer, to form a model of the desired shape. Figure 2.19b) shows examples of possible units produced in this way.

Exercises

Try to design the kinematic layout of a:

1. Sewing machine.
2. Machine for producing the chain shown in Figure 2.1 in accordance with the production layout given in Figure 2.2.
3. Internal combustion engine.
4. Domestic dough mixer (dough kneader).
5. Typewriter.
6. Mechanical toy, spring or electrically driven.
7. Machine gun.
8. Automatic record player.
9. Photocopying machine.

3

Dynamic Analysis of Drives

In this chapter we shall discuss examples illustrating the operation time computation techniques for drives of different physical natures. We begin with the simplest—a purely mechanical drive.

3.1 Mechanically Driven Bodies

The first case we shall consider in this section may be classified as a free-fall phenomenon. This is the situation which occurs, for instance, when a stack of parts moves vertically downwards in a magazine-type hopper (or dispenser). The simplest example is presented in Figure 3.1, which shows a body falling from level I to level II through a distance L. Assuming that there is no resistance of any kind, we can write the following expression for the time t required for this process:

$$t = \sqrt{2L/g}. \tag{3.1}$$

Figure 3.2 shows a mechanism used in automatic machines (lathes) for feeding rod-like material during processing. The weight M acts on the slider 2 via a cable I

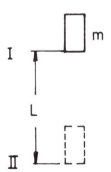

FIGURE 3.1 Model of a free-falling body.

which passes over a roller with moment of inertia I. The slider 2 pushes the rod m which is supported by frictional guide 3. Thus, the acting force $F = Mg$ must overcome the friction F_1 in the guides; F_1 may be expressed as:

$$F_1 = fmg, \qquad [3.2]$$

where $f =$ the dry friction coefficient and m is the mass of the rod.

In addition, the force F rotates the roller with moment of inertia I. Therefore, the equilibrium equation of forces takes the form

$$Mg = fmg + (M - m)a + I\alpha / r, \qquad [3.3]$$

where $a =$ the linear acceleration of the weight (or rod),
  $r =$ the radius of the roller, and
  $\alpha =$ the angular acceleration of the roller.

Since

$$a = \alpha r, \qquad [3.4]$$

from Equation (3.3) we can derive an expression for a in the form

$$a = g \frac{N - fm}{M + m + I / r^2}. \qquad [3.5]$$

The time t needed to displace the rod through distance L can be calculated from the formula

$$t = \sqrt{2L \frac{M + m + I / r^2}{g(M - fm)}} \qquad [3.6]$$

Obviously, for $I/r^2 \ll (M + m)$ (i.e., the influence of the roller is negligible in comparison with that of the moving masses), Equation (3.6) can be rewritten in the form

$$t = \sqrt{2L \frac{M + m}{g(M - fm)}}. \qquad [3.7]$$

In this case we analyze movement along an inclined plane. This is the case that occurs when, for instance, parts slide along a tray from a feeder, as is shown in Figure 3.3. Here ϕ is the inclination angle of the tray. The friction between the parts and the tray is described by the force $F_1 = fmg \cos \phi$ (here again, $f =$ the dry friction coefficient

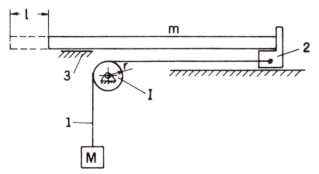

FIGURE 3.2 Layout of a rod-feeding mechanism driven by the force of gravity.

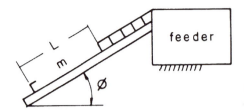

FIGURE 3.3   Model of gravitation drive on an inclined tray.

which resists the movement along the tray). The driving force F in this case can be found from the known formula

$$F = mg \sin \phi.$$ [3.8]

The equilibrium equation thus has the form

$$mg \sin \phi = fmg \cos \phi + ma.$$ [3.9]

From Equation (3.9) we obtain

$$a = g(\sin \phi - f \cos \phi).$$ [3.10]

The time t required to displace a part through a distance L equals

$$t = \sqrt{\frac{2L}{g(\sin \phi - f \cos \phi)}}.$$ [3.11]

[Note: When $\sin \phi = f \cos \phi$ or $f = \tan \phi$, no movement will occur. The time tends to infinitely long values.]

Here we analyze the movement of a mass driven by a previously deformed spring. The layout of such a mechanism is shown in Figure 3.4a). A spring as a driving source is described by its characteristic shown in Figure 3.5. This characteristic shows the dependence of the force P developed by the spring on the values of the deformation x (in both the stretched and compressed modes). When this dependence is linear, as shown in Figure 3.5, parameter c, which is the stiffness of the spring, is constant for this case. In other words, stiffness of the spring is a proportionality coefficient tying the deformation of the spring to the force P it develops. It also defines the value of the slope of the characteristic and can be described as

$$c = \tan \phi$$ [3.12]

and

$$P = -cx.$$ [3.13]

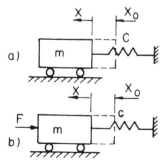

FIGURE 3.4   Spring-driven body: a) Without and b) With resisting force F.

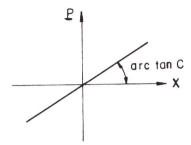

FIGURE 3.5   Characteristic of the spring: force versus deformation.

The force P always acts in the direction opposite to x.

Thus, the movement of the mass m is described by the following equation based on the Dalamber principle:

$$m\ddot{x} + cx = 0. \qquad [3.14]$$

This differential homogeneous equation has a simple solution:

$$x = A\cos\omega t + B\sin\omega t, \qquad [3.15]$$

where the unknown parameters A, B, and ω must be determined. Substituting Expression (3.15) into Equation (3.14), we obtain

$$(c - m\omega^2)(A\cos\omega t + B\sin\omega t) = 0 \qquad [3.16]$$

and

$$w^2 = c/m.$$

The parameter ω is known as the natural frequency of the system. To find the unknown parameters A and B, we have to use the initial conditions of the system. Say, at the moment $t = 0$ the deformation of the spring $x = x_0$ and $\dot{x} = 0$. We then substitute these data into expression (3.15) and obtain directly $A = x_0$ and $B = 0$. Thus the complete solution of Equation (3.14) is

$$x = x_0 \cos\sqrt{c/m}\, t. \qquad [3.17]$$

Expression (3.17) is interpreted graphically in Figure 3.6.

To find the time needed to move the mass from the point x_0 to any other point x_1 located at a distance L from x_0, we rewrite Expression (3.17) in the following way:

$$t_1 = \sqrt{m/c}\, \mathrm{arc\,cos}\,(x_1/x_0). \qquad [3.18]$$

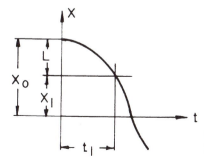

FIGURE 3.6   Displacement versus time for a body driven by a spring.

In a more realistic approach, we must consider a resisting force acting on the mass m during its motion, as shown in Figure 3.4b). Since the nature of the force can vary, so can its analytic description. For example, if it is caused by dry friction, the force may be described analytically in the form

$$F = |F| \operatorname{sgn} \dot{x}. \qquad [3.19]$$

This graphic interpretation of Equation (3.19) is given in Figure 3.7. The movement of mass m can be described by

$$m\ddot{x} + |F|\operatorname{sgn}\dot{x} + cx = 0. \qquad [3.20]$$

which can be replaced by a system of equations in the form

$$\ddot{x} + \mu + \omega^2 x = 0 \quad \text{for } \dot{x} > 0,$$
$$\ddot{x} - \mu + \omega^2 x = 0 \quad \text{for } \dot{x} < 0. \qquad [3.21]$$

Here $\mu = |F|/m$ and $\omega^2 = c/m$.

Substituting $k = \mu/\omega^2$, in Equations (3.21), we obtain

$$\ddot{x} + \omega^2 (x + k) = 0 \quad \text{for } \dot{x} > 0,$$
$$\ddot{x} + \omega^2 (x - k) = 0 \quad \text{for } \dot{x} < 0. \qquad [3.22]$$

It is convenient to transform these equations multiplying them by $2\dot{x}$ and integrating them into the following form:

$$\dot{x}^2 + \omega^2 (x + k)^2 = R_{i-1}^2,$$
$$\dot{x}^2 + \omega^2 (x - k)^2 = R_i^2, \quad i = 1, 2, \ldots \qquad [3.23]$$

The value R is an integration constant which must be defined for every change of $\operatorname{sgn}\dot{x}$.

This form of interpretation permits us to express the behavior of the mass in the terms of the phase plane which is shown in Figure 3.8. The oscillating movement of the mass ceases at the moment when $R_n \leq 2k\omega$. In our case, the spring moves the mass from a point $x = x_0$ through a distance L to a point $x = x_1$. In accordance with the diagram given in Figure 3.8, the value R equals $\omega x_0 - \omega k$. This enables us to rewrite the first of the two Equations (3.23) in the following way:

$$\dot{x} = \sqrt{2k\omega^2 (x_0 - x)^2 + \omega^2 (x_0^2 - x^2)} \qquad [3.24]$$

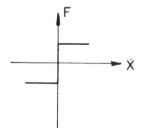

FIGURE 3.7 Force developed by dry friction versus speed of the body.

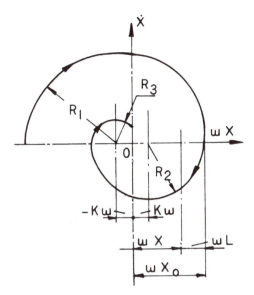

FIGURE 3.8 Speed versus displacement for an oscillating body influenced by dry friction.

and

$$\int_{x_0}^{x_1} \frac{dx}{\sqrt{2k\omega^2 x_0 + \omega^2 x_0^2 - 2k\omega^2 x - \omega^2 x^2}} = \int_0^{t_1} dt,$$

[3.25]

$$t_1 = \frac{1}{\omega}\left[\text{arc sin}\frac{x_0 + k}{\sqrt{2kx_0 + x_0^2 - k^2}} - \text{arc sin}\frac{x_1 + k}{\sqrt{2kx_0 + x_0^2 - k^2}}\right].$$

For the case $k = 0$, it follows from Equation (3.25) that

$$t_1 = \frac{1}{\omega}\left[\frac{\pi}{2} - \text{arc sin}\frac{x_1}{x_0}\right] = \frac{1}{\omega}\,\text{arc cos}\frac{x_1}{x_0}.$$

[3.26]

Thus, Equation (3.26) coincides with the Formula (3.18) which describes the same situation $k = 0$ (as a result of $\mu = 0$ in this case).

Let us consider the case in which the resisting force F shown in Figure 3.4b) is constant. Such a case can arise when, for instance, the force of gravity acts on the mass driven by the spring. This is the situation which occurs in Figure 2.16 where the spring 9 lifts the lever 6, the cutter 7, and the armature of the magnet 5. The following equation describes this situation:

$$m\ddot{x} + cx = -F.$$

[3.27]

The solution of this equation consists of two components:

$$x = x' + x'',$$

[3.28]

where, x' is the solution of the homogeneous equation and therefore has the form given in Equation (3.15), while the partial solution x'' must be some constant value $x'' = D = \text{const.}$

Substituting $x'' = D$ into Equation (3.27), we obtain

$$D = -\frac{F}{c}.$$ [3.29]

Thus, Equation (3.28) can be rewritten in the form

$$x = A \cos \omega t - \frac{F}{c}.$$ [3.30]

For initial conditions $t = 0$, $x = x_0$, and $\dot{x} = 0$, Solution (3.30) gives

$$x = \left(x_0 + \frac{F}{c} \right) \cos \omega t - \frac{F}{c}.$$ [3.31]

Naturally, the initial deformation of the spring A at the beginning of the motion must include the deformation caused by the force F which entails the appearance of the initial coordinate x_0 of the mass location.

The above-considered spring-driven mechanisms can also be rotating in nature, as in Figure 3.9. Equation (3.27), rewritten in terms of angular motion, takes the form

$$I\ddot{\phi} + c\phi = -T,$$ [3.32]

where

      I = the moment of inertia of the rotating body,
      c_ϕ = stiffness of the spring lumped to the angular displacement,
      T = resisting torque $T = Fr$,
      r = the radius on which the force F acts, and
      ϕ = angle of rotation.

We should not forget that the dimensions here are different from those in Equation (3.27). The solution of Equation (3.32) has a form analogous to that of Equation (3.30), as follows:

$$\phi = \phi \cos \omega t - \frac{T}{c_\phi}.$$ [3.33]

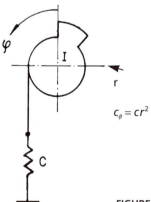

FIGURE 3.9 Rotating motion caused by a spring.

Here ϕ is calculated from the initial conditions $t = 0$, $\phi = \phi_0$, and $\dot{\phi} = 0$, and has the form

$$\phi = \left(\phi_0 + \frac{T}{c} \right) \cos \omega t - \frac{T}{c_\phi}. \qquad [3.34]$$

3.2   Electromagnetic Drive

An electromagnet used in machine design for drive purposes usually consists of a housing 1, a coil 2, an armature 3, and a spring 4, as shown schematically in Figure 3.10. The armature slides in guides 5, which are made of a nonmagnetic material so as to create minimal friction between the armature and the guides. The housing, armature, guide bushing, and air gap constitute the magnetic circuit. The coil carries the electric circuit.

In this case the computation includes the determination of the current changes in the coil, the magnetic flux in cross sections of the magnetic circuit, the pulling force developed by the magnet, the influence of the air gap, and the speed of motion of the armature. The initial data for these computations include the geometric dimensions, the structure and materials of the magnet, and the parameters of the voltage. These computations cannot be completed analytically. The equations are cumbersome, and there are no analytical methods of solution for such nonlinear differential equations.

However, to simplify the equations, the following assumptions may be made:

- There are no energy losses in the form of eddy current and hysteresis;
- Reluctance of the air gap is proportional to its length;
- There is no leakage flux or reluctance in the magnetic circuit.

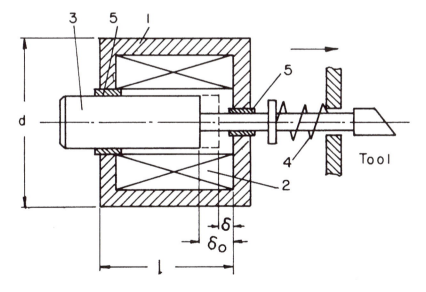

FIGURE 3.10   Layout of an electromagnet.

Then the equation of the behavior of the armature has the form

$$C_1 m \frac{d^3\delta}{dt^3} + C_1 C_5 \frac{d\delta}{dt} - C_2 \sin \omega t \sqrt{m \frac{d^2\delta}{dt^2} + C_5\delta + F_c} + (C_3 + mC_4\delta) \cdot$$

$$\left(m \frac{d^2\delta}{dt^2} + C_5\delta + F_c\right) = 0,$$

[3.35]

where

$$C_1 = 4\pi \ W^2 \ 10^{-7} S,$$
$$C_2 = 2u \sqrt{2\pi W^2 10^{-7} S},$$
$$C_3 = 2R\delta_0,$$
$$C_4 = 2R/m,$$

C_5 = stiffness of the spring the magnet has to overcome,
W = the number of turns in the coils of the electromagnet,
S = cross-sectional area of the air gap,
u = amplitude of the voltage,
R = ohmic resistance of the coil,
δ_0 = initial air gap,
δ = air gap,
m = mass of the moving parts (armature plus the driven bodies),
ω = frequency of the voltage,
t = time,
F_c = friction (considered to be constant), and
V_s = speed of the moving parts.

In addition we denote:

i = current in the coil,
l = linear longitudinal dimensions of the magnet,
d = linear transverse (diametrical) dimensions of the magnet,
A = kinetic energy of the moving armature, and
γ = specific weight of the material.

Even after these simplifications, Equation (3.35) still looks frightening. Attempts to tackle its solution by a computer will give us only approximate results because of the above-made assumptions, which make this approach almost worthless.

We therefore consider another, more practical approach based on the use of the similarity principle. It can be proved that the following similarity criteria Π_1, Π_2, and Π_3 exist:

$$\Pi_1 = \frac{S_0 W^2}{Rt\delta_0} = \text{idem,}$$

[3.36]

$$\Pi_2 = \frac{uW\sqrt{S}}{R\delta_0\sqrt{F_e}} = \text{idem,}$$

[3.37]

$$\Pi_3 = \frac{(F_e - F_r)^{t^2}}{m\delta_0} = \text{idem},$$ [3.38]

where

$$F_r = F_c + C_5\delta \quad \text{and} \quad F_e = m\frac{d^2\delta}{dt^2} + F_c + C_5\delta,$$

and

    F_e = force developed by the moving armature,

    F_r = resisting force made up of friction with the spring deformation force.

The main idea underlying this approach is that, for every magnet, Equations (3.36), (3.37), and (3.38) are equal. [The symbol idem indicates this fact.] It means that if we know the behavior of at least one electromagnet, we can derive the behavior of another one of the same design pattern if its parameters are known. Let us say we have a device about which all the information is known; then for any other device under design we can determine the time, speed, and force describing the movement of the armature of the new device. Denoting the similarity coefficients by indexes s, we can write

$$l_s = l_2 / l_1; \quad \delta_s = \delta_2 / \delta_1; \quad d_s = d_2 / d_1;$$

$$t_s = t_2 / t_1 = \sqrt{m_s \delta_s / F_s};$$

$$V_s = \delta_s / t_s; \quad S_s = S_2 / S_1 = d^2 s;$$

$$m_s = m_s / m_1 = l_s \cdot d^2 s \quad \text{for } \gamma_s = \gamma_1 / \gamma_2 = 1;$$ [3.39]

$$F_s = i^2 {}_s W^2 s; \quad W_s = t^2 s u_s / \sqrt{m_s \delta_s S_s};$$

$$i_s = l_s / W_s; \quad R_s = S_s W_s^2 / t_s \delta_s = u_s^2 / l_s^2;$$

$$A_s = m_s V_s^2; \quad C_{5s} = m_s / t_s^2.$$

Let us consider some examples.

a. An electromagnet is given and its parameters are known; for instance, operation time $t_1 = 0.01$ sec and dimensions l_1 are given. We then wish to find the operation time t_2 of a new design which will have proportionally enlarged dimensions. Say the coefficients have the values

$$d_s = 2, \quad l_s = 2, \quad \text{and} \quad \delta_2 = 2.$$

We can use the above-given dependencies to obtain

$$t_s = \sqrt{\frac{m_s \delta_s}{F_s}} = \sqrt{\frac{m_s \delta_s}{i_s^2 W_s^2}} = \sqrt{\frac{m_s \delta_s}{l_s^2}} = \sqrt{\frac{m_s}{l_s}}.$$ [3.40]

From Equations (3.39), it follows that $m_s = l_s d_s^2 = 8$ and we then obtain

$$t_s = 2 \quad \text{or} \quad t_2 = t_s t_1 = 2 \cdot 0.01 = 0.02 \text{ sec.}$$

In addition, we can calculate that the energy A_2, which the armature will develop, will change eight times. To preserve these conditions, the stiffness of the spring must be taken to be twice as high as that in the pattern:

$$A_s = m_s V^2_s = \frac{l_s d^2_s \delta^2_s}{t^2_s} = \frac{2 \cdot 4 \cdot 4}{4} = 8,$$

$$C_{5s} = \frac{m_s}{t^2_s} = \frac{8}{4} = 2.$$

The force F_2 developed by the magnet will be four times higher than that of the pattern:

$$F_s = i^2_s W^2_s = \delta^2_s = 4$$

   b. The same pattern of electromagnet is given. We wish to design another magnet, but this time with a different length. Thus,

$$\delta_s = 2, \quad d_s = 1, \quad l_s = 2.$$

We now wish to know the value of the force developed by the new magnet or, in other words, the value of F_s:

$$F_s = i^2_s W^2_s = l^2_s = 4.$$

   We have therefore derived the conclusion that this magnet will be four times stronger than the pattern. The operation time, however, will not change. Indeed, as was shown in the previous example,

$$t_s = \sqrt{\frac{m_s \delta_s}{l^2_s}} = \sqrt{\frac{d^2_s \delta_s}{l_s}} = 1.$$

   c. The same pattern of an electromagnet is given. The new design is two times "thicker," i.e.,

$$\delta_s = 1, \quad l_s = 1, \quad d_s = 2.$$

We can obtain the following answers by means of the above-derived formulas:

$$F_s = \delta^2_s = 1 \quad \text{and} \quad t_s = \sqrt{\frac{d^2_s \delta_s}{l_s}} = 2.$$

The calculation approach described above is not absolute. It cannot, for instance, take into account a situation in which the diameter of the housing is not reduced to the same extent as the diameter of the armature in comparison with the pattern design, or, when the geometry of the design is not the same as that of the pattern— for example, if an opening is drilled in the armature, whereas the armature of the pattern is solid. The calculation method does, however, enable us to obtain, cheaply and quickly, satisfactory estimations of the behavior of the new electromagnet or of the values of the dimensions (or other design parameters) of the device required to provide the required behavior.

3.3 Electric Drives

In this section we consider the behavior of electrically driven mechanical systems. The most widely used drive of this kind is an electromotor drive of the type illustrated schematically in Figure 3.11. It consists of a motor 1, a clutch 2, and a driven mass I. The movement of this mass can be described by the general equation

$$I\ddot{\phi} + T_r = T_d \qquad\qquad [3.41]$$

where T_r and T_d are the resistance and driving torques, respectively.

The question the designer must usually answer is how long it will take to reach a specific speed of the movement or a specific displacement of a specific link. If the values I, T_r and T_d were constant, we would obtain, instead of Expression (3.41), the solutions

$$\ddot{\phi} = \frac{T_d - T_r}{I} \qquad\qquad [3.42]$$

and

$$\dot{\phi} = \frac{T_d - T_r}{I} t \qquad\qquad [3.43]$$

for the initial condition $t = 0$, $\dot{\phi} = 0$ and

$$\phi = \frac{T_d - T_r}{2I} t^2, \qquad\qquad [3.44]$$

which brings us to the simple formula expressing the time t_1 needed to reach speed $\dot{\phi}_1$ or angle ϕ_1, respectively:

$$t_1 = \dot{\phi}_1 \frac{I}{T_d - T_r} \quad \text{and} \quad t_1 = \sqrt{\frac{2\phi_1 I}{T_d - T_r}}. \qquad\qquad [3.45]$$

In reality, however, none of these parameters is constant.

The torque T_d developed by electromotors is usually a function of the rotation speed ω, and the nature of this function, in turn, depends on the type of motor. We consider here some of these dependencies or characteristics.

DC motors with compound or independent excitation (shunt motors) have a so-called flat characteristic, which has the general form shown in Figure 3.12. This characteristic can be usually approximated by a straight line and can therefore be described analytically in a corresponding manner:

$$\omega = b_1 - b_2 T_d,$$
$$T_d = a_1 - a_2 \omega, \qquad\qquad [3.46]$$

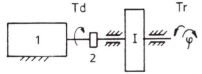

FIGURE 3.11 Layout of an electrically driven machine.

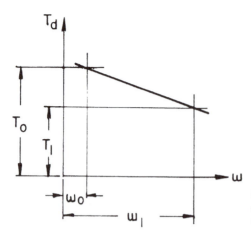

FIGURE 3.12 Characteristic of a DC motor with compound or independent excitation: torque versus speed of the rotor. Flat characteristic.

where a_1, a_2, b_1, and b_2 are constants defining the location and inclination of the line in the coordinate plane.

DC motors with series excitation have a different type of characteristic, known as a drooping characteristic. Figure 3.13 shows an example of such a characteristic. Analytically, this function may be described as

$$T_d = a_1 \left(\frac{a_2}{\omega} \right)^q ,$$ [3.47]

where a_1, a_2, and q are constants describing the location and the shape of the curve in the coordinate plane.

For AC induction motors a typical characteristic is shown in Figure 3.14.

After some assumptions (which we will not discuss in depth) have been made, an analytical approximation of this characteristic may be given in the following form:

$$T_d = \frac{2 T_m s_m s}{s_m^2 + s^2} ,$$ [3.48]

where

$\quad T_d$ = torque developed by the motor,
$\quad T_m$ = the maximal torque value,
$\quad\ s$ = slip, $s = (\omega_0 - \omega) / \omega_0$,
$\quad \omega_0$ = the rotational speed of the flux of the motor,

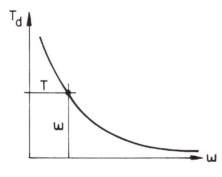

FIGURE 3.13 Characteristic of a DC motor with series excitation: torque versus speed of the rotor. Drooping characteristic.

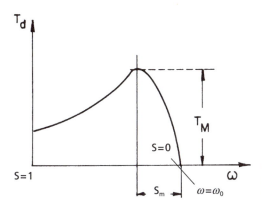

FIGURE 3.14   Characteristic of an AC induction motor: torque versus speed of the rotor.

s_m = the slip value at the point of maximal torque, and
ω = speed of the rotor.

For example, we consider an electromotor with parameters:

$T_m = 1\,\mathrm{Nm}$, $\omega_0 = 314\,\mathrm{rad/sec}$, $s_m = 0.15$.

The characteristics of this motor will have the form shown in Figure 3.14a). (We created the figure using the MATHEMATICA language as follows:

z0=Plot[471 (314−w)/(2218+(314−w)^2),{w,0,314},AxesLabel−>{"w","T"}]).

Then we can answer the natural question: how will the process of acceleration of a system driven by such a motor develop?

Assuming the moment of inertia of the system $I = 0.075$ kg m² and neglecting, for this example, the resistance, we formulate the equation of motion in terms of MATHEMATICA as:

f20=.075 w'[t]− 471 (314− w[t])/(2218+(314−w[t])^2)
j20=NDSolve[{f20==0,w[0]==0},{w[t]},{t,0,10},MaxSteps−>1500]
y20=Plot[Evaluate[w[t]/.j20],{t,0,10},AxesLabel−>{"t","w"},
Frame−>True,GridLines−>Automatic]

One can see in Figure 3.14b) that in 10 seconds the system reaches the maximal rotation speed of about 300 rad/sec.

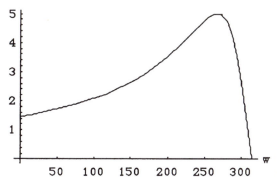

FIGURE 3.14a)   Characteristics of an asynchronous motor.

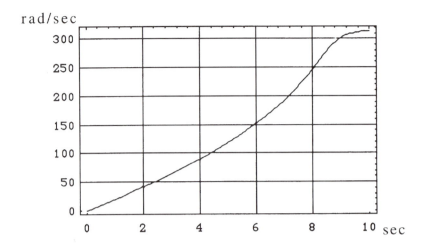

FIGURE 3.14b)   Rotation speed of the system driven by a synchronous motor.

For synchronous induction motors, the characteristic is a straight line for a certain range of torques, as is shown in Figure 3.15. After the maximal value T_m of the torque is reached, the motor stops and is not able to work. This maximal torque occurs when the angle θ between the vector of the rotating flux of the stator and the geometric axis of the rotor's poles equals θ_m which is about half of the angular pitch of the poles. The dependence $T_d(\theta)$ is very important in the theory of synchronous inductive motors and usually has the form shown in Figure 3.16.

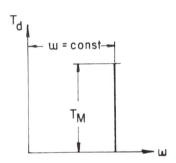

FIGURE 3.15   Torque versus the constant rotation speed of a synchronous induction motor.

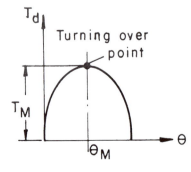

FIGURE 3.16   Characteristic of a synchronous induction motor: torque versus the angle between the rotating flux and the geometric axis of the rotor's poles.

A special type of electromotor—a stepper motor—is widely used in robotics of different kinds. It is therefore necessary to explain the principle of operation of such motors and to compare them with the other motors mentioned above.

Figure 3.17 shows, in principle, the layout of a stepper motor (sometimes called a step or stepping motor). Here, the stator 1 has a N (multiple of three) number of poles and three corresponding systems of coils A, B, and C. The rotor is also provided with poles, the number of poles N_1 being a multiple of any factor other than that of N. (In Figure 3.17 $N = 3 \times 4 = 12$ and $N_1 = 8$.) Let us follow the behavior of the rotor:

1. When the coils A are energized, the poles designated a will be pulled by the magnetic field, thus moving the rotor into the position shown in Figure 3.17a).
2. When the coils C are energized, the poles designated c will be pulled by the field, thus moving the rotor against the C poles of the stator and slewing the rotor by one-third of the poles' angular pitch (Figure 3.17b)).
3. When the poles B are energized, the rotor poles b will be pulled against the stator's B poles, again slewing the rotor for one-third of the angular pitch. Thus, energizing the stator in the sequence A, C, B causes clockwise rotation of the rotor. Obviously, reversing sequence A, B, C will reverse the direction of rotation of the rotor. In this example, the pitch of the stator poles is 30°, which corresponds to a pitch of the rotor poles of 45°, and after each recommutation of the coils the rotor moves through 15°. However, there are motors with 1.8° rotation for every step and others with 45° and even 90° per step.

A different design for a stepper motor consists of several stators (three or five) that are offset one from the other. In these motors, the magnetic coupling between phases is eliminated and they therefore provide excellent slew capabilities. Such motors are called variable reluctance motors. In other designs, the rotor is a permanent magnet. The different types of stepper motor are characterized by the load torque T_L they develop, the inertia of the rotor I_r, the maximum pulse rates, and the accuracy of the rotor's tracing of the magnetic field. The ranges and values of these parameters depend

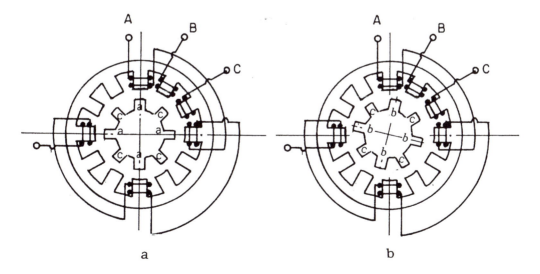

a   b

FIGURE 3.17   Layout of a stepper motor: a) The first step; b) The second step.

on the design and dimensions of the motors. These parameters usually change in the following range:

$$\text{Torque: } T_L = 22.5 \div 1125 \text{ kg m,}$$

$$\text{Inertia of the rotor: } I_r = 1.2 \div 10000 \text{ g cm}^2,$$

$$\text{Maximum pulse rate: } S = 150 \div 50000 \text{ pps.}$$

A typical torque-versus-time characteristic for stepper motors is presented in Figure 3.18. The point that should be stressed is that changes in the torque are different for different pulse rates. The lower the value of the pulse rate (i.e., the duration of one pulse is longer), the higher the torque at the beginning of the switching and the lower the torque at the end of it. In more detail, the dependence torque versus pulse rate is shown in Figure 3.19. (These data are taken from *Machine Design*, April 29, 1976, p. 36.) Point A represents the conditions that ensure the maximum speed at which a load can be run bi-directionally without losing a step. This condition occurs by a speed of about five steps per second. Point B indicates the so-called stall torque. At this point, the stator windings being energized, for all kinds of motors, resist movement. Point C represents the detent-like torque which is typical only for motors with permanent magnet rotors. At this point even a nonenergized stator resists the movement of the rotor to move. The motor "remembers" its position. Curves 1 and 2 represent the behav-

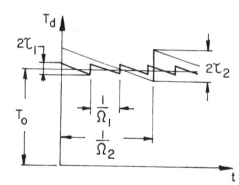

FIGURE 3.18 Torque-versus-time dependence for a stepper motor.

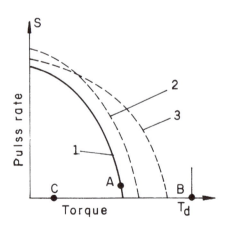

FIGURE 3.19 Torque-versus-pulse-rate dependence for stepper motors.

ior of motors provided with permanent magnet rotors, while curve 3 represents that of variable reluctance motors.

Now that we have learned the characteristics of the most frequently used electromotors as drives for automatic machines and systems, some other comparative features of these electromotors must be discussed.

The advantage of DC motors lies in the ease of speed control, whereas speed control in AC motors requires the installation of sophisticated equipment (frequency transformers). The advantage of AC motors (both one- and three-phase) is that they operate on the standard voltage available at any industrial site. In addition, a three-phase induction motor with a squirrel-cage rotor is cheaper than any other type of motor of the same power. For accurate positioning both DC and AC motors require feedbacks. In contrast, stepper motors, although more expensive, are suitable for accurate positioning (almost always without any feedback) and speed control. Such motors are convenient for engagement with digital means (computers).

Let us now analyze Equation (3.41) for the case when T_d is described by Equations (3.46). Let us suppose that the resistance torque T_r is also described by a linear expression which is proportional to the speed of rotation of the machine. Thus we can write for T_r:

$$T_r = \alpha_1 + \alpha_2 \omega. \qquad [3.49]$$

Obviously, α_1 and α_2 are constants. The physical meaning of the value α_1 is the initial resistance of the driven system. Until the drive has developed this value of the driving torque, the system will not move. The value of α_2 controls the rate of the resistance torque during the speed increase of the accelerated system. The problem is how to estimate the values of α_1 and α_2. We feel that the only way to do this is to measure the resistance torques of existing machines and interpolate or extrapolate the results to the case under design.

Substituting Equations (3.46) and (3.49) into Equation (3.41) we obtain a linear equation in the form

$$I\ddot{\phi} + \alpha_1 + \alpha_2\dot{\phi} = a_1 - a_2\dot{\phi}. \qquad [3.50]$$

After simplification we obtain

$$\ddot{\phi} + C\dot{\phi} = B, \qquad [3.51]$$

where $C = (\alpha_2 + a_2)/I$ and $B = (a_1 - \alpha_1)/I$.

Remembering that $\dot{\phi} = \omega$, we can rewrite Equation (3.51) to obtain

$$\dot{\omega} + C\omega = B. \qquad [3.52]$$

The solution of this equation has the form

$$\omega = \omega_1 + \omega_2, \qquad [3.53]$$

where ω_1 is the solution of the homogeneous equation

$$\dot{\omega} + C\omega = 0. \qquad [3.54]$$

For ω_1 we have

$$\omega_1 = Ae^{at}. \qquad [3.55]$$

Substituting Equation (3.55) into Equation (3.54) we obtain

$$Aae^{at} + CAe^{at} = 0 \qquad [3.56]$$

or

$$a = -C = -\frac{\alpha_2 + a_2}{I}. \qquad [3.57]$$

For ω_2 we seek a solution in the form

$$\omega_2 = \Omega. \qquad [3.58]$$

Substituting Equation (3.58) into Equation (3.52) we obtain

$$C\Omega = B \qquad [3.59]$$

or

$$\Omega = \frac{B}{C} = \frac{a_1 - \alpha_1}{\alpha_2 + a_2}. \qquad [3.60]$$

From Equation (3.53) we derive

$$\omega = A \exp\left[-\frac{\alpha_2 + a_2}{I}t\right] + \frac{a_1 - \alpha_1}{\alpha_2 + a_2}. \qquad [3.61]$$

For the initial conditions at time $t = 0$ and speed $\omega = 0$, we can rewrite Equation (3.61) and extract the unknown constant A in the following way:

$$A + \frac{a_1 - \alpha_1}{\alpha_2 + a_2} = 0 \qquad [3.62]$$

or

$$A = \frac{\alpha_1 - a_1}{\alpha_2 + a_2}. \qquad [3.63]$$

And finally we obtain the solution:

$$\omega = \frac{\alpha_1 - a_1}{\alpha_2 + a_2}\left[\exp\left(-\frac{\alpha_2 + a_2}{I}t\right) - 1\right]. \qquad [3.64]$$

To calculate the time needed to achieve some speed of rotation ω_1 we derive the following equation from Equation (3.64):

$$t = \frac{I}{\alpha_2 + a_2}\ln\left[\frac{\alpha_1 - a_1}{\omega_1(\alpha_2 + a_2) + \alpha_1 - a_1}\right]. \qquad [3.65]$$

From Expression (3.65) some particular cases can be obtained. For a constant resistance torque $\alpha_2 = 0$ we obtain

$$t = \frac{I}{a_2}\ln\left[\frac{\alpha_1 - a_1}{\omega_1 a_2 + \alpha_1 - a_1}\right]. \qquad [3.66]$$

When there is no resistance torque, i.e., $\alpha_1 = \alpha_2 = 0$ we obtain

$$t = \frac{I}{a_2} \ln \left[\frac{a_1}{a_1 - \omega_1 a_2} \right].$$ [3.67]

For the case when $\alpha_1 = 0$ we derive from Expression (3.65) the following formula:

$$t = \frac{I}{\alpha_2 + a_2} \ln \left[\frac{a_1}{a_1 - \omega_1(\alpha_2 + a_2)} \right].$$ [3.68]

Expression (3.64) allows us to find the dependence of the rotation angle ϕ on time. For this purpose we must rewrite this expression as follows:

$$\frac{d\phi}{dt} = \frac{\alpha_1 - a_1}{\alpha_2 + a_2} \left[\exp\left(-\frac{\alpha_2 + a_2}{I} t \right) - 1 \right].$$ [3.69]

Integrating this dependence termwise, we obtain:

$$\int_0^\phi d\phi = \frac{\alpha_1 - a_1}{\alpha_2 + a_2} \int_0^t \left[\exp\left(-\frac{\alpha_2 + a_2}{I} t \right) - 1 \right] dt$$ [3.70]

and

$$\phi = I \frac{\alpha_1 - a_1}{(\alpha_2 + a_2)^2} \left[1 - \exp\left(-\frac{\alpha_2 + a_2}{I} t \right) \right] + t \frac{a_1 - \alpha_1}{\alpha_2 - a_2}.$$ [3.71]

Now for the particular case when $\alpha_2 = 0$:

$$\phi = I \frac{\alpha_1 - a_1}{a_2^2} \left[1 - \exp\left(-\frac{a_2}{I} t \right) \right] + t \frac{a_1 - \alpha_1}{a_2}.$$ [3.72]

For $\alpha_1 = \alpha_2 = 0$,

$$\phi = I \frac{-a_1}{a_2^2} \left[1 - \exp\left(-\frac{a_2}{I} t \right) \right] + t \frac{a_1}{a_2}.$$ [3.73]

For $\alpha_1 = 0$,

$$\phi = I \frac{-a_1}{(\alpha_2 + a_2)^2} \left[1 - \exp\left(-\frac{\alpha_2 + a_2}{I} t \right) \right] + t \frac{a_1}{\alpha_2 - a_2}.$$ [3.74]

In Figure 3.20 the graphic representation of this dependence is shown to be composed of three components. This graph provides us with a tool for determining the time t_1 needed to achieve rotation for the angle ϕ_1.

Let us consider the case where the drive is supplied by a *series DC* motor with the characteristics given by Expression (3.47). Because of analytical difficulties, we will

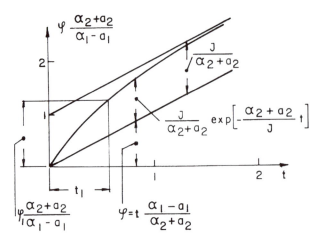

FIGURE 3.20 Angular displacement of the DC motor's rotor (composed of three components) versus time as a solution of Equation 3.50.

discuss here the simplest case, i.e., when in Equation (3.41) $T_r = 0$, which means that resistance is negligible. Thus, we obtain

$$I\ddot{\phi} = a_1 \left(\frac{a_2}{\omega} \right)^q \qquad [3.75]$$

or

$$\dot{\omega} = \frac{a_1}{I} \left(\frac{a_2}{\omega} \right)^q .$$

From Equation (3.75) it follows that

$$\int_0^\omega \omega^q \, d\omega = \frac{a_1 a_2^q}{I} \int_0^t dt, \qquad [3.76]$$

and finally that

$$\omega = \frac{d\phi}{dt} = \left[(q+1) \frac{a_1 a_2^q}{I} t \right]^{1/q+1} . \qquad [3.77]$$

Thus, for the dependence $\phi(t)$ we obtain

$$\int_0^\phi d\phi = \left[\frac{a_1 a_2^q}{I} (q+1) \right]^{1/q+1} \int_0^t t^{1/q+1} dt \qquad [3.78]$$

or

$$\phi = \left[\frac{a_1 a_2^q}{I} (q+1) \right]^{1/q+1} \cdot \frac{q+1}{q+2} t^{(q+2)/q+1} .$$

When the drive is supplied by an asynchronous induction motor, we substitute Equation (3.48) into Equation (3.41). Here again, we will discuss the simplest case when $T_r = 0$. Thus, we rewrite Equation (3.41) in the form

$$I\dot{\omega} = \frac{2T_m s_m s}{s_m^2 + s^2}. \tag{3.79}$$

Remembering the definition of slip given above, we obtain, instead of Equation (3.79), the expression

$$\dot{\omega} = \frac{2T_m s_m \omega_0 (\omega_0 - \omega)}{I\left[s_m^2 \omega_0^2 + (\omega_0 - \omega)^2\right]}. \tag{3.80}$$

Denoting $(2T_m s_m \omega_0)/I = A$ and $s_m^2 \omega_0^2 = B$, we rewrite Equation (3.80) in the form

$$\int_0^\omega \frac{B + (\omega_0 - \omega)^2}{\omega_0 - \omega} d\omega = A \int_0^t dt. \tag{3.81}$$

After obvious transformation, the final result can be obtained in the form:

$$\frac{A}{B}t = \ln\frac{1}{\omega_0 - \omega} - \ln\frac{1}{\omega_0} - \frac{1}{2B}(\omega_0 - \omega)^2 + \frac{1}{2B}\omega_0^2. \tag{3.82}$$

For a synchronous motor the driving speed (as was explained above) remains constant over a certain range of torques until the motor stops. Thus,

$$\omega = \omega_0 = \text{constant.}$$

To reach the speed ω_0 from a state of rest when $\omega = 0$, an infinitely large acceleration must be developed. To overcome this difficulty, synchronous motors are started in the same way as are asynchronous motors. Therefore, the calculations are of the same sort, and they may be described by Equations (3.79–3.82), which were previously applied to asynchronous drives.

For the drive means of stepper motors, we must make two levels of assumption. First, we assume that the stepper motor develops a constant driving torque, $T_d = T_0 = \text{constant}$ (the higher the pulse rate, the more valid the assumption), which is the average value of the torque for the "saw"–like form of the characteristic. Then, from the basic Equations (3.41) and (3.49), we obtain for the given torque characteristic the following equation of the movement of the machine:

$$I\dot{\omega} + \alpha_1 + \alpha_2 \omega = T_0.$$

Rewriting this expression, we obtain

$$\dot{\omega} + \frac{\alpha_2}{I}\omega = \frac{T_0 - \alpha_1}{I}. \tag{3.83}$$

The solution consists of two components, $\omega = \omega_1 + \omega_2$. For the solution of the homogeneous equation we have

$$\omega_1 = Ae^{-Bt}, \tag{3.84}$$

and for the particular solution we have

$$\omega_2 = C. \tag{3.85}$$

Substituting these solutions in the homogeneous form of Equation (3.83) and in its complete form, respectively, we obtain

$$B = \frac{\alpha_2}{I} \quad \text{and} \quad C = \frac{T_0 + \alpha_1}{\alpha_2}. \tag{3.86}$$

Using the initial conditions that for $t = 0$ the speed $\omega = 0$, we obtain for the constant A

$$A = -\frac{T_0 + \alpha_1}{\alpha_2}. \tag{3.87}$$

Thus, the complete solution has the form

$$\omega = \frac{T_0 + \alpha_1}{\alpha_2}\left[1 - \exp\left(-\frac{\alpha_2}{I}t\right)\right]. \tag{3.88}$$

The next step is to calculate the $\phi(t)$ dependence. This can obviously be achieved by direct integration of solution (3.88):

$$\int_0^\phi d\phi = \frac{T_0 + \alpha_1}{\alpha_2}\int_0^t\left[1 - \exp\left(-\frac{\alpha_2}{I}t\right)\right]dt$$

or

$$\phi = \frac{T_0 + \alpha_1}{\alpha_2}\left[t + \frac{I}{\alpha_2}\left\{\exp\left(-\frac{\alpha_2}{I}t\right) - 1\right\}\right]. \tag{3.89}$$

For the second assumption, we introduce into the excitation torque a "saw"-like periodic component. To do so we must express this "saw" in a convenient form, i.e., describe it in terms of a Fourier series. Let us approximate this "saw" by inclined straight lines, as shown in Figure 3.21 (the reader can make another choice for the approximation form). Then, this periodic torque component T_p can be described analytically by the expression

$$T_p = \tau\frac{\pi_1 - \Omega t}{2} \quad \text{for} \quad 0 < \Omega t < 2\pi, \tag{3.90}$$

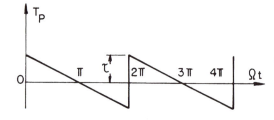

FIGURE 3.21 Approximation of the "saw"-like characteristic (see Figure 3.18) of a stepper motor by inclined straight lines.

and its expansion into a Fourier series becomes

$$T_p = \tau \sum_{K=1}^{\infty} \frac{\sin K\Omega t}{K} \cdot (-1)^K, \qquad [3.91]$$

where τ is the torque amplitude.

Thus, Equation (3.83) for this case can be rewritten in the form

$$\dot{\omega} + \frac{\alpha_2}{I}\omega = \frac{T_0 + \alpha_1}{I} + \tau \sum_{K=1}^{\infty} \frac{\sin K\Omega t}{K} \cdot (-1)^K, \qquad [3.92]$$

and its solution will be composed of three components:

$$\omega = \omega_1 + \omega_2 + \omega_3. \qquad [3.93]$$

The solutions ω_1 and ω_2 are found as in the previous case for the corresponding forms and may be expressed as

$$\omega_1 = A\exp\left[-\frac{\alpha_2}{I}t\right]; \quad \omega_2 = \frac{T_0 + \alpha_1}{\alpha_2}.$$

Here we show the solution ω_3 only for one first term of the series, namely,

$$\omega_3 = C\sin(\Omega t + \theta). \qquad [3.94]$$

Substituting it into Equation (3.92) for the corresponding case, we obtain the form

$$C\Omega\cos(\Omega t + \theta) + C\frac{\alpha_2}{I}\sin(\Omega t + \theta) = \tau\sin\Omega t. \qquad [3.95]$$

After rearrangement of the members and comparison between the left and right sides of this equation, we obtain

$$\theta = \arctan\left[-\frac{I\Omega}{\alpha_2}\right] \quad \text{and} \quad C = \frac{\tau}{\frac{\alpha_2}{I}\cos\theta - \Omega\sin\Theta}. \qquad [3.96]$$

Introducing the initial conditions, namely for $t = 0$, $\omega = 0$, we derive

$$A = -\left[\frac{\tau + \alpha_1}{\alpha_2} + C\sin\theta\right].$$

Finally, the solution of the full Equation (3.92) is

$$\omega = \left[\frac{\tau + \alpha_1}{\alpha_2} + \frac{\tau\sin\theta}{(\alpha_2/I)\cos\theta - \Omega\sin\theta}\right]\left[1 - \exp\left(-\frac{\alpha_2}{I}t\right)\right]. \qquad [3.97]$$

From Equation (3.97) we obtain the dependence

$$\phi = \int_0^t \left[\frac{\tau + \alpha_1}{\alpha_2} - \frac{\tau\sin\theta}{(\alpha_2/I)\cos\theta - \Omega\sin\theta}\right]\left[1 - \exp\left(-\frac{\alpha_2}{I}t\right)\right]dt. \qquad [3.98]$$

Obviously, the calculations shown above answer the following questions:

- How long does it take for the drive to reach the desired angle or (using corresponding transmission) displacement?
- How long does it take for the drive to reach the desired speed?
- What angle, displacement, or speed can be reached during a specific time interval?
- Which parameters of the motor must be taken into account to reach the desired angle, displacement, or speed in a specific time?

3.4 Hydraulic Drive

Let us now learn how to estimate the displacement time of a mass driven by a hydromechanism.

Let us consider the hydromechanism presented schematically in Figure 3.22. This device consists of a cylinder 1, a piston 2, a piston rod 3 with a driven mass M, and a piping system 4 for pressure supply. We can describe the movement of the mass M by the differential equation

$$M\ddot{s} + (\text{sgn}\,\dot{s})\psi\dot{s}^2 = pF - Q, \tag{3.99}$$

where

s = the displacement of the driven mass,
p = the pressure at the input of the cylinder,
F = the area of the piston,
Q = the useful and detrimental forces,
$\psi = F^3\rho/2a^2f^2$ = the coefficient of hydraulic friction of the liquid flow in the cylinder,

where
ρ = density of the liquid,
f = the area of the inlet-pipe cross section,
a = the coefficient of the inlet hydraulic resistance.

For movement of the piston to the right, the hydraulic friction is directed to the left and thus sgn $\dot{s} = 1$.

Denoting

$$m = \frac{2\psi}{M} \quad \text{and} \quad A = \frac{pF - Q}{M},$$

we can rewrite Equation (3.99) in the form

$$\ddot{s} + \frac{m}{2}\dot{s}^2 = A. \tag{3.100}$$

The excitation A causes the movement of the mass M.

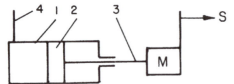

FIGURE 3.22 Layout of a hydraulic drive.

Let us now try to define the operation time of the piston in the hydromechanism under consideration. For this purpose, we will rewrite Equation (3.100) in the form

$$\dot{V} + \frac{m}{2}V^2 = A,$$ [3.101]

where V is the speed of the mass.

Let us assume that A can be taken as a constant value. Then Equation (3.101) can be rearranged as

$$\frac{dV}{A - (m/2)V^2} = dt.$$ [3.102]

Integrating Equation (3.102), we obtain

$$t = \frac{1}{\sqrt{2mA}} \ln \frac{1 + V\sqrt{m/2A}}{1 - V\sqrt{m/2A}} + C,$$

where C is the constant of the integration.

The initial conditions are that when $t = 0$, $V = 0$; thus, $C = 0$, and we can finally write

$$t = \frac{1}{\sqrt{2mA}} \ln \frac{1 + V\sqrt{m/2A}}{1 - V\sqrt{m/2A}} \quad \text{or} \quad e^{\beta t} = \frac{1 + V\sqrt{m/2A}}{1 - V\sqrt{m/2A}},$$ [3.103]

where

$$\beta = \sqrt{2mA}.$$

From Equation (3.103) we obtain the following expression for the speed:

$$V = \frac{e^{\beta t} - 1}{\sqrt{m/2A}(e^{\beta t} + 1)}.$$ [3.104]

From equation (3.104) we derive an expression describing the dependence $s(t)$. We rewrite correspondingly:

$$\begin{aligned} e^{\beta t} - 1 &= e^{\beta t/2} - e^{-\beta t/2} \\ e^{\beta t} + 1 &= e^{\beta t/2} + e^{-\beta t/2} \end{aligned}$$ [3.104a]

and

$$\frac{e^{\beta t/2} - e^{-\beta t/2}}{e^{\beta t/2} + e^{-\beta t/2}} = th(\beta t/2).$$ [3.104b]

Now, finally, we obtain

$$V = \frac{ds}{dt}; \quad \sqrt{m/2A} \int_0^s ds = \frac{2}{\beta} \int_0^t th(\beta t/2)\, dt.$$ [3.104c]

Integrating this, latter expression we obtain the required formula in the following form:

$$s = \frac{2}{m} \ln ch\,(\beta\, t\,/\,2).$$ [3.104d]

An example in MATHEMATICA language is given. Let us suppose that a device corresponding to Figure 3.22a) is described by the following parameters:

M = 1000 kg, p = 700 N/cm^2, Q = 5000 N, F = 75 cm^2,
Ψ = 100 N sec^2/m^2.
Then β = 4.36 1/sec, m = 0.2 1/m, A = 47.5 m/sec^2.
The solution for this specific example is:
s[t] = = 2/.2 Log[Cosh[4.36/2 t]]
j = Plot[2/.2 Log[Cosh[4.36/2 t]],{t,0,.1},AxesLabel−>{"t","s"},
PlotRange−>All,Frame−>True,GridLines−>Automatic]

It is more difficult to solve the problem for a case in which the value A varies, say, a function of the piston's displacement. Thus: $A(s)$. For this purpose we rearrange Equation (3.100) and substitute $y = s^2$ in that expression. We can then rewrite the equation in the form

$$dy\,/\,ds + my = 2A(s).$$ [3.105]

(*Note:* If $s^2 = y$, then $dy/dt = 2\dot{s}\ddot{s}$, which gives $\ddot{s} = dy/2ds$.)

Equation (3.105) is linear with respect to y, and thus, in accordance with the superposition principle, the solution must be expressed as the following sum:

$$y = y_0 + y_1,$$

where

y_0 = the solution of the homogeneous equation,
y_1 = the partial solution for A in the right-hand side of the equation.

We seek y_0 in the form

$$y_0 = Ye^{-Bs}.$$ [3.106]

Substituting Equation (3.106) into Equation (3.105), we find that

$$B = m.$$

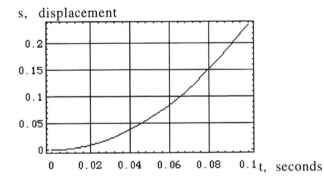

FIGURE 3.22a) Solution: piston displacement versus
time for the above-given mechanism.

Supposing, for example, that $A(s) = a_1 + a_2 s$; we seek the solution y_1 in the form

$$y_1 = b_1 + b_2 s.$$

Substituting this expression into Equation (3.105) and comparing the coefficients on both sides of the equation, we find that

$$b_1 = \frac{m a_1 - a_2}{m^2} \quad \text{and} \quad b_2 = \frac{a_2}{m}. \qquad [3.107]$$

For initial conditions $t = 0$ and $V = 0$, we also have $y = 0$. (Remember: $y = 2V.\dot{V}$). This condition gives the following expression for Y:

$$Y = \frac{a_2 - a_1 m}{m^2}. \qquad [3.108]$$

Finally, the complete solution can be written as

$$y = \frac{a_2 - a_1 m}{m^2} [e^{-ms} - 1] + \frac{a_2}{m} s. \qquad [3.109]$$

Substituting back the meaning of y we obtain

$$t = \int_0^s \frac{ds}{\sqrt{\dfrac{a_2 - a_1 m}{m^2}[e^{-ms} - 1] + \dfrac{a_2}{m} s}}. \qquad [3.110]$$

3.5   Pneumodrive

In general, the dynamics of a pneumomechanism may be described by a system of differential equations which depict the movement of the pneumatically driven mass and the changes in the air parameters in the working volume. The work of a pneumomechanism differs from that of a hydraulic mechanism in the nature of the outflow of the air through the orifices and the process of filling up the cylinder volume. Let us consider the mechanism for which the layout is given in Figure 3.23. Let us suppose the processes of outflow and filling up are adiabatic, and the pressure p_r in the receiver 1 is constant. From thermodynamics we know that the rate of flow of the air through the pipeline 2 may be described by the formula

$$G = \alpha F_p p_r \sqrt{\frac{2g}{RT_r} \cdot \frac{k}{k-1} \left[\beta^{2/k} - \beta^{(k+1)/k} \right]}, \qquad [3.111]$$

where

> G = the rate of flow,
> α = coefficient of aerodynamic resistance,
> F_p = cross-sectional area of pipe 2 (m^2),
> p_r = air pressure in the receiver 1,
> T_r = absolute temperature of the air in the receiver,

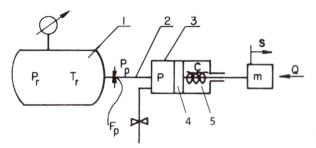

FIGURE 3.23 Layout of a pneumatic drive.

R = gas constant,
$\beta = p/p_r$ = ratio of the pressure in cylinder 3 to that in the receiver,
k = adiabatic exponent ($k = 1.41$).

At this stage, we must distinguish between supercritical and subcritical regimes. If we denote p_{cr} as the critical pressure, then

$$\beta_{\mathrm{cr}} = \frac{p_{cr}}{p_r} = \left[2/(k+1)\right]^{k(k+1)}$$ [3.112]

(for air $\beta_{cr} = 0.528$) for the supercritical regime $p_r > p/\beta_{cr}$, and for the subcritical regime, $p_r < p/\beta_{cr}$, where p = pressure in the cylinder.

To explain the meaning of the concept of the critical regime, let us make use of the example shown in Figure 3.24. In the left volume a pressure p_r, which is much higher than pressure p, is created and maintained permanently. At some moment of time the valve is opened and the gas from the left volume begins to flow into the right volume. Because we stated in the beginning that $p < p_r$, we have a situation where the ratio $p/p_r = \beta$ grows theoretically from zero to one (when p becomes equal to p_r). The air consumption through the connecting pipe develops as shown in Figure 3.25: from $\beta = 0$ up to $\beta = \beta_{cr}$ it stays constant until $\beta_{cr} = 0.528$, while for the range $\beta_{cr} < \beta \leq 1$ the consumption changes from G_{cr} to zero as the pressure in the right volume becomes equal to p_r.

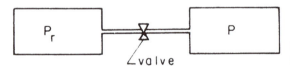

FIGURE 3.24 Layout of gas transfer from one volume into another.

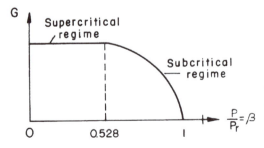

FIGURE 3.25 Air consumption in the connecting pipe of the layout shown in Figure 3.24 versus the pressure ratio in the connected volumes.

For the supercritical regime, the pressure p_1 in the cylinder input orifice is constant, i.e.,

$$p_1 = \beta_{cr} p_r = 0.528 p_r > p. \tag{3.113}$$

Substituting p_1 from Equation (3.113) into Equation (3.111), we obtain, for the supercritical flow rate G_{cr}

$$G_{cr} = \alpha F_p p_r \left[\frac{2}{k+1} \right]^{1/(k-1)} \sqrt{\frac{2g}{RT_r} \cdot \frac{k}{k+1}}. \tag{3.114}$$

Under standard conditions ($T_r = 293°$K) we have

$$G_{cr} = 0.023 \alpha F_p p_r. \tag{3.115}$$

Now let us find the time required to fill the cylinder from the initial value of the pressure p_0 to the final value p_1 ($p_r < 0.528 p_1$) while the air temperature remains constant. For a cylinder volume V_c we have

$$pV_c = G_c RT \tag{3.116}$$

where

G_c = instantaneous value of the air weight, and
T = the air temperature.

During an infinitesimal time period dt the pressure will change over a value dp, and the amount of air will change by $G_{cr} \, dt$. Thus,

$$(p + dp)V_c = RT(G_c + G_{cr} \, dt). \tag{3.117}$$

Substituting Equation (3.116) from Equation (3.117), we obtain

$$dt = \frac{V_c}{RT} \frac{DP}{G_{cr}}. \tag{3.118}$$

By integration we obtain

$$t_1 = \frac{V_c}{RT} \frac{p_1 - p_0}{G_{cr}}. \tag{3.119}$$

By substituting Equation (3.115) into Equation (3.119), we finally reach

$$t_1 = \frac{V_c}{0.67T} \frac{p_1 - p_0}{\alpha F_p p_r}. \tag{3.120}$$

When time t_1 has passed, two outcomes are possible:

1. The piston begins its movement earlier than t_1 (i.e., the movement begins before pressure p_1 appears in the cylinder volume).
2. The pressure in the cylinder is still not sufficient to initiate the movement of the piston.

For the first case we have to integrate Equation (3.118) in the limits from p_0 to some pressure p^* which is less than that at t_1. Thus, in place of Equation (3.120) we obtain

$$t_1 = \frac{V_c}{0.67T} \cdot \frac{p^* - p_0}{\alpha F_p p_r}.$$ [3.121]

For the second case, we have to continue our investigation for the subcritical regimes. For a subcritical regime,

$$p_1 = p > \beta_{cr} p_r = 0.528 p_r.$$

Now, in Equation (3.111) the value of β varies from the initial value of 0.528 to 1. In this case we must substitute in the differential Equation (3.118) G, which is not constant and is defined by Equation (3.111). Thus,

$$dt = \frac{V_c \sqrt{\dfrac{RT_r}{2g} \cdot \dfrac{k-1}{k}} \, dp}{RT \, \alpha F_p p_r \sqrt{\beta^{2/k} - \beta^{(k+1)/k}}}.$$ [3.122]

Since $p = \beta p_r$,

$$dp = p_r \, d\beta.$$

Therefore,

$$dt = \frac{0.022 V_c}{\alpha F_p T} \sqrt{T_r} \, \beta^{-1/k} \left[1 - \beta^{(k-1)/k} \right]^{-1/2} d\beta.$$ [3.123]

To integrate this equation, we introduce an auxiliary function,

$$\lambda^2 = 1 - \beta^{(k-1)/k},$$ [3.124]

which gives

$$2\lambda \, d\lambda = -\frac{k-1}{k} \beta^{-1/k} d\beta.$$ [3.125]

After substituting Equations (3.124) and (3.125) in Equation (3.123), we obtain

$$dt = \frac{0.022 V_c}{\alpha F_p T} \sqrt{T_r} \, \frac{2k}{k-1} d\lambda.$$ [3.126]

The limits of integration are determined by the initial value of β_0 and the critical value of β_{cr}. Thus,

$$t_2 = \frac{0.154 V_c}{\alpha F_p T} \sqrt{T_r} \left(0.41 - \sqrt{1 - \beta_0^{0.29}} \right) \text{sec.}$$ [3.127]

In the general case, the time t^* required to reach a pressure sufficient to move the piston and overcome the load and the forces of resistance may be written in the form

$$t^* = t_1 + t_2.$$

There is an additional time component t_0, which is the time needed by the pressure wave to travel from the valve to the orifice. This time can be estimated as follows:

$$t_0 = \frac{L}{V_s}, \qquad [3.128]$$

where L is the length of the pipe from the valve to the input to the cylinder, and V_s is the sound speed $V_s \cong 340 \text{ m/sec}$.

In Figure 3.26 we show the pressure development versus time.

To calculate the movement of the piston, we must deduce the differential equation for its displacement. This requires some intermediate steps. The thermodynamic equation for the air in the volume of the cylinder has the following form:

$$p_c V_c = G_c RT. \qquad [3.129]$$

(The subscript c indicates values belonging to the cylinder volume.)

For the volume V_c, we can substitute the obvious expression

$$V_c = F_c s,$$

where

$\quad s =$ the displacement of the piston, and
$\quad F_c =$ the cylinder's cross-sectional area

After differentiating Equation (3.129), we obtain

$$p_c dV_c + V_c dp_c = RT \, dG_c = RTG \, dt \qquad [3.130]$$

or

$$p_c \dot{s} + s \frac{dp_c}{dt} = \frac{RT}{F_c} G.$$

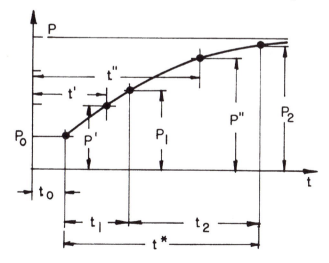

FIGURE 3.26 Pressure development versus time in a real pneumatic drive. Here: t_0 is determined from (3.128); t_1 from (3.120); t_2 from (3.127); p' and p'' are intermediate pressure values.

For the supercritical airflow regime, the piston moves in such a way that G_{cr} = constant (see Equation (3.115)). Thus, after integration of Equation (3.130), we obtain

$$sp_c = s_0 p_0 + \frac{RT}{F_c} G_{cr} t. \qquad [3.131]$$

(This expression is written for the initial conditions: when $t = 0$, then $s = s_0$ and $p = p_0$.)

For the layout shown in Figure 3.23 (where 3 = cylinder, 4 = piston, 5 = spring with stiffness c), the differential equation of the movement of the piston 4 may be written as

$$m \frac{dV}{dt} - p_c F_c + cs = Q, \qquad [3.132]$$

where

V = speed of the piston,
Q = the load, which includes the useful and harmful forces.

Now, by expressing p_c in terms of Equation (3.131) and substituting it into Equation (3.132), we obtain

$$m\ddot{s} - \frac{s_0}{s} p_0 F_c - \frac{RTG_{cr}}{s} t + cs = Q. \qquad [3.133]$$

This equation is essentially nonlinear even for the supercritical regime when G_{cr} = constant. It becomes more complicated for the subcritical regime of the air flow when we have to substitute for the value of G from Equation (3.111) for this regime.

We give here a numerical example in MATHEMATICA language for Equation (3.133), which also takes Expression (3.115) into consideration. The data for a device corresponding to that shown in Figure 3.23 are:

m = 400 kg,
c = 100 N/m,
α = 0.5,
F_p = 0.0002 m²,
p_r = 500000 N/m²,
p_0 = 100000 N/m²,
s_0 = 0.05 m,
F_c = 0.01 m²,
R = 28.7 Nm/kg °K.

The following three equations differ in the temperatures of the acting air. In Equations (f1), (f2), and (f3), the absolute temperatures are taken as T_1 = 293° K, T_2 = 340° K and T_3 = 400° K, respectively. The higher the value of the temperature, the faster the piston moves, and the longer the distance s it travels during equal time intervals. For instance (see Figure 3.26a)), in the considered example during 0.8 seconds for given temperatures, displacements of the piston correspondingly are

T_1 = 293⁰K $s_1 \cong$ 2.12 meters,   [f1]
T_2 = 340⁰K $s_2 \cong$ 2.40 meters,   [f2]
T_3 = 400⁰K $s_3 \cong$ 2.76 meters.   [f3]

s, displacement

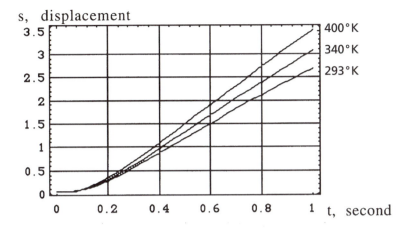

FIGURE 3.26a)   Displacement of the piston *s* versus time. For a mechanism shown in Figure 3.23, for different air temperatures: 293, 340, and 400°K.

```
f1 = 400 s"[t]−(.023 .5 28.7 293 .0002 500000 t +.05
100000 .01)/s[t]+ 4000+100 s[t]
j1 = NDSolve[{f1 = = 0,s[0] = = .05,s'[0] = = 0},{s[t]},{t,0,2}]
b1 = Plot[Evaluate[s[t]/.j1],{t,0,1},AxesLabel−>{"t","s"},
PlotRange−>All,Frame−>True,GridLines−>Automatic]
f2 = 400 s"[t]−(.023 .5 28.7 340 .0002 500000 t +.05
100000 .01)/s[t]+ 4000+100 s[t]
j2 = NDSolve[{f2 = = 0,s[0] = = .05,s'[0] = = 0},{s[t]},{t,0,1}]
b2 = Plot[Evaluate[s[t]/.j2],{t,0,1},AxesLabel−>{"t","s"},
PlotRange−>All]
f3 = 400 s"[t]−(.023 .5 28.7 400 .0002 500000 t +.05
100000 .01)/s[t]+ 4000 +100 s[t]
j3 = NDSolve[{f3 = = 0,s[0] = = .05,s'[0] = = 0},{s[t]},{t,0,1}]
b3 = Plot[Evaluate[s[t]/.j3],{t,0,1},AxesLabel−>{"t","s"},
PlotRange−>All]
s11 = Show[b1,b2,b3]
```

Let us now consider some simplified cases when Equation (3.133) can be made linear. As an example, we consider the situation in which the pressure p_c in the cylinder can be taken as constant during the movement of the piston. For such a simplified case, when the process can be assumed to be subcritical for most of the period of the piston's movement (which is the case for mechanisms with relatively long cylinders, low resistance of the manifold, and a relatively high load), we can approximate the description of the piston's movement by a linear differential equation. For instance, the mechanism shown in Figure 3.23 can be described by an equation which follows from Equation (3.133):

$$m\ddot{s} + cs = Q + p$$

or

$$\ddot{s} + q^2 s = A, \qquad\qquad\qquad\qquad [3.134]$$

where

$$q^2 = \frac{c}{m}; \quad A = \frac{Q+p}{m}; \quad p = p_c F_c.$$

The complete solution has the following form:

$$s = s_0 + s_1.$$

The solution's component s_0, corresponds to the homogeneous case of Equation (3.134) and is sought in the harmonic form,

$$s_0 = b_1 \cos \omega t + b_2 \sin \omega t. \qquad [3.135]$$

Substituting this solution into the corresponding form of Equation (3.134), we obtain

$$\omega = q.$$

The component s_1 is the partial solution of the same equation and its shape depends on the function A. Assuming that the external forces acting on the driven mass are a linear function of time in the form

$$A = a_1 t + a_2,$$

we must seek s_1 in an analogous form. Thus,

$$s_1 = C_1 t + C_2. \qquad [3.136]$$

Substituting Equation (3.136) into Equation (3.134), we obtain

$$C_1 = \frac{a_1}{q^2} \quad \text{and} \quad C_2 = \frac{a_2}{q^2};$$

and the complete solution then looks as follows:

$$s = b_1 \cos qt + b_2 \sin qt + \frac{a_1}{q^2} t + \frac{a_2}{q^2}. \qquad [3.137]$$

For initial conditions, when $t = 0$, the position of the driven mass $s = 0$, and the initial speed of the mass $\dot{s} = 0$, we have

$$b_1 = -\frac{a_2}{q^2} \quad \text{and} \quad b_2 = -\frac{a_1}{q^3}. \qquad [3.138]$$

Finally, we have

$$s = \frac{a_2}{q^2}[1 - \cos qt] - \frac{a_1}{q^3} \sin qt + \frac{a_1}{q^2} t. \qquad [3.139]$$

For the particular cases when $a_1 = 0$ or $a_2 = 0$, we obtain from Equation (3.139), respectively,

$$s = \frac{a_2}{q^2}[1 - \cos qt]$$

or

$$s = \frac{a_1}{q^2}\left[t - \frac{1}{q}\sin qt\right].$$

The second simplified case we consider here is that which occurs when Equation (3.133) can be reduced to the form

$$m\ddot{s} = \frac{RTG_{cr}}{s}t \qquad [3.140]$$

(assuming that the spring is extracted from the mechanism and $s_0 = 0$). We carry out linearization by substituting

$$\frac{s}{t} \cong V.$$

(Here: V = speed of the driven mass.) Thus, from Equation (3.140) we obtain

$$\dot{V} = \frac{RTG_{cr}}{mV} \qquad [3.141]$$

and for $T \approx$ constant

$$\int_0^v V\, dV = \frac{RTG_{cr}}{m}\int_0^t dt$$

or

$$\frac{V^2 m}{2RTG_{cr}} = t. \qquad [3.142]$$

Finally, we rewrite Equation (3.142) in the form

$$\frac{ds}{dt} = \sqrt{2RTG_{cr}t/m}$$

or

$$\int_0^s ds = \sqrt{2RTG_{cr}/m}\int_0^t \sqrt{t}\,dt, \qquad [3.143]$$

which gives

$$t = \sqrt[3]{9ms^2/8RTG_{cr}}. \qquad [3.144]$$

3.6 Brakes

In this section we consider a special type of drive, one which must reduce the speed of a moving element until complete cessation of movement of the element is achieved, i.e., a brake system. Such a mechanism must be able to facilitate speed reduction in the shortest possible time followed by locking of the drive as soon as the moving element has come to a stop. A braking mechanism is shown schematically in Figure 3.27. Here 1 is the driving motor, 2 the driven machine, 3 the drum of the brake, and 4 brake shoes.

The type of brakes we consider here can be classified according to the analytical approximation used to characterize the dependence of the brake torque on the variables of the system under consideration. Thus, the following kinds of brake torque T_b will be analyzed:

1. Constant torque: $T_b = \text{const} = T$.
2. Proportional-to-time torque: $T_b = T_0 + T_1 t$.   [3.145]
3. Proportional-to-displacement torque: $T_b = T_0 + T_2 \phi$.   [3.146]
4. Proportional-to-speed torque: $T_b = T_0 + T_3 \dot{\phi}$.   [3.147]

To simplify the consideration we assume that the resistance torque T_r for all the cases mentioned above is constant: $T_r = \text{const}$.

The general brake equation is

$$I\ddot{\phi} + T_b + T_r = 0,$$ [3.148]

All the solutions we seek here must answer the question: How long will the braking take? In other words, we need to know the amount of time needed for the moving part to reduce its speed from a value ω to a complete stop $\omega = 0$, and the value of the displacement executed by the element in that time.

For constant torque,

$$I\ddot{\phi} = -T_r - T;$$

hence

$$\dot{\omega} = -\frac{T_r + T}{I} \quad \text{and} \quad \int_{\omega}^{0} d\omega = -\int_{0}^{t} \frac{T_r + T}{I} dt$$

or

$$\omega = \frac{T_r + T}{I} t$$ [3.149]

and

$$\phi = \frac{T_r + T}{2I} t^2.$$ [3.150]

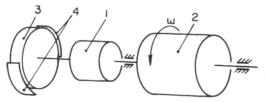

FIGURE 3.27   Layout of a braking mechanism.

For torque proportional to time,

$$I\ddot{\phi} = -(T_0 + T_r) - T_1 t \qquad [3.151]$$

or

$$\dot{\omega} = -\frac{T_0 + T_r + T_1 t}{I};$$

hence

$$\int_{\omega}^{0} d\omega = -\int_{0}^{t} \frac{T_0 + T_r + T_1 t}{I} dt$$

or

$$\omega = \frac{T_0 + T_r}{I} t + \frac{T_1}{2I} t^2 \qquad [3.152]$$

and

$$\phi = \frac{T_0 + T_r}{2I} t^2 + \frac{T_1}{6I} t^3. \qquad [3.153]$$

For torque proportional to displacement,

$$I\ddot{\phi} + T_r + T_0 + T_2 \phi = 0$$

or

$$\ddot{\phi} + \frac{T_2}{I} \phi = -\frac{T_r + T_0}{I}. \qquad [3.154]$$

In this case the solution is composed of two components, and therefore,

$$\phi = \phi_0 + \phi_1;$$

the homogeneous solution ϕ_0 may be sought in the harmonic form. Thus,

$$\phi_0 = a_1 \cos\Omega t + a_2 \sin\Omega t.$$

Substituting the harmonic form in the homogeneous variant of Equation (3.154) we obtain

$$\Omega = \sqrt{\frac{T_2}{I}}. \qquad [3.155]$$

The partial solution can then be sought in the form

$$\phi_1 = b = \text{const.}$$

Substituting this solution into Equation (3.154), we find, for ϕ_1:

$$\phi_1 = -\frac{T_r + T_0}{T_2}. \qquad [3.156]$$

Hence,

$$\phi = a_1 \cos\sqrt{\frac{T_2}{I}}\,t + a_2 \sin\sqrt{\frac{T_2}{I}}\,t - \frac{T_r + T_0}{T_2}.$$

From the initial conditions, for $t = 0$, $\phi = 0$, and $\dot{\phi} = \omega_0$, it follows that

$$a_1 = \frac{T_r + T_0}{T_2},$$

$$a_2 = \omega_0 \sqrt{\frac{I}{T_2}}.$$

[3.157]

Thus,

$$\phi = \frac{T_r + T_0}{T_2} \cos\sqrt{\frac{T_2}{I}}\,t + \omega_0 \sqrt{\frac{I}{T_2}} \sin\sqrt{\frac{T_2}{I}}\,t - \frac{T_r + T_0}{T_2}.$$

[3.158]

Differentiating Equation (3.158) we obtain for $\omega(t)$:

$$\omega = \omega_0 \cos\sqrt{\frac{t_2}{I}}\,t - \frac{T_r + T_0}{T_2}\sqrt{\frac{T_2}{I}} \sin\sqrt{\frac{T_2}{I}}\,t.$$

[3.159]

For torque proportional to speed,

$$I\ddot{\phi} + T_r + T_0 + T_3\dot{\phi} = 0.$$

After transformation, this equation can be rewritten as

$$\ddot{\phi} + \frac{T_3}{I}\dot{\phi} = -\frac{T_r + T_0}{I}$$

or

$$\dot{\omega} + \frac{T_3}{I}\omega = -\frac{T_r + T_0}{I},$$

[3.160]

where ω is a composite solution of two components: $\omega = \omega_0 + \omega_1$.

The homogeneous solution ω_0 may be found in an exponential form. Thus,

$$\omega_0 = a\exp[bt].$$

Substituting the exponential form into the homogeneous variant of Equation (3.160), we find for b

$$b = -\frac{T_3}{I}.$$

[3.161]

The partial solution ω_1 is sought in the form

$$\omega_1 = C = \text{const.}$$

Substituting ω_1 into Equation (3.160), we obtain

$$\omega_1 = -\frac{T_r + T_0}{T_3}.$$
[3.162]

Hence,

$$\omega = a\exp\left[-\frac{T_3}{I}t\right] - \frac{T_r + T_0}{T_3}.$$

For the initial condition $t = 0$, $\omega = \omega_0$, we find the value of a:

$$a = \omega_0 + \frac{T_r + T_0}{T_3}.$$

Finally, the solution of Equation (3.160) has the form

$$\omega = \left(\omega_0 + \frac{T_r + T_0}{T_3}\right)\exp\left(-\frac{T_3}{I}t\right) - \frac{T_r + T_0}{T_3}.$$
[3.163]

Integrating this expression we will find the function $\phi(t)$. Indeed, from Equation (3.163) it follows that

$$\int_0^\phi d\phi = \int_0^t \left\{\left(\omega_0 + \frac{T_r + T_0}{T_3}\right)\cdot\exp\left(-\frac{T_3}{I}t\right) - \frac{T_r + T_0}{T_3}\right\}dt$$

or

$$\phi = \left[\omega_0 + \frac{T_r + T_0}{T_3}\right]\left[\frac{I}{T_3}\right]\left[1 - \exp\left(-\frac{T_3}{I}t\right)\right] - \frac{T_r + T_0}{T_3}t.$$
[3.164]

Equations (3.150), (3.153), (3.158), and (3.164) answer the question formulated at the beginning of this section by showing the dependencies $\phi(t)$ for all four cases under consideration. Equations (3.149), (3.152), (3.159), and (3.163) express the changes in speed in terms of time, thus describing the form of the function $\omega(t)$.

3.7   Drive with a Variable Moment of Inertia

We can imagine a number of systems in which the moment of inertia changes during rotation—for instance, the situation represented in Figure 3.28a). In this example a vessel rotating around a vertical axis during its rotation is filled with some granular material (say, sand). The rotation is achieved by a specific drive means. Obviously, the rotating mass changes in a certain manner and, thus, its moment of inertia changes. An analogous example is given in Figure 3.28b). Here the vessels are driven translationally and filled by some material while being accelerated.

Other examples are shown in Figure 3.29a) and b). In Figure 3.29a) the column rotates around the vertical axis carrying a horizontal beam on which a sliding body with an inertial mass m is mounted. The distance r between the mass center of the

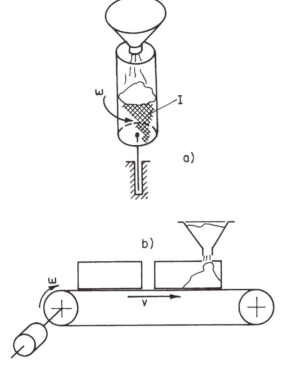

FIGURE 3.28 Models of systems with variable masses: a) Changing moment of inertia in a rotating motion; b) Changing mass in a translational motion.

body and the geometrical axis of the column may be varied, thereby changing the moment of inertia of the whole system. The case given in Figure 3.29b) is another illustration of the same case. The drives providing the rotating speed ω in both cases are influenced by the variable moments of inertia of these systems. We will now consider some simplified calculation examples corresponding to the cases described above.

Case 1 (Figure 3.28a))

The general equation of motion has the following form:

$$d(I\omega)/dt = T(\omega). \tag{3.165}$$

Here, we consider particular forms of the driving torque $T(\omega)$ and the moment of inertia $I(t)$. We assume that these forms appear as follows:

$$I(t) = I_0 + I_1 t \tag{3.166}$$

and

$$T(\omega) = T_0 - \tau_1 \omega. \tag{3.167}$$

Thus, substituting Equations (3.166) and (3.167) into Equation (3.165) we obtain

$$\int_0^\omega \frac{d\omega}{T_0 - T_1\omega} = \int_0^t \frac{dt}{I_0 + I_1 t} \quad (\text{here } T_1 = \tau_1 + I_1). \tag{3.168}$$

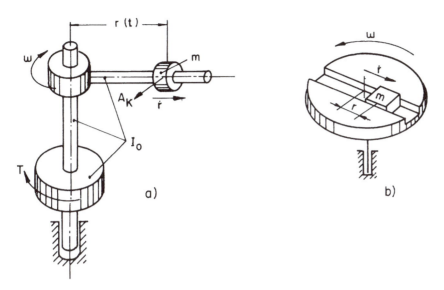

FIGURE 3.29 Variable moment of inertia caused by changing geometry of the system: a) Cylindrical manipulator; b) Indexing table with a movable body. (Here Coriolis acceleration occurs.)

Thus,

$$\frac{1}{T_1}\ln\frac{T_0-T_1\omega}{T_0}=\frac{1}{I_1}\ln\frac{I_0+I_1t}{I_0} \qquad [3.169]$$

and

$$T_1\omega=-T_0\left\{\left[\frac{I_0+I_1t}{I_0}\right]^{T_1/I_1}+1\right\}. \qquad [3.170]$$

Integrating this expression further with respect to time, we obtain the relationship between the angle of rotation and time:

$$\int_0^\phi d\phi=-\frac{T_0}{T_1}\int_0^t\left\{\left[\frac{I_0+I_1t}{I_0}\right]^{T_1/I_1}+1\right\}dt. \qquad [3.171]$$

From Equation (3.171) it follows that

$$\phi=\frac{T_0}{T_1}I_0^{-T_1/I_1}\left[T_1+I_1\right]^{-1}\left\{I_0^{((T_1/I_1)+1)}-\left(I_0+I_1t\right)^{((T_1/I_1)+1)}\right\}+\frac{T_0}{T_1}t. \qquad [3.172]$$

Obviously, for the layout shown in Figure 3.28b an expression of the same form can be derived, the only difference being that in place of moments of inertia we substitute masses; in place of angular speeds, translational speed; in place of angle of rotation, translational displacement; and in place of torques, forces. Thus, the assumed dependencies describing the changes in the mass $m(t)$ and the driving force $F(V)$ are obtained in a form analogous to that of Equation (3.166) and we have

$$m(t)=m_0+m_1t \quad \text{and} \quad F(V)=F_0+f_1V, \qquad [3.173]$$

and the final result for displacement $s(t)$ will have the form

$$s = \frac{F_0}{F_1} m_0^{-F_1/m_1} \left[F_1 + m_1 \right]^{-1} \left\{ m_0^{((F_1/m_1)+1)} - \left[m_0 + m_1 t \right]^{((F_1/m_1)+1)} \right\} + \frac{F_0}{F_1} t. \qquad [3.174]$$

Here $F_1 = f_1 + m_1$.

Case 2 (Figure 3.29a) and b))

Analyzing these systems, we must also take into consideration Expression (3.165). Thus the equation of motion describing the rotation of the column (Figure 3.29a) appears as follows:

$$d \left\{ m \left[r(t, \phi, \dot{\phi}) \right]^2 \omega \right\} / dt = T(\omega), \qquad [3.175]$$

where

  m = the mass moving along the beam, during the rotation of the column,
  $\omega(t)$ = the angular speed of the column,
  $r(t)$ = the position of the mass m along the beam,
  $T(\omega)$ = the column driving torque.

Let us suppose that:

$$r(t) = r_0 + r_1 t; \quad r_1 = \dot{r}; \quad \text{and} \quad T(\omega) = T_0 - T_1 \omega. \qquad [3.176]$$

Substituting the chosen functions (3.176) into Equation (3.175) and rearranging the equation, we obtain

$$\dot{\omega} + \frac{2\omega r_1}{r_0 + r_1 t} = \frac{T_0 - T_1 \omega}{m(r_0 + r_1 t)^2} \qquad [3.177]$$

or, rewriting this equation, we arrive at the form

$$\dot{\omega} + \frac{\omega}{r_0 + r_1 t} \left[2 r_1 + \frac{T_1}{m(r_0 + r_1 t)} \right] = \frac{T_0}{m(r_0 + r_1 t)^2}. \qquad [3.178]$$

Denoting:

$$A(t) = \frac{2 r_1}{r_0 + r_1 t} + \frac{T_1}{m(r_0 + r_1 t)^2} \quad \text{and} \quad B(t) = \frac{T_0}{m(r_0 + r_1 t)^2},$$

we can rewrite Equation (3.178) in the form

$$\dot{\omega} + \omega A(t) = B(t). \qquad [3.179]$$

For this equation the solution is sought as

$$\omega = e^{-\int_0^t A(t) dt} \left[\int_0^t B(t) e^{\int_0^t A(t) dt} dt \right].$$

The integral in the latter expression cannot be solved in a nonnumerical way.

Therefore, we show here some computation examples of Equation (3.178) made with the MATHEMATICA program. For this purpose, let us decide about the values of the parameters constituting this equation, as follows:

$$r_0 = 0.1\,\mathrm{m},\; r_1 = 0.1\,\mathrm{m/sec},\; T_0 = 5\,\mathrm{Nm},\; T_1 = 1\,\mathrm{N\,sec}.$$

In addition, we take three different values for the moving mass: $m = 2\,\mathrm{kg}$, $3\,\mathrm{kg}$, and $4\,\mathrm{kg}$.

In keeping with these parameters and for each of the chosen mass values, we write the needed expressions and obtain the solutions in graphic form as follows.

The calculations, as was mentioned above, were carried out for three different mass values. This fact is reflected in the three curves on each graph shown in Figure 3.30a) and b). The upper ones belong to the smallest mass value (in our case $m = 2\,\mathrm{kg}$), and the lowest curves to the largest mass ($m = 4\,\mathrm{kg}$).

```
f1=w'[t]+w[t]*(.2+.5/(.1+.1*t))/(.1+.1*t)−2.5/(.1+.1*t)^2
y1=NDSolve[{f1==0,w[0]==0},w,{t,0,.1}]
z1=Plot[Evaluate[w[t]/.y1],{t,0,.1}, AxesLabel-
>{"t,time","w,speed"}]

f2=w'[t]+w[t]*(.2+.333/(.1+.1*t))/(.1+.1*t)−1.7/(.1+.1*t)^2
y2=NDSolve[{f2==0,w[0]==0},w,{t,0,.1}]
z2=Plot[Evaluate[w[t]/.y2],{t,0,.1}, AxesLabel-
>{"t,time","w,speed"}]

f3=w'[t]+w[t]*(.2+.25/(.1+.1*t))/(.1+.1*t)−1.25/(.1+.1*t)^2
y3=NDSolve[{f3==0,w[0]==0},w,{t,0,.1}]
z3=Plot[Evaluate[w[t]/.y3],{t,0,.1}, AxesLabel-
>{"t,time","w,speed"}]

x1=Show[z1,z2,z3]
```

First we compute the behavior of the column at the very beginning of the motion during 0.1 sec (Figure 3.30a)), afterwards for the rest of the action time which is taken, in this case about 1 sec (Figure 3.30b)).

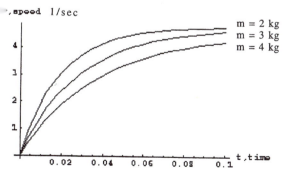

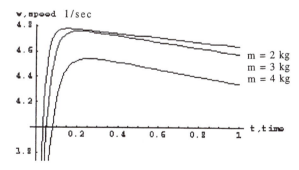

ɔURE 3.30a)   Angular speed of the column ɪveloped during one-tenth of a second according Equation (3.178) versus time.

FIGURE 3.30b)   Angular speed of the column developed during one second according to Equation (3.178) versus time.

To complete this brief discussion we show another, simplified case of the same mechanism when the driving torque can be assumed to be constant, i.e., $T_1 = 0$. The equations written in MATHEMATICA also follow the solution given in Figure 3.31.

f5=w'[t]+w[t]*.2/(.1+.1*t)−1.255/(.1+.1*t)^2
y5=NDSolve[{f5==0,w[0]==0},w,{t,0,1}]
z5=Plot[Evaluate[w[t]/.y5],{t,0,1},AxesLabel->{"t,time","w,speed"}]

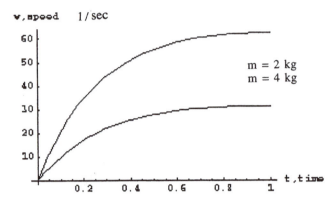

FIGURE 3.31 Angular speed of the column developed during one second when the driving torque is constant according to Equation (3.178) versus time.

Exercise 3E-1

The mechanism shown in Figure 3E-1 consists of two drums, 1 and 2, with moments of inertia $I_1 = 0.01$ kg m² and $I_2 = 0.045$ kg m², respectively. The drums are connected by a gear transmission with a ratio of 1:3 so that drum 1 rotates faster than drum 2. Drum 1, with a radius $R = 0.05$ m, is driven by a spring via a rope, while drum 2 is braked by torque $T_r = 5$ Nm. The stiffness of the spring $c = 500$ N/m. The drum was initially rotated for one revolution, stretching the spring; thereafter, at a particular time, the system was freed. Calculate the time needed by the drum 1 to complete 0.5 of a revolution under the influence of the spring overcoming the torque T_r.

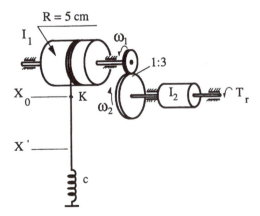

FIGURE 3E-1

Exercise 3E-2

A blade with mass $m = 1$ kg driven by a spring is shown in Figure 3E-2. In the beginning, the spring is compressed by a distance $L_0 = 0.2$ m. When freed, the blade descends for a distance $L_1 = 0.1$ m until it comes into contact with a wire having a thickness $h = 0.004$ m. The required cutting force $P = 800$ N. The spring has a linear characteristic shown in the figure with a constant $c = 5000$ N/m. Calculate the time needed by the blade to complete cutting the wire or, in other words, to travel the distance $L = L_1 + h$. At initial time, $t = 0$, the blade is at rest.

Exercise 3E-3

The DC electromotor shown in Figure 3E-3, provided with a drum, lifts a mass $m = 10$ kg by means of a rope wound on the drum with a radius $r = 0.035$ m. The rotating part of this system (rotor of the motor, shaft, and drum) has a moment of inertia $I_0 = 0.005$ kg m². The motor has a linear characteristic $T = 5 - 0.05\,\omega$ Nm (where $T = $ torque, $\omega = $ angular speed). Calculate the time needed to obtain a rotational speed $\omega = 10$ 1/sec, and the height reached by mass m at this moment in time. At the beginning of the process the motor is at rest.

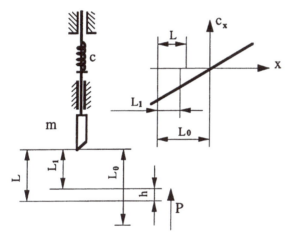

FIGURE 3E-2

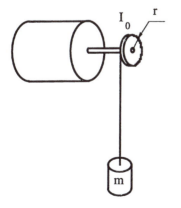

FIGURE 3E-3

Exercise 3E-3a)

The given mechanism (in Figure 3E-3a) consists of a DC electromotor with a characteristic $T = 4 - 0.1\,\omega$ Nm, a speed reducer carrying out a ratio 1:4, and a driven pinion with a radius $R = 2$ cm and moment of inertia together with the attached wheel of $I_1 = 0.01$ kg m². The pinion of radius 2 cm is engaged with a rack of mass $m_1 = 5$ kg. The rack lifts a mass $m_2 = 5$ kg by means of a flexible cable. The rotor of the motor, its shaft, and the wheel fastened to it have a moment of inertia $I_0 = 0.001$ kg m². Find the speed of the mass m_1 after the motor has been working for 0.1 second; find the distance the mass m_1 travels in this time. At the beginning of the process the motor is at rest.

Answer the same questions for the case in which the drive is performed by an AC electromotor. The characteristic of the latter is (see Expression (3.48)):

$$T = 2 \cdot T_m \frac{s_m \cdot s}{s_m^{\,2} + s^2} = 2 \cdot 1 \frac{0.1 \cdot s}{0.01 + s^2}, \qquad \text{[3.180]}$$

where

$T_m = 1$ Nm,
$s_m = 0.1$, and
$\omega_0 = 157$ 1/sec.

Exercise 3E-3b)

A screw jack driven by a DC electromotor is shown in Figure 3E-3b). The characteristic of the motor is $T = 4 - 0.1\,\omega$ Nm. The speed ω developed by the motor is reduced by a transmission with a ratio 1:3. The "screw-nut" device lifts a mass $m = 200$ kg. The moment of inertia of the rotor and the wheel attached to it $I_0 = 0.001$ kg m²; the moment of inertia of the screw and its driving wheel $I_1 = 0.01$ kg m²; and the pitch of the screw is $h = 10$ mm. Find the height the mass will travel during time $t = 0.5$ sec. At the beginning of the process the motor is at rest.

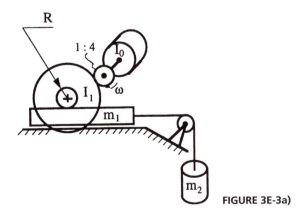

FIGURE 3E-3a)

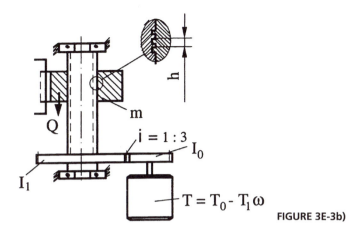

FIGURE 3E-3b)

Exercise 3E-4

The hydraulic cylinder shown in Figure 3E-4 is described by the following parameters:

Pressure of the working liquid $p = 500$ N/cm^2,
Force of resistance $Q = 5000$ N,
Cross-area of the piston $= 50$ cm^2,
Moving mass $M = 200$ kg, and
Coefficient of hydraulic resistance $\psi = 150$ Nsec^2/m^2.

Calculate the time needed to develop a piston speed $V = 5$ m/sec; Estimate the time needed to obtain a displacement $s = 0.1$ m.

Exercise 3E-4a)

A hydraulic drive is shown in Figure 3E-4a. The cylinder with an inner diameter $D_0 = 0.08$ m is used to move a piston rod with mass $m_1 = 100$ kg. The piston rod ($D = 0.02$ m) serves as a rack engaged with a gear wheel block with a ratio of radii $R/r = 2.5$ and $r = 0.04$ m. The moment of inertia of the block $I = 0.2$ kg m^2. The block drives a mass $m_2 = 50$ kg. The hydraulic pressure on the input of the device $p = 200$ N/cm^2. Coefficient of hydraulic resistance in the piping $\psi = 120$ Nsec^2/m^2. Find

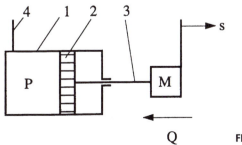

FIGURE 3E-4

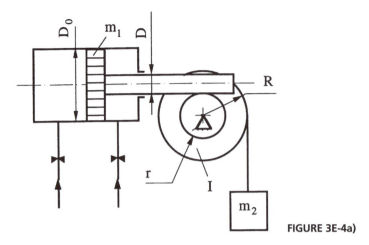

FIGURE 3E-4a)

the time needed to achieve a speed of the piston $V = 2\,\text{m/sec}$ when the height of the mass increases; find the distance travelled by the piston. At the beginning of the process the piston is at rest.

Exercise 3E-5

Figure 3E-5 shows a pneumatic cylinder serving as an elevator. Pressure $P_r = 50\,\text{N/cm}^2$ to this elevator is supplied from an air receiver 2 located about $L = 10\,\text{m}$ away. The initial position of the piston $1 = 0.1\,\text{m}$. The mass of the elevator handles $m = 400\,\text{kg}$ for case a), and $m = 550\,\text{kg}$ for case b). The stroke s_{max} is about 1.5 m. Other pertinent data are:

Inner diameter of the cylinder $D = 0.15\,\text{m}$,
Inner diameter of the pipe $d = 0.012\,\text{m}$,
Absolute temperature of the air in the receiver $T_r = 293°\text{K}$, and
Coefficient of aerodynamic resistance $\alpha = 0.5\,\text{sec/m}$.

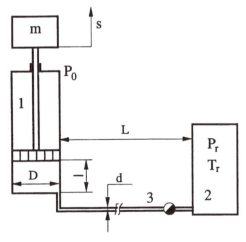

FIGURE 3E-5

Calculate the time needed to lift the mass (for both cases separately) from the moment in time that the valve 3 is actuated until the time the mass reaches point s_{max}.

Exercise 3E-5a)

The pneumatic system in Figure 3E-5a) consists of a volume $V_c = 0.2$ m², a pipe with a diameter $d = 0.5''$ and length $L = 20$ m, and a coefficient of resistance $\alpha \cong 0.5$ sec/m. The provided pressure $P_r = 50$ N/cm². The absolute air temperature in the system $T_r = 293°$ K. Find the time needed to bring the pressure P_c in the volume V_c to the value $P_r = 50$ N/cm².

Exercise 3E-5b)

The pneumatically actuated jig in Figure 3E-5b) is used to support a weight $Q = 5,000$ N. The designations are clear from the figure. The inner diameter of the cylinder $D = 0.125$ m, the initial volume of the cylinder $V_c = 0.002$ m³, the diameter of the piping $d = 0.5''$, the constant air pressure in the system $P_r \cong 60$ N/cm², the air temperature $T_c = 293°$ K, and the coefficient of aerodynamic resistance in the piping $\alpha = 0.5$ sec/m. The distance from the valve to the cylinder $L = 20$ m. Find the time needed to close the jig from the time the valve is actuated (the real travelling distance of the piston $s \cong 0$).

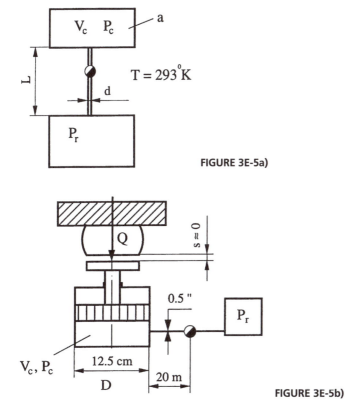

FIGURE 3E-5a)

FIGURE 3E-5b)

Exercise 3E-5c)

The machine shown in Figure 3E-5c) must be stopped by a brake. The initial speed of the drum with a moment of inertia $I_1 = 0.1$ kg m^2 is $n_0 = 1,500$ rpm. The ratio of the speed reducer connecting the drum to the brakes is $z_1/z_2 = 3.16$. The moment of inertia of the brake's drum is $I_2 = 0.01$ kg m^2. Find the time required for the machine to stop completely when the braking torque $T = 5 + 4\,\phi$ Nm (where ϕ is the rotation angle); find the time needed for the machine to stop when the braking torque $T = 5 + 4\,\omega$ Nm (the final speed ω_f is about 0.5% of that in the beginning).

Exercise 3E-6

Figure 3E-6 shows a mechanism consisting of a rotating column with a moment of inertia I_0, to which a lever is connected by means of a hinge. The hinge moves so that the angle θ changes according to the law $\theta = \alpha t$, where α = constant. A concentrated mass m is fastened to the end of the lever, whose length is r. Write the equation of motion of this system when a constant torque T is applied to the column. Solve the equation for the following data:

$I_0 = 0.01$ kg m^2,
$m = 2$ kg,
$T = 0.1$ Nm,
$\alpha = 0.5$ 1/sec, and
$r = 0.5$ m.

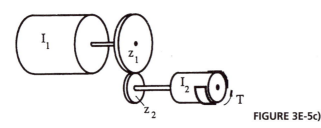

FIGURE 3E-5c)

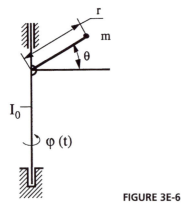

FIGURE 3E-6

Exercise 3E-6a)

A mass m is moving along the diameter of a disc-like rotating body with a moment of inertia I_0 (see Figure 3E-6a)). The law of motion $r(t)$ of this mass relative to the center of the disc is $r = R_0 \cos \alpha t$ where R_0 and ω are constant values. The mechanism is driven by a DC electric motor with a characteristic $T = T_1 - T_0 \omega$. Write the equation of motion for the disc in this case. At the beginning of the process the motor is at rest.

Exercise 3E-7

Consider the electromagnet shown in Figure 3E-7. How will its response time change if:

The number W of winds on its coil is doubled? The voltage U is doubled? The mass m of the armature is doubled? (In each of the above-mentioned cases the rest of the parameters stay unchanged.)

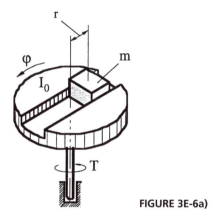

FIGURE 3E-6a)

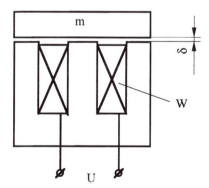

FIGURE 3E-7

4

Kinematics and Control of Automatic Machines

4.1 Position Function

We begin our discussion with the study of case 4, described in Chapter 1 and shown in Figure 1.5. The drive, whatever its nature, imparts the required movement to the tools through a mechanical system that controls the sequence and regularity of the displacements. Every mechanism has a driving link and a driven link. The first question in kinematics is that of the relationship between the input (driving motion) and the output (driven motion). Let us denote:

x = the input motion, which can be linear or angular,
s = the output motion, which also can be linear or angular.

Thus, we can express the relationship between these two values as:

$$s = \Pi(x). \tag{4.1}$$

We call $\Pi(x)$ the position function.
From Equation (4.1), it follows that

$$\dot{s} = \Pi'(x) \cdot \dot{x} \tag{4.2}$$

and

$$\ddot{s} = \Pi''(x)\dot{x}^2 + \ddot{x}\Pi'(x). \tag{4.3}$$

The importance of Equation (4.2) is that it expresses the interplay of the forces: by multiplying both sides of Equation (4.2) by the force (or torque, when the motion is angular), we obtain an equation for the power on the driving and driven sides of the mechanism (at this stage frictional losses of power can be neglected). Hence,

$$\dot{s}\,F_{output} = \dot{x}\,F_{input}\,.$$

From Equation (4.2),

$$\dot{s}/\dot{x} = \Pi'(x);$$

then

$$F_{input}/F_{output} = \Pi'(x). \qquad [4.4]$$

Obviously, $\Pi'(x)$ is the ratio between the driving and driven links. In the particular case where the input motion can be considered uniform (i.e., $\dot{x} = $ constant and $\ddot{x} = 0$), it follows, from Expression (4.3), that

$$\ddot{s} = \Pi''(x)\dot{x}^2. \qquad [4.5]$$

The designer often has to deal with a chain of n mechanisms, for which

$$s_n = \Pi_n\Big\{\Pi_{n-1}\Big[\Pi_{n-2}\big(...\Pi_1[x]\big)\Big]\Big\}. \qquad [4.6]$$

To illustrate this, let us take the Geneva mechanism as an example for calculation of a Π function. The diagram shown in Figure 4.1 will aid us in this task. It is obvious that this mechanism can be analyzed only in motion, that is, when the driving link is engaged with the driven one. For the four-slot Geneva cross shown on the right side of the figure, this occurs only for 90° of the rotation of the driving link; during the remainder of the rotation angle (270°) the driving link is idle. To avoid impact between the links at the moment of engagement, the mechanism is usually designed so that, at that very moment, there is a right angle between 0_1A and 0_2A.

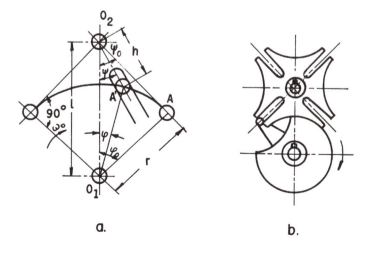

FIGURE 4.1   Layout of a four-slot Geneva mechanism.

A scheme of this mechanism is presented on the left side of the figure. Here $0_1A = r$ for the driving link at the very moment of engagement, and $0_1A' = r$ is constant at all intermediate times. The number of slots n in the cross determines the angle ψ_0, i.e.,

$$\psi_0 = 180°/n,$$ [4.7]

and, obviously,

$$\phi_0 = 90° - \frac{180°}{n}.$$

From the triangle 0_10_2A we obtain

$$l = \frac{r}{\cos\phi_0}.$$ [4.8]

Applying Equation (4.8) to the triangle $0_10_2A'$ we can express

$$\overline{0_10_2} = l = r\cos\phi + h\cos\psi,$$ [4.9]

where the value of ψ is unknown, and the length of $0_2A' = h$.
    From the sine law, we obtain

$$\frac{h}{\sin\phi} = \frac{r}{\sin\psi}.$$ [4.10]

And thus from Equations (4.9) and (4.10):

$$\frac{r}{\cos\phi_0} = r\cos\phi + r\frac{\cos\psi\sin\phi}{\sin\psi}.$$ [4.11]

Denoting $\lambda = r/l$ and simplifying Equation (4.11), we obtain

$$\tan\psi = \frac{\lambda\sin\phi}{1 - \lambda\cos\phi}$$ [4.12]

or

$$\psi = \Pi(\phi) = \arctan\frac{\lambda\sin\phi}{1 - \lambda\cos\phi}.$$ [4.13]

From Equation (4.13) we obtain the following expression for the velocity of the driven link $\dot{\psi}$:

$$\dot{\psi} = \frac{\lambda(\cos\phi - \lambda)}{1 - 2\lambda\cos\phi + \lambda^2}\frac{d\phi}{dt}$$ [4.14]

or

$$\Pi'(\phi) = \frac{\lambda(\cos\phi - \lambda)}{1 - 2\lambda\cos\phi + \lambda^2}.$$

When $d\phi/dt = \omega_0 = $ constant, we obtain

$$\dot{\psi} = \frac{\lambda(\cos\phi - \lambda)\omega_0}{1 - 2\lambda\cos\phi + \lambda^2}.$$ [4.15]

For acceleration of the driven link we obtain

$$\varepsilon = \ddot{\psi} = \frac{-\lambda(1-\lambda^2)\sin\phi(\dot\phi)^2}{(1-2\lambda\cos\phi+\lambda^2)^2} + \frac{\lambda(\cos\phi-\lambda)\ddot\phi}{1-2\lambda\cos\phi+\lambda^2}.$$ [4.16]

When $\dot\phi = \omega_0$ we can simplify the expression to the form

$$\varepsilon = \frac{-\lambda(1-\lambda^2)\sin\phi}{(1-2\lambda\cos\phi+\lambda^2)^2}\omega_0^2$$

and

$$\Pi''(\phi) = \frac{-\lambda(1-\lambda^2)\sin\phi}{(1-2\lambda\cos\phi+\lambda^2)^2}.$$ [4.17]

Graphical interpretations of Expressions (4.15) and (4.17) are shown in Figure 4.2.

This mechanism is very convenient whenever interrupted rotation of a tool is necessary. Naturally, various modifications of these mechanisms are possible. For instance, two driving pins can drive the cross, as in Figure 4.3. The resting time and the time of rotation for this mechanism are equal. More than four slots (the minimum number of slots is three) can be used. Figure 4.4 shows a Geneva mechanism with eight slots. One driver can actuate four (or some other number of) mechanisms, as in Figure 4.5. The durations of the resting times can be made unequal by mounting the driving pins at angle λ (see Figure 4.6). One stop will then correspond to an angle $(\lambda - 90°)$, the other to an angle $(270° - \lambda)$. Another modification of such a mechanism is shown in Figure 4.7.

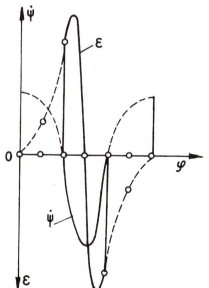

FIGURE 4.2   Speed and acceleration of the driven link of the Geneva mechanism shown in Figure 4.1.

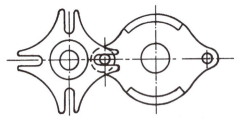

FIGURE 4.3   Geneva mechanism with two driving pins.

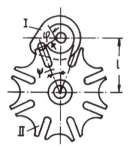

FIGURE 4.4   Eight-slot Geneva mechanism.

FIGURE 4.5   Geneva mechanism with multiple driven links.

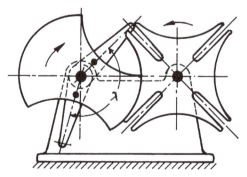

FIGURE 4.6   Asymmetrical Geneva mechanism with two driving pins.

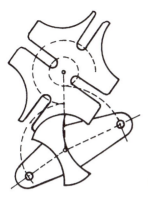

FIGURE 4.7   Modified Geneva mechanism.

Here, the slots have different lengths and the driving pins are located on different radii. These changes also produce different stop durations. Figure 4.8 shows a Geneva mechanism with internal engagement. The driven body a rotates for an angle $\psi = 2\pi/4$ while the driving link b passes through an angle $\phi = \pi + \psi$. We will return to these mechanisms later.

Another example of the derivation of the position function is shown for a crankshaft mechanism (Figure 4.9). Omitting the intermediate strokes, we obtain for the coordinates y and x of point M on the connecting rod the following expressions:

$$y = r\cos\phi + m\cos\left(\delta - \arcsin\frac{r\sin\phi - l}{\overline{BC}}\right),$$

$$x = r\sin\phi + m\sin\left(\delta - \arcsin\frac{r\sin\phi - l}{\overline{BC}}\right),$$ [4.18]

where $\phi = \omega t$, and ω = angular speed of the driving crank.

Differentiating y and x, with respect to time, we obtain the components of the speed in the corresponding directions. The components of the acceleration can be obtained

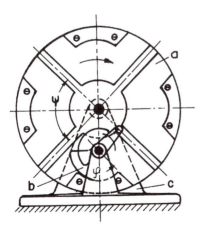

FIGURE 4.8   Internal engagement Geneva mechanism.

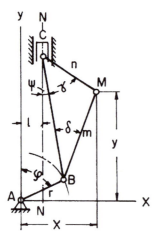

FIGURE 4.9   Crankshaft mechanism.

in the same manner. Thus, the position function is expressed separately for each coordinate. Obviously, this analytical approach can be useful for analyzing other specific kinds of trajectories. Figure 4.10 shows the different trajectories of several points belonging to the same link: the connecting rod of a four-bar mechanism.

[*Note:* The reader may, of course, use the modern vector loop method to make the kinematic analyses of designed or chosen mechanisms. This approach is especially useful for computerized calculations or animation of the mechanism on the computer's screen. However, the author's opinion remains that the choice of mathematical approach is a private affair depending on personal taste, predilection, etc. In the author's opinion the offered approach gives a better physical understanding of the kinematic events. His duty was to show that such-and-such things at this-and-this design stage must be calculated.]

In all the examples we have discussed so far, neither the position function nor the kinematic properties (except speeds) can be modified after the dimensions and shapes are established. However, this lack of flexibility can be overcome by altering the design. Take, for instance, the mechanism for contour grinding in Figure 4.11. Grinding tool (grinding wheel) 4, with its motor, is mounted on the connecting rod of the crankshaft mechanism. The mechanism is adjusted by moving joint 3 and guide 1 and securing them in the new position by means of set screws 2. The radius of crank 5 can also be changed by moving it along the slot in rotating table 6.

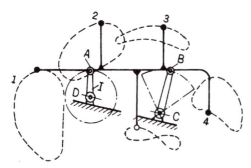

FIGURE 4.10   Trajectories of different points of a connecting rod of a four-link mechanism.

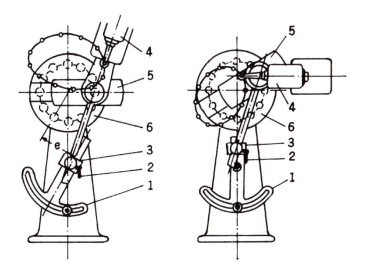

FIGURE 4.11 Mechanism for contour grinding. Views of contours produced when settings of parts 1 and 3 are changed.

4.2 Camshafts

It is not always possible to satisfy the desired position function by means of the mechanisms discussed in the previous section. The requirements dictated by the timing diagram (see Chapter 2) vary, but they can often be met by using cam mechanisms. The idea underlying such mechanisms is clear from Figure 4.12a), in which a disc cam is presented schematically. A linearly moving follower has a roller to improve friction and contact stresses at the cam profile–follower contact joint. It is easy to see that, by rotating the cam from positions 0 to 11, the follower will be forced to move vertically in accordance with the radii of the profile. Graphical interpretation of the position function has the form shown in Figure 4.12b). During cam rotation through the angle ϕ_1 (positions 0-1-2-3-4-5) the follower climbs to the highest point; during the angle ϕ_2 (6-7-8-9-10-11) it goes down, and during the angle ϕ_3 the follower dwells (because this angle corresponds to that part of the profile where the radius is constant). Changing the profile radii and angles yields various position functions, which in turn produce different speeds and acceleration laws for the follower movements. Figure 4.13 illustrates the cosine acceleration law of follower movement. The analytical description of this law is given by the following formulas:

$$s = \Pi(\phi) = \frac{h}{2}\left(1 - \cos\pi\frac{\phi}{\phi_1}\right),$$

$$\frac{ds}{d\phi} = \Pi'(\phi) = \frac{h}{2}\ \frac{\pi}{\phi_1}\sin\frac{\phi}{\phi_1}\ , \qquad\qquad [4.19]$$

$$\frac{d^2s}{d\phi^2} = \Pi''(\phi) = \frac{h}{2}\ \frac{\pi^2}{\phi^2_1}\cos\pi\frac{\phi}{\phi_1}\ .$$

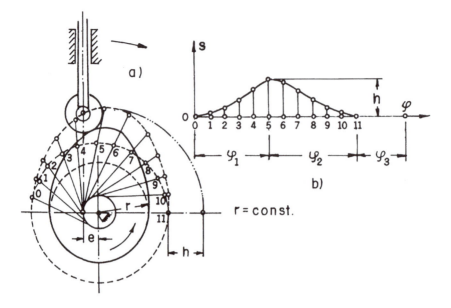

FIGURE 4.12 a) Disc cam mechanism; b) The follower motion law.

To provide the desired sequence and timing of actions, it is convenient to mount all cams needed for the machine being designed on one shaft, thus creating a camshaft, for example, as shown in Figure 4.14. One rotation of this shaft corresponds to one cycle of the machine, and thus one revolution lasts one period or T seconds. As can be seen from Figure 4.14, a camshaft can drive some mechanisms by means of cams (mechanisms A, B, C, and D), some by cranks (mechanism E), and some by gears (mechanism F). Sometimes a single straight shaft is not optimum for a given task. Then the solution shown in Figure 4.15 can be useful. Here, motor 1, by means of belt drive 2, drives camshaft 3, which is supported by bearings 4. A pair of bevel gears 5 drive shaft 6. The ratio of transmission of the bevel gears is 1:1; thus, both shafts complete one revolution in the same time and all cams and cranks (7 and 8) complete their tasks at the same time.

(Figure 4.14a) shows a photograph of a specific camshaft controlling an automative assembly machine serving the process of dripping irrigation devices production).

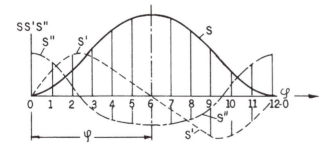

FIGURE 4.13 Cosine acceleration law carried out by link driven by cam mechanism.

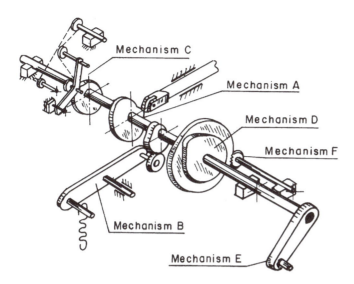

Mechanism C

Mechanism A

Mechanism D

Mechanism F

Mechanism B

Mechanism E

FIGURE 4.14 Camshaft as a mechanical program carrier. Main camshaft.

This solution is convenient when the tools handle the product from more than one side. However, this example also illustrates the disadvantage of mechanical types of kinematic solutions, namely, difficulties in transmission of motion from one point to another. For instance, Figure 4.16 shows what must be done to transform the rotational motion of shaft 1 into the translational motion of rack 2. (Note that the guides of follower 3, bearings of intermittent shaft 4, and guides of rack 2 are not shown, although they belong to the design and contribute to its cost.)

FIGURE 4.14a) General view of a camshaft serving a specific production machine for automative assembly of dripping irrigation units. (Netafim, Kibbutz Hatzerim, Israel.)

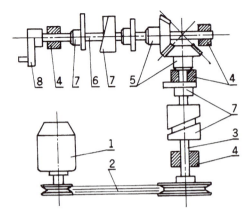

FIGURE 4.15 Generalized concept of a main camshaft.

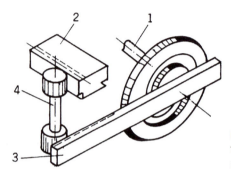

FIGURE 4.16 Complexity of motion transformations carried out by purely mechanical means; see text for explanation.

To make the structure more flexible, some kinds of transmission can be adopted.

Figure 4.17a) presents a ball transmission for very lightly loaded mechanisms. The action of this transmission is obvious—it is a rough model of a hydraulic transmission. For larger loads it is recommended that cylindrical inserts 1 be used between the balls. A purely hydraulic transmission (Figure 4.17b)) can also be used. Bellows 1 on the cam-follower's side transmits pressure through connecting pipe 3 to bellows 2, located on the tool side, and actuates the latter. A third possibility for transmitting motion in a flexible manner by mechanical means is shown in Figure 4.18. This device consists of guide 1 made of some flexible metal in which plastic ribbon 2, which possesses openings, is borne. The friction between the ribbon and the guide is reduced by proper

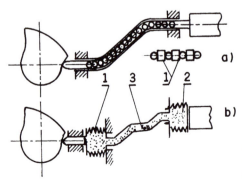

FIGURE 4.17 Models of a "flexible" transmission of the camfollower's motion: a) Ball transmission; b) Hydraulic transmission.

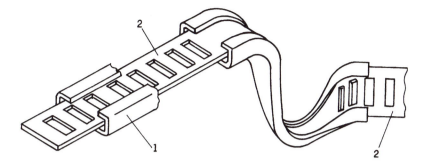

FIGURE 4.18 Flexible toothed rack for motion transmission.

choice of materials and coatings, so that the guide can be folded in various ways and still provide satisfactory transmission of the motion.

Cam mechanisms, being a kind of mechanical program carrier, operate under certain restrictions which must be known to the designer, together with the means to reduce the harm these restrictions cause. The main restriction is the pressure angle. This is the angle between the direction of follower movement and a line normal to the profile point in contact with the follower at a given moment. Figure 4.19 illustrates the situation at the follower-cam meeting point. The profile radius $r = \overline{0A}$ makes angle γ with the direction of follower motion KA. Angle β, between the tangent at contact point A and speed vector V_{a1} (perpendicular to the radius vector r) may be termed the profile slope. The same angle β appears between radius vector r and the normal N at contact point A. Thus, we can express the pressure angle α as follows:

$$\alpha = \beta - \gamma. \qquad [4.20]$$

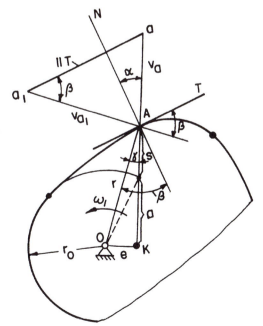

FIGURE 4.19 Pressure angle in cam mechanism.

Calling the follower speed V_a, we obtain from the sine law

$$\frac{V_a}{\sin\beta} = \frac{V_{a_1}}{\cos\alpha}.$$ [4.21]

From Figure 4.19 we have

$$a = \sqrt{r_0^2 - e^2},$$

where r_0 = the constant radius of the dwelling profile arc. Obviously,

$$V_{a1} = \omega r$$

and

$$r = \frac{a+s}{\cos\gamma}.$$ [4.22]

Substituting (4.22) into (4.21) and taking into account (4.20), we can rewrite Expression (4.21) as follows:

$$\frac{V_a}{\sin(\alpha+\gamma)} = \frac{(a+s)\omega}{\cos\alpha\cos\gamma}$$

or

$$\frac{1}{\cos\alpha} = \frac{V_a\cos\gamma}{(a+s)(\sin\alpha\cos\gamma+\cos\alpha\sin\gamma)\omega},$$

which gives

$$\tan\alpha + \tan\gamma = \frac{V_a}{(a+s)\omega}.$$

From Figure 4.19 it follows that

$$\tan\gamma = \frac{e}{a+s}.$$

Thus, we obtain

$$\tan\alpha = \frac{(V_a/\omega)+e}{a+s}.$$ [4.23]

Remembering that

$$s = \Pi(\phi) \quad \text{and} \quad \frac{V_a}{\omega} = \Pi'(\phi),$$

we finally obtain, from (4.23)

$$\tan\alpha = \frac{\Pi'(\phi)-e}{a+\Pi(\phi)}.$$ [4.24]

For the central mechanism, where $e = 0$, we obtain a simpler expression for (4.23), i.e.:

$$\tan\alpha = \frac{V_a}{\omega(r_0 + s)}$$

or

$$\tan\alpha = \frac{\Pi'(\phi)}{r_0 + \Pi(\phi)}.$$

The larger the pressure angle α, the lower the efficiency of the mechanism. When this angle reaches a critical value, the mechanism can jam. The critical value of the pressure angle depends on the friction conditions of the follower in its guides, on the geometry of the guides, on the design of the follower (a flat follower always yields $\alpha = 0$ but causes other restrictions), and on the geometry of the mechanism. To reduce the pressure angle, we must analyze Expressions (4.23) and (4.24). It follows from them that the pressure angle decreases as:

1. The value of a or r_0 increases;
2. The $\Pi'(\phi)$ function that describes the slope of the profile decreases.

Taking advantage of the first conclusion is impractical since it involves enlarging the dimensions of the mechanism. Thus, we usually recommend use of the second conclusion, that is, to "spread" the profile over a wider profile angle. However, to stay within the limits determined by the timing diagram, we must increase the rotating speed of the cam. This can be done by introducing the concept of an auxiliary camshaft. (See Figure 4.20a)) The main camshaft 1 is driven by a worm reducer and controls three mechanisms by means of cams I, II, and III. Cam III has a special function, namely, to actuate the auxiliary camshaft. This shaft is driven by a separate motor 4 and belt drive 5. The latter brings into rotation one-revolution mechanism 6, which is controlled by

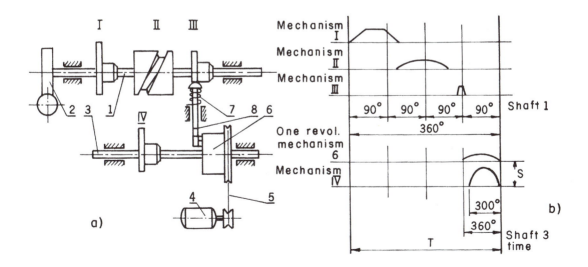

FIGURE 4.20 Concept of an auxiliary camshaft. a) Mechanical layout; b) Timing diagram.

cam III. This cam actuates follower 8 by pressing against spring 7. The one-revolution mechanism is then switched on, carries out one revolution, and stops. Obviously, the speed of shaft 3 can be considerably higher than the rotational speed of camshaft 1. Thus, cam IV, which is mounted on shaft 3, completes its revolution much faster than those on shaft 1.

The timing diagram shown in Figure 4.20b) is helpful here. According to this diagram, auxiliary shaft 3 rotates four times faster than main camshaft 1; that is, during one-quarter of a revolution of the main shaft, the auxiliary shaft completes a full revolution, then rests until shaft 1 finishes its revolution. This makes it possible to design the profile of cam IV with a more gradual slope, thus reducing the pressure angle. In our example, the profile of cam IV is extended over an angle of 300°, providing the needed displacement s for the follower. Obviously, without the auxiliary shaft the same profile must extend over an angle less than 90°, and the pressure angle would be much larger.

It is worthwhile to study the operation of the one-revolution mechanism. In Figure 4.21 we show a possible design of this mechanism, consisting of: permanently rotating part 1 (in Figure 4.20 motor 4 and belt 5 drive this part); driven part 2, key 3, and stop 4. The rotating part is provided with a number of semicircular slots (say 6). The driven part 2 has one slot. As is clear from cross section A-A, in a certain position of key 3 (frontal view a)), part 1 can rotate freely around driven part 2 (i.e., key 3 does not hinder this rotation). However, when key 3 takes the position shown in frontal view b), parts 1 and 2 are connected and rotate together as one body. When not actuated, key 3 is usually kept in the disconnecting position by stop 4, which presses lever 3a of the key. When the command to actuate the mechanism is given (cam III actuates follower 8 in Figure 4.20), stop 4 is removed from its position, freeing lever 3a. Thus, key 3 rotates

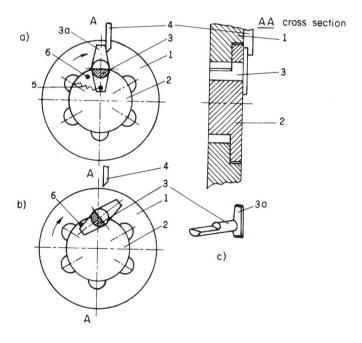

FIGURE 4.21   One-revolution mechanism: a) Disengaged state; b) Engaged state; c) Key.

into the connecting position, due to spring 5, when the slots of part 1 permit this. From this moment on, parts 1 and 2 move together, as mentioned earlier. However, if, during this first revolution, stop 4 returns to its previous position (cam III in Figure 4.20 ensures this), then before the revolution is completed lever 3a meets the stop, rotates key 3 into the disconnecting position, and frees part 1 from part 2. Pins 6 restrict the angular motion of the key. This kind of one-revolution mechanism has extensive applications.

At this point it is interesting to consider the problem of the flexibility of camshafts and other mechanisms for carrying out position functions in hardware. Indeed, from our description of the action of main and auxiliary camshafts, it would appear that, once designed, manufactured, and assembled, these mechanisms cannot be changed. This is, of course, true, and is completely adequate for "bang-bang" robotic systems (type 5 in Figure 1.5). However, there are ways of introducing some flexibility even into this seemingly stiff, purely mechanical approach. Figure 4.22 shows the design of cam 1 and shaft 2 together with special lock 3 which permits rapid cam change on the shaft. To some extent, such cam change is like reprogramming a programmable machine. Another design with the same purpose is shown in Figure 4.23a). Here, cam 1 is fixed on shaft 2 by means of nut 3. In Figure 4.23b) cam 1 is fixed by means of tooth-like coupling 2.

There are other ways in which the position function may be realized in a relatively flexible way by mechanical means. For instance, the cam shown in Figure 4.24 is built so that profile piece 2 can be fastened by bolts 3 and 4 at any angle on the circular base 1, which has a circular slot (it can be moved along this slot), thus yielding a wide range of *s* values. Another example, shown in Figure 4.25, allows easy adjustment of the cam

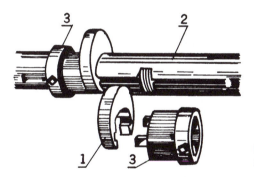

FIGURE 4.22   Arrangement for rapid cam exchange on a camshaft.

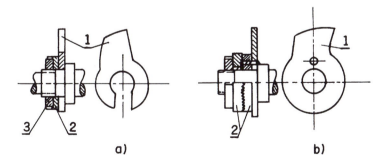

FIGURE 4.23   Another arrangement for a) Cam exchange; b) Change of fixation angle.

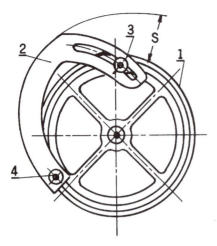

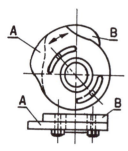

FIGURE 4.24   Arrangement for rapid cam profile exchange.

FIGURE 4.25   Arrangement for rapid timing change.

profile. Here the cam is composed of parts A and B, which can be fixed by bolts in different relative positions.

Use of a spatial cam introduces flexibility, as in Figure 4.26. Here, drum 1, having a system of openings, is fastened onto the shaft, while pieces 2 of profiles of appropriate shapes and sizes are mounted on the drum by bolts. A diagram of how the profiles are located on the drum surface is also shown. The spatial approach to cams offers additional solutions to the flexibility problem. The common feature of the examples illustrated in Figures 4.27–4.29 is the use of the third dimension, namely the z-axis, for adjusting the position function. To change the s value, one can use the solutions shown

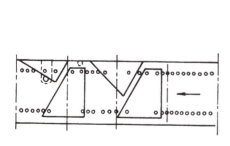

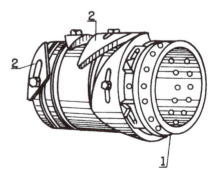

FIGURE 4.26   Spatial cam with an arrangement for rapid profile exchange.

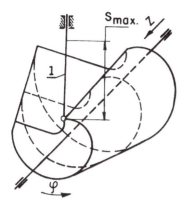

FIGURE 4.27 Spatial cam mechanism with a possibility of changing the follower's motion law by moving it along the z-axis.

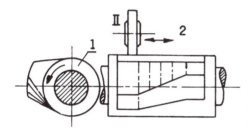

FIGURE 4.28 Spatial cam mechanism with a possibility of changing the timing by moving the follower along the z-axis.

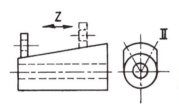

FIGURE 4.29 Spatial cam mechanism with a possibility of changing the follower's stroke by moving it along the z-axis.

in Figure 4.27 or Figure 4.29; to change the profile angle (the duration of the follower's dwell in its upper position), it is convenient to use the design shown in Figure 4.28.

Lastly, we can imagine and realize a design in which each cam or crank is driven by a separate drive, say a DC or stepper motor. An example is the automatic assembly machine presented schematically in Figure 4.30. An eight-position indexing table is driven by spatial cam 1 mounted on shaft 2 and driven by motor 3. The cam is engaged with rotating follower 4 and, through bevel gear 5, moves table 6. Two other mechanisms are shown around the table: automatic arm 7 (for manipulation with two degrees of freedom) driven by motors 8 and 9. Motor 8 rotates screw 10 to raise and lower arm 7. This is done with the aid of nut 11, which in turn is driven (through transmission 12) by motor 9. By controlling these two motors we can achieve simultaneous displacement of the arm according to the angular and linear coordinates of the system. The next mechanism carries out the final assembly by pressing one part into another: Motor 13 drives cam 14 which stretches spring 15 (through lever 16 which serves as a follower of cam 14); lever 16 moves pressing punch 17. Other mechanisms are in their rest positions and are not shown here.

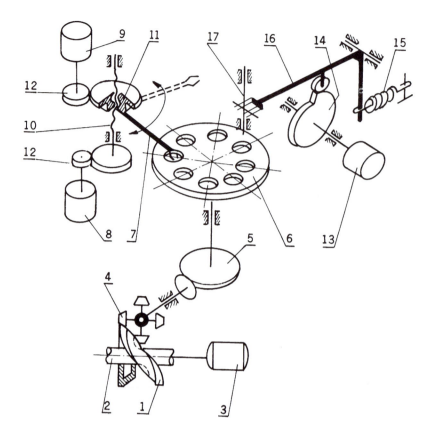

FIGURE 4.30 Layout of an automatic machine with autonomous, independent drive of mechanisms.

This kinematic solution allows flexibility in location of the mechanisms, because the motors need only wiring and, of course, wires can be extended and bent as desired. This solution is flexible also with respect to control. Indeed, the motors can be actuated in any sequence, for any time period, and with almost any speed and acceleration, by electrical commands.

In this section we discussed the important case where the cam (or camshaft) must rotate faster than the main shaft, and for this purpose we introduced the concept of an auxiliary shaft and explained the action of the one-revolution mechanism. At this point it is profitable to discuss the opposite case, where the cam must carry out a much longer cycle than the main shaft. An example of such a system is shown in Figure 4.31. This mechanism is usually called a differential cam drive. On main shaft 1, cam 5 is permanently fixed and freely rotating sleeve 2 is driven by shaft 1 through a transmission which includes four wheels, z_1, z_2, z_3, and z_4. The speeds of the cam and sleeve are not similar. Thus, roller 4, which is attached to the sleeve, moves the latter along the shaft according to the profile of cam 5. The time t of one cycle (one relative revolution between cam and sleeve) can be calculated from the following formula:

$$t = \frac{z_1 z_3}{n(z_2 z_4 - z_1 z_3)},$$ [4.25]

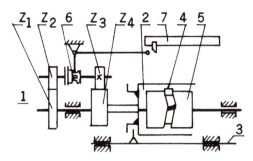

FIGURE 4.31   Layout of a cam drive for considerable reduction of the cam's rotation speed.

where n is the rotation speed of shaft 1. Sleeve 2 is engaged with follower 3. Coupling 6 and stop 7 serve to disconnect this mechanism.

4.3   Master Controller, Amplifiers

Let us go to case 6 in Figure 1.5. Here, control is effected through an amplifier. There are several standard solutions for this type of kinematic layout. One is the so-called master controller, which can be considered the simplest program carrier. The amplifying energy is usually electricity, compressed air, or liquid. However, purely mechanical solutions are also possible, and we will discuss them below. An example of a classical master controller that regulates electrical contacts is presented in Figure 4.32. Here, a low-power motor drives small camshaft 1. Cams 2 actuate contacts 3, connecting and disconnecting their circuits. The cams can be mounted on the shaft at different angles, according to the timing diagram. In the figure, every even-numbered cam turns on specific circuits (motor, magnet, etc.), while the odd cams turn them off. By adjusting the angles between even and odd cams according to the desired sequence and duration of action of every mechanism or device, we can make the entire system work as we wish. Setting up the cams is relatively simple and, thus, such systems are flexible to a certain extent. The power of the master cam device is obviously limited; however, the power of the mechanisms controlled by it can be much higher.

We can imagine a case where the cams actuate hydraulic or pneumatic valves instead of electrical contacts. The amplifying energy will then be the energy of compressed liquid or air. One difference between this particular use of cams and the applications discussed is striking, namely, that here the cams are able to produce only "on-off" commands, and the transient processes depend completely on the nature of the controlled system. Master camshafts do not control the manner in which the piston of the hydraulic or pneumatic cylinder develops its motion, nor how the electric drive accelerates its rotation. It only determines the precise timing of the starts and stops. There are cases where this kind of control is enough, but, of course, in other situations such behavior is not sufficient, and refinement of the movement of the controlled item is essential. For instance, a winding mechanism must be provided with a cam that ensures uniform distribution of the turns of the reel; the cam that throws the shuttle of a loom must develop an acceleration high enough to ensure travel of the shuttle from one side of the produced fabric to the other.

We have seen that the cams in the mechanisms described above transmit practically no power and work on an "on-off" regime, turning real power transmitters on or

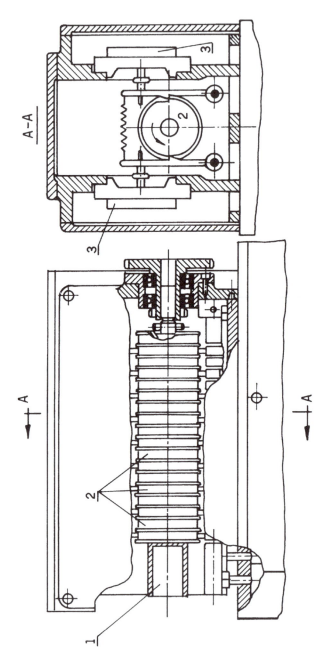

FIGURE 4.32   General view of a master controller: left, side view; right, cross section.

off. As one might expect, other solutions for the same purpose exist in several conceptual forms. We will first briefly consider the family of devices using punched cards and perforated tapes. These are used in concert with specific readout devices. The latter can be of various types, e.g., electrical, pneumatic, photoelectric, and mechanical (although the latter are rarely used). Let us consider them in the listed order.

Figure 4.33a) shows the layout of an electrical readout device consisting of base 1 on which perforated card 2 is placed. Contacts 3 are fastened onto a moving block 4 and lowered. Those contacts that meet the card are bent and no connection is made (because of the insulating properties of the material of which the card is made); those which meet an opening in the punched card connect with the corresponding contact 5 in base 1. Thus, the output represents a combination of electric connections.

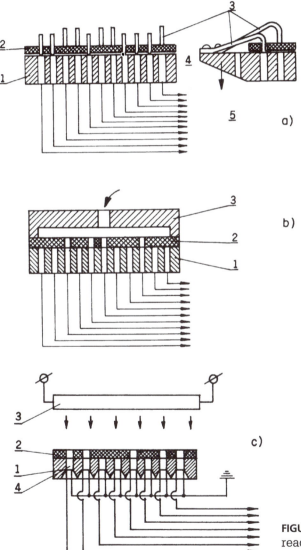

FIGURE 4.33 Layout of a punched-card readout device: a) Electrical; b) Pneumatic; c) Photoelectric.

Analogously, a pneumatic readout system can be devised. Figure 4.33b) shows a diagram of a pneumatic readout device that can work with either vacuum or pressure. Perforated card 2 is placed on base 1 and is sealed by hollow clamp 3. The clamp is connected to a pressure or vacuum source so that pressure or vacuum is transmitted through the openings in the card to the piping system. Here, the readout of the device is a combination of pressures or vacuums.

The fastest readout device is based on photoelectric sensors, a scheme of which is shown in Figure 4.33c). This device consists of base 1 on which punched card 2 is placed. The perforated card is exposed to light source 3. Thus, those photosensors 4 that are protected by the card are not actuated, while those exposed to light entering the perforations in the card are actuated, yielding a combination of electrical connections in the output wiring. The high speed of response of photoelectric devices makes it possible to use continuously running perforated tapes, as opposed to the devices discussed earlier, which require discontinuous (discrete) reading of information because of their slow response. Some electrical devices constitute an exception to this rule. However, the pressure of contacts sliding along the tape causes significant wear of the tape and the contacts, and therefore discontinuous readouts are preferable. This is not to mention the lower speed of the tape (because of the time response) than in photoelectric devices. The latter devices also have the advantage of no mechanical contact, so that wear due to friction does not occur. Mechanical readout devices for perforated cards will be discussed after amplifiers are considered. (A combination of such a readout with a purely mechanical amplifier will be shown in Figure 4.37.)

Figure 4.34a) shows the layout of a hydraulic amplifier consisting of cylinder A. piston B, and slide valve C; piston B does not move. The cylinder and housing of the

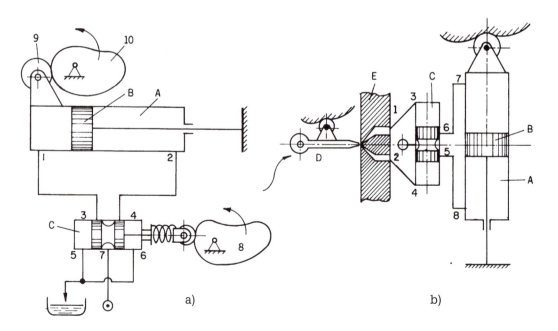

FIGURE 4.34 Layout of a hydraulic amplifier: a) Mechanical input;
b) Pneumatic or hydraulic input.

valve are made as one unit. Thus, when for some reason the piston of the valve is displaced (leftwards, say), the pressure from port 7 passes through outlet 3 to cylinder port 1, while in this situation the idle volume of the cylinder is connected to outlet 6 and liquid tank through its outlet 2 and port 4 of the valve. The pressure entering the left volume of the cylinder causes a leftward movement and equivalent displacement of the slide valve housing. The movement of this housing relative to the valve piston closes all ports and therefore stops the cylinder. To continue the movement of the cylinder, the valve piston must again be displaced leftward, and so on. To reverse the motion of the cylinder, the piston of the valve must be moved rightward; ports 7, 4, and 2 then connect and the right volume of the cylinder is under pressure, while the liquid is freed to flow into the tank through ports 1, 3, and 5. The valve's piston can be moved by a cam 8 and thus the cylinder will almost copy the cam profile. This is a good way of making a tracer: as the gauge fastened to the valve piston rod follows, say, a wooden model, the cylinder drives a milling head which processes a metal blank 10.

An additional amplifying stage can be introduced, as shown in Figure 4.34b. Here, a pneumatic gauge connected to the slide valve (as in the previous example) controls its movement. The pneumatic stage consists of nozzle D which is installed opposite partition E, which has two channels 1 and 2. The pressure difference between these channels depends upon the position of the edge of the nozzle D relative to the inlets of the channels. (The diameter of the nozzle output is about 0.5 mm.) The air flow from the nozzle is divided by the partition dividing the channel's input and brought to valve ports 3 and 4, through ports 1 and 2, moving the piston in accordance with the pressure difference. The subsequent action of the system is as described above.

The ideas applied in the above amplifiers can also be used to design an electrohydraulic stepping motor. An example is the layout presented in Figure 4.35, a solution implemented by the Fujitsu company. The device is controlled by a valve that regulates liquid flow through a number of channels. Oil pressure is applied to inlet 1 and can be directed to outlets 2 or 4, while ports 3 and 5 return the oil to its reservoir. Outlet ports 2 and 4 are connected to ports 6 and 7, respectively, of the rotary hydraulic motor 8, which consists of rotor 9 provided with (in our case) 11 holes that serve as cylinders for plungers 10. The rotor is pressed against oil distributing plate 11. The contact surfaces of both rotor and distributor are processed so as to provide perfect sealing (to prevent oil leakage) and free relative rotation. Figure 4.35c) shows the cross section of the mechanism through that contact surface. Here arched oil-distribution slots 18 are made in part 11. Plungers 10 are axially supported by inclined thrust bearing 12. The rotor is fastened on motor shaft 13, the tail part of which is shaped as nut 14. The latter engages with piston 15 of the valve by means of threaded end 16. Stepping motor 17 drives piston 15, so that, due to its rotation, it moves axially relative to the inlet and outlet ports of the valve, because of the threaded joint with shaft 13. In Figure 4.35a) the situation of the valve corresponds to the resting state of the hydraulic motor. When, due to rotation of motor 17, the piston begins to move rightward (see arrow in Figure 4.35a)) and thereby connects port 1 with port 4 (see Figure 4.35b)), the pressure reaches port 6 of the hydraulic motor, while port 7 connects with ports 2, 3, and 5, permitting drainage of the oil into the reservoir. The oil flow causes rotation of the motor in a certain direction (say, counter-clockwise, as in Figure 4.35c)). The rotation of shaft 13 moves the piston leftward (arrow in Figure 4.35b)) and thus the piston 15 locks the oil-conducting channels, stopping the hydraulic motor. This system, as follows from the

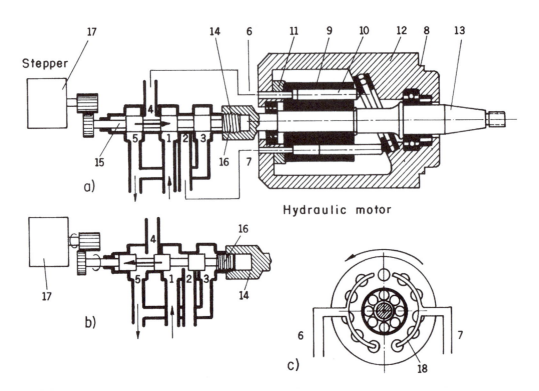

FIGURE 4.35 Layout of Fujitsu hydraulic pulse motor-stepmotor combination with a hydraulic servomotor. a) General view of the device; b) Leftward movement of the valve's piston; c) Cross section of the oil distributor.

above description, responds to each pulse or step of electric motor 17 by a step or pulse of the hydraulic motor. However, the power of stepping motors is usually modest, making them unsuitable for heavy-duty work, while hydraulic motors can develop practically unlimited power and torque.

Another electromechanical amplifier is presented in Figure 4.36. It consists of two drums 1 and 2 freely rotating around shafts 3 and 4, respectively. The drums are permanently driven by electromotors 5 and 6, which rotate the drums in opposite directions. Shafts 3 and 4 are specially shaped and connected by means of elastic metallic ribbons 7 and 8 wound around the drums in opposite directions. One of the shafts, say 3, is designed to be the input. When rotated in a given direction (say clockwise), it causes ribbon 7 to stretch around drum 1, resulting in a high frictional torque. Thus, drum 1 acts as a friction clutch, connecting motor 5 with ribbon 7 and shaft 4 (the driven shaft). This torque can be much larger than that of shaft 3. When the motion at the input is counter-clockwise, the action develops in the following manner. Shaft 3 stretches ribbon 8 around drum 2, connecting motor 6 by friction with output shaft 4. The torque created between the ribbons and the corresponding drums can be estimated by the Euler formula:

$$T_{output} = T_{input}e^{f\phi}. \qquad [4.26]$$

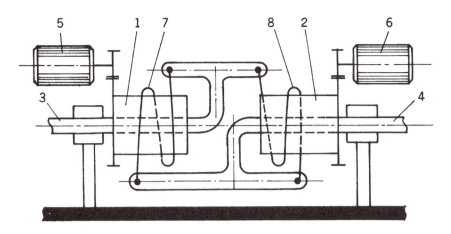

FIGURE 4.36   Electromechanical amplifier.

Here,

T_{output}, T_{input} = output, input torques, respectively;
f = frictional coefficient describing the friction between the ribbons and drum surfaces;
ϕ = the wrapping angle of the ribbon around the drums; and
e = natural logarithm base.

The following example seems to be of great technical and educational importance. On the one hand, it is the first known technical application of a perforated device and program carrier; on the other, it is the first purely mechanical amplifier. The invention in question is the Jacquard loom (see Chapter 1). The Jacquard mechanism controls every thread of the warp of the fabric being produced. Individual control of each thread makes it possible to produce fabrics with very complicated woven patterns in an automatic manner. The mechanism consists of drum 1 (in Figure 4.37 it is octagonal) which drives chain 2 of punched cards. Every face of drum 1 has a set of holes 3. A set of needles 4 is placed in frame 7 in an order corresponding to the holes on the drum's sides. Frame 7 rests on lever 5, which is driven in a periodical manner by cam 6. Frame 7 is raised during rotation of the drum for one-eighth of a revolution and lowered when the drum stops in its next position (the discontinuous drive of the drum is not shown in Figure 4.37). When it is lowered, needles 4 meet the card. Where it is not punched the needles remain in a raised position, but where the needles meet a hole they fall through. Each needle 4 supports a lever 8 which, by means of rods 9, moves a lever 10 so that the state of needles 4 and levers 8 is reproduced by levers 10. By means of an angular lever 13, the levers 10 are connected with the heddles of the loom. The heddles are hung on frames 14. Amplifying energy comes through a connecting rod 11 which drives specially shaped lever 12. This lever oscillates each time the loom's shuttle is thrown from one side to the other. When tooth 15 of lever 12 moves leftward, it engages those levers (10) that are lowered because of the position of the corresponding needles 4, and due to this engagement the corresponding heddles and therefore threads of the

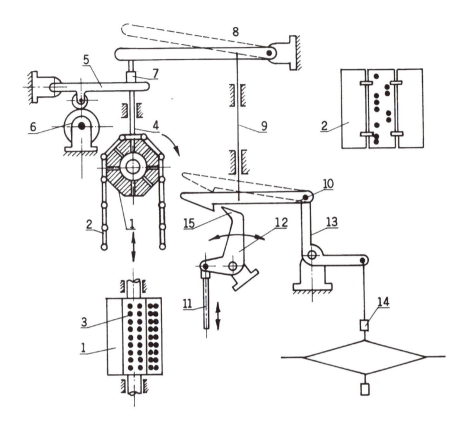

FIGURE 4.37 Mechanical readout for a punched-card controller. The Jacquard loom mechanism.

warp are lifted, creating the required weave of the cloth. Those levers (10) that are in a raised position at any given moment obviously fail to engage lever 12 and thus perform no positive action, and respective heddles and threads stay in place.

Our discussion about master camshafts led us to consider the use of punched cards (or tapes) for control purposes. It is natural to consider additional ways of accomplishing the same task. Figure 4.38 shows a rotating hydraulic (or pneumatic) valve consisting of housing 1 with curved grooves 2, and rotating slider 3 with radial slot 4. Slider 3 is driven by shaft 5 and, say, gear wheel 6. The slider is pressed towards the housing by spring 7. To ensure rotation, the spring acts through thrust bearing 8. Thus as it rotates, slider 3 connects pressure inlet port 9 with grooves 2, each of which is connected by a pipe to a separate cylinder. Obviously, the speed of the slider's rotation determines the time during which the cylinders are under pressure, and this time must be longer than that needed for the piston to be displaced as required. (Opening 10 in the slider connects the cylinders with the atmosphere to permit the pistons to return to their initial positions, assuming that there are springs for this purpose in the cylinders.) It is clear that the number of cylinders controlled by such a device cannot be large, because of the limited dimensions of the valve.

An electric analog of such a controller (electric distributor) is presented in Figure 4.39a). Arclike contacts 2 are fastened on the housing. Rotating slider contact 3 connects these immovable contacts 2 with the power source in a certain order, the dura-

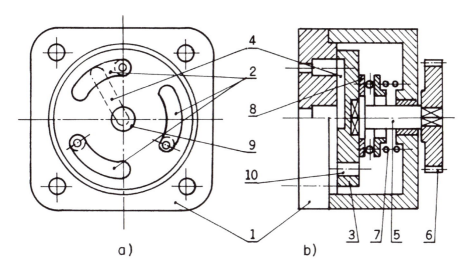

FIGURE 4.38   Hydraulic valve as a master controller: a) Front view; b) Cross section.

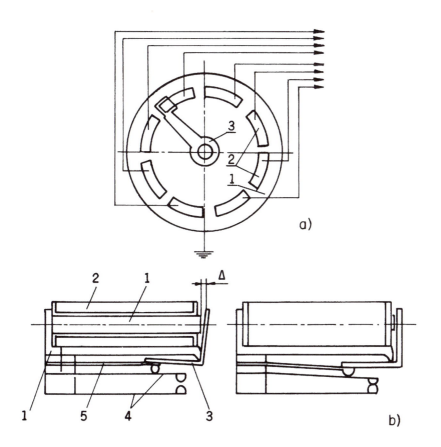

FIGURE 4.39   a) Electric distributor as a master controller;
b) Electromagnetic relay.

tion of these connections obviously depending on the arc lengths. This device can, for instance, actuate electromagnetic valves.

We should also mention the electromagnetic relay. Originally devised for amplifying current in the first telegraph lines, it has found wide use in the automation of machines and tools. The structure and action of this device, shown in Figure 4.39b, is as follows. Core 1 also serves as a base and holds coil 2. Armature 3 has the form of an angular lever. One arm of this lever actuates a group of electric contacts 4. Sometimes a spring 5 is used to keep the armature at a certain distance from the core. When energized, the coil induces a magnetic flux and the armature overpowers the spring, closing the gap and simultaneously actuating the contacts. In the general case, some of the contacts can be normally open while the others are normally closed. Thus, when the coil is energized the contacts change their state: the normally opened contacts close and the normally closed contacts open. The power needed for energizing the coil is much smaller than that controlled by the contacts. This kind of amplifier belongs to the so-called "on-off" kind.

Figure 4.40 shows a schematic layout of an electromagnetic valve. When electromagnet 1 is energized, it compresses spring 2 and connects channel 3 with outlet port 4; when the magnet is switched off, channel 3 is connected with port 5. Here too the common feature of these controllers is that they are "on-off" type devices; after the controlled element is turned on, they do not affect its behavior until it finishes its function.

Special devices must be designed to control the movement of, say, a hydraulic piston in the time interval between application of pressure to the cylinder and its removal. Figure 4.41 illustrates a hydraulic device for keeping the speed of a piston constant. Here the pressure difference $p_1 - p_2$ in the first throttle 1 can be made constant by changing the cross-sectional area in the second throttle 2. Thus, the flow rate through the throttles is kept constant, ensuring constant speed of the piston even for a variable resisting force p.

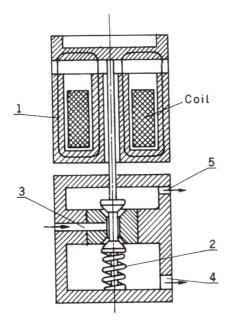

FIGURE 4.40 Layout of an electromagnetic valve.

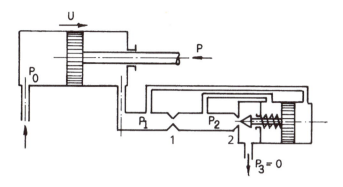

FIGURE 4.41   Layout of a hydraulic circuit for uniform piston speed control.

As we have seen, the control devices discussed so far dictate a certain pace for the machine's movements, thereby determining the time interval for every tool action and the duration of the whole cycle (remember, we are mainly considering periodically working systems). This approach, although very widespread, suffers from one serious disadvantage: spare time must be allowed between every operation to avoid collisions between tools, blanks, and parts of the machine, to ensure completion of each operation, and to provide the required level of reliability. The way to overcome this disadvantage is to use limit switches or valves (or even purely mechanical solutions). Figure 4.42a) shows a pneumatic limit valve. The example presented here consists of cylinder 1 (pneumatic or hydraulic) and piston 2, with its rod 3 on which specially shaped bushing 4 is mounted. This bushing actuates limit valves 5 and 6 by means of levers 7 and 8, respectively. Pressure is applied to inlet ports 9 and 10, respectively. This pressure is supplied to the inlet ports valve 11 and actuates piston 12, which in turn connects the inlet and outlet ports of cylinder 1 in accordance with the location of piston 2. This device operates a shutoff mechanism consisting of magazine 13 and slide 14. Thus, parts 15 are removed from the magazine individually. In the situation illustrated in the figure, bushing 4 actuates lever 7, depressing spring 16 and opening the valve; thus the pressure (say, of air) reaches the left side of valve 11, moving piston 12 rightward, connecting pipe 17 with pipe 18, and causing piston 2 to move leftward (at this point the left volume of cylinder 1 is connected with outlet ports 20 through pipe 19). When the bushing reaches lever 8, it moves piston 12 leftward, connecting 17 with 19, and 18 with 20, resulting in a backward action of the whole device, and so on.

When we discussed the advantages and disadvantages of pneumatic drives, we mentioned the difficulties that arise in achieving accurate displacements (mechanical stops being practically the only means for achieving this aim). A pneumatic device was recently introduced for this purpose (Figure 4.42b)). It can also be controlled by limit valves. In the figure, two states of the locking arrangement are shown:

I. The piston rod 1 is free to move;
II. The piston rod 1 is jammed.

The jamming device can be built as an integral part of pneumocylinder 2 (where piston 3 drives rod 1) and consists of auxiliary piston 4, lock 5, and housing 6. There is

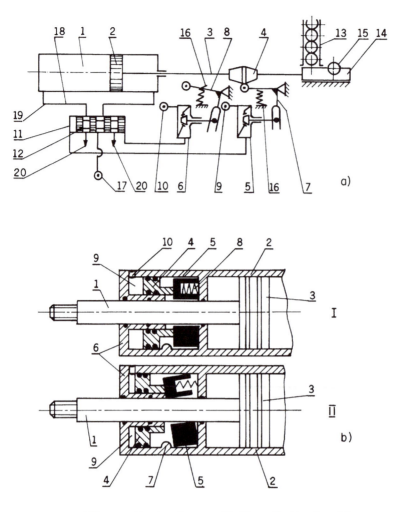

FIGURE 4.42   a) Pneumatic circuit controlled by a limit valve; b) Pneumatic jam for stopping the piston rod at any point in its stroke.

a protuberance 7 on the inner wall of the housing. Spring 8 applies asymmetrical pressure on lock 5. When volume 9 of the device is pressurized via inlet port 10 to auxiliary piston 4, the latter moves lock 5 rightward, compressing spring 8 and freeing piston rod 1 for movement. However, when the volume 9 is exhausted, spring 8 shifts auxiliary piston 4 leftward and, due to protuberance 7, skews lock 5, creating between it and rod 1 a friction high enough to stop and lock the rod at that position. Obviously, at this moment the main piston 3 is also freed of pressure by the control circuit.

A purely mechanical limit switch is shown in Figure 4.43 for an automatic machine for cutting metal bars into sections of a certain length L. Support 1 must be fastened onto rod 2 at a certain position. Rotating feeding rollers 3 advance material 4 until it touches support 1, compressing spring 5. This action causes rotation of shaft 6 and frees key 7 of the one-revolution device (see Figure 4.21). Engine 8 drives flywheel 9 and the latter drives the one-revolution device when allowed. This in turn drives an eccentric press that drives cutter 10. After the section of material 4 is cut, spring 5

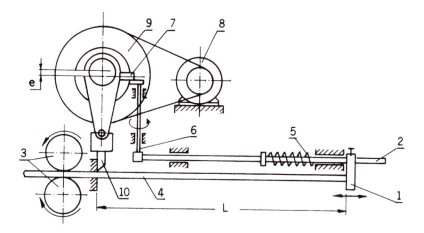

FIGURE 4.43   Layout of an automatic machine for measuring and cutting bars.

immediately returns rod 2 to its initial position, stopping the one-revolution device and cutter 10 in a raised position and making it possible for a new section of material to be fed and measured.

Obviously, in this case electric limit switches could have been used. It is possible to short an electric circuit through the material being processed when the part, material, or blank is metallic, which gives more design possibilities. A simple example of this is shown in Figure 4.44a). A metal cutting device is again shown. Wire 1 is fed by rollers 2 driven by motor 3. When the desired length L of wire is drawn, it reaches contact 4. The latter is insulated from the machine by its support 5. Thus, the wire itself shorts circuit 6 through coil 7 of relay R. This relay, when energized, actuates corresponding contacts. Contact K_1 is then closed and energizes magnet 8, which brings cutter 9 into action. In addition, normally closed contacts K_2 and K_3 open, thereby switching off motor 3; to bring the motor and the wire to a fast stop. Electromagnet 10 is also switched off, thus activating spring 11, which operates brake 12 by stretching ribbon 13 around the brake drum.

The subject of limit switches brings us closer to the concept of feedback control. Indeed, feedback devices monitor information about the location of specific parts or links at definite instants in time (i.e., not continuously). Thus a system including limit switches belongs to group 8 of our list of automatic machines (Figure 1.5). There are many different limit-switch designs. We will confine ourselves to a few special cases. Figure 4.44b) shows a reed-relay in combination with a permanent magnet. Here 1 is a moving part, whose displacement the limit switch must determine. Magnet 2 is fastened on this part. The reed-relay consists of glass ampoule 3 and a pair of elastic steel contacts 4 (gold-coated for better conductivity and prevention of corrosion). When magnetic flux flows through the contacts, it closes them; otherwise they stay open. The ampoule is evacuated. Compared with the usual electromagnetic relay, the reed-relay has a much shorter response time and a longer lifetime. This is because of the small inertial mass of the moving contacts—no armature need be moved—and because of the vacuum—no sparks occur and thus no contact erosion takes place.

Another kind of limit switch also consists of a glass ampoule (Figure 4.44c): in ampoule 1 are placed two contacts 2 and a drop of mercury. In certain positions the

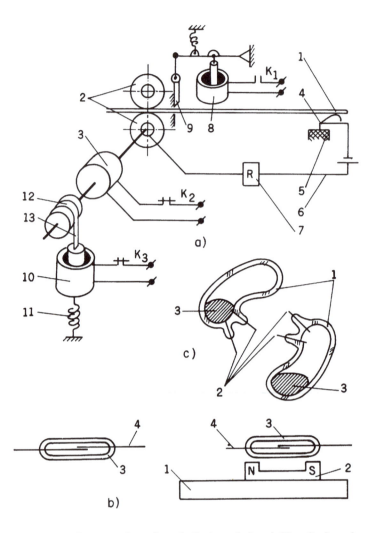

FIGURE 4.44 Layout of an electric limit switch: a) Circuit closed by the fed wire 1; b) Reed relay: left—free of magnetic field, right—under magnetic field; c) Mercury-actuated switch.

mercury covers both contacts, closing the electric circuit. Otherwise the contacts are disconnected. This switch is convenient for control of angular displacement.

4.4 Dynamic Accuracy

Up to now, we have discussed the problem of motion at the level of kinematics, without taking into account elastic deformation of the links in the mechanism. This section deals with the problem of motion errors caused by dynamics. It is convenient to use, for this example, the cam mechanisms we discussed above as an illustration. We have already mentioned that the position function of these mechanisms can be chosen from among a very wide range of possibilities, and Figure 4.13 shows a partial graphical interpretation of a specific function with the displacement s, velocity $\dot{s}$, and acceleration $\ddot{s}$. While the graph shows the ideal shapes of these kinematic characteristics, one can never actually achieve such curves, whatever the effort made to attain accuracy in manufacturing. And the reason for this pessimistic note is the limited stiffness of the links constituting the mechanism, with their deformation by external and inertial forces. These deformations are usually too small to significantly alter the shape of the displacement $s(t)$. However, the first- and second-order derivatives, namely, the velocity $\dot{s}$ and especially the acceleration $\ddot{s}$, can (and usually do) acquire significant deflections or errors. Sometimes the errors in acceleration reach the order of magnitude of the nominal acceleration.

An example of such disturbed results (Figure 4.45) was obtained with an experimental device in the Mechanical Engineering Department of Ben-Gurion University. Here, a section of a cosine-type acceleration carried out by a cam-driven follower is presented. The thicker line shows the calculated curve, while the thin line shows the real data for the acceleration of the follower. Obviously, the higher the rotational speed of the cam, the larger are the disturbances in the accelerations.

These errors should be estimated during the design process. It is worthwhile to see how these disturbances appear in the dynamic model of the mechanism under consideration. How can we foresee these errors? What are the analytical means of obtaining an estimation of their values?

To clarify the discussion we use the mechanism shown in Figure 4.46 as an example. This mechanism consists of a flywheel with a moment of inertia J_0, which drives gear wheel z_1 through a shaft with stiffness c_1. Gear wheel z_1 is engaged with another wheel z_2 which drives a cam (the eccentricity of which is e), and the latter, in turn, drives a follower. The motion $s(t)$ of the follower is described by the motion function and equals $\Pi(\phi)$. By means of a connecting rod, the follower drives mass m and overcomes external force F. The rod connecting the follower to the mass has stiffness c_2. The damping effects in the system are described by damping coefficients b_1 and b_2. These coefficients will help us to take into consideration the energy losses due to internal and

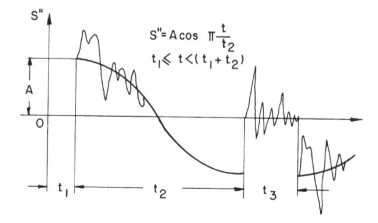

FIGURE 4.45   Comparison between the ideal and measured follower motion of a cosine cam mechanism.

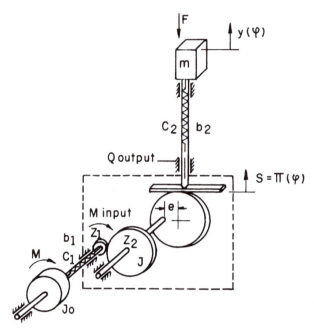

FIGURE 4.46 Dynamic model of a mechanism.

external friction in and between the links of the mechanism. To simplify calculations, we use the following relationships for the damping forces and torques, respectively:

$$F_d = b_1 \dot{s} \quad \text{and} \quad T_d = b_2 \dot{\phi}. \tag{4.27}$$

The dissipated forces and torques are proportional to the linear and angular speeds, respectively.

For the mechanism described, the motion function relating the rotation of the drive shaft to the motion s of the follower link has the following form:

$$s = \Pi(\phi) = e \cos\phi \frac{z_1}{z_2}.$$

Here e is the eccentricity of the cam. (To simplify the example, the cam is circular and rotates around an axis eccentric to its geometric center.)

The simplest approach to analyzing this mechanism is, of course, the kinematic one. In this case, the mechanism can be represented by the model shown in Figure 4.47a) and described analytically as follows:

$$T_{\text{input}} = -\Pi'(\phi^*)Q_{\text{output}} = e\frac{z_1}{z_2}\left(\sin\phi^*\frac{z_1}{z_2}\right)Q_{\text{output}}, \tag{4.28}$$

$$Q_{\text{output}} = F + m\ddot{s}^*.$$

These equations as well as the following ones are written by taking into account Equations (4.1), (4.2), and (4.4). The symbols denoted by asterisks are used for ideal values (not disturbed, without errors or dynamic deflections). The deflections (or what we

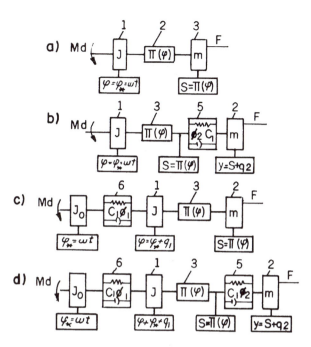

FIGURE 4.47   Mathematical models of different dynamic approaches to mechanisms: a) Kinematic approach to mechanism in Figure 4.46; b) Stiffness of the follower taken into account; c) Stiffness of the drive shaft taken into account; d) Both stiffnesses taken into account.

call errors) are denoted by q_1 and q_2 corresponding to the place where these errors appear. Here, all the designations are clear from Figures 4.46 and 4.47, and Q is the force at the output of the kinematic layout.

Next, we take the stiffness c_2 of the connecting rod into account. This model is presented in Figure 4.47b) and described as follows:

$$T_{\text{input}} = -\Pi'(\phi^*)(b_2\dot{q}_2 + c_2 q_2) = e\frac{z_1}{z_2}\left(\sin\phi * \frac{z_1}{z_2}\right)(b_2\dot{q}_2 + c_2 q_2),$$

$$m\ddot{q}_2 + b_2\dot{q}_2 + c_2 q_2 = -F - m\ddot{s}*,$$

[4.29]

where q_2 is the additional motion caused by the dynamic effect of c_2. Thus, the real motion of mass m is

$$y = s* + q_2,$$

where $s*$ is the ideal motion at the output of the mechanism on the assumption that $\phi = \phi^* = \omega t$, where ω is constant.

Let us consider a different approach by taking the limited stiffness of the drive shaft into account and assuming the connecting rod to be absolutely stiff. This model is shown in Figure 4.47c) and is described analytically as

$$J\ddot{q}_1 + b_1\dot{q}_1 + c_1 q_1 = -\Pi(\phi^* + q_1)\cdot Q_{\text{output}} = e\frac{z_1}{z_2}\left[\sin(\phi^* + q_1)\frac{z_1}{z_2}\right]Q_{\text{output}}, \qquad [4.30]$$

$$Q_{\text{output}} = F + m\ddot{s},$$

where q_1 is the additional output due to the limited stiffness of the shaft, so that

$$\phi = \phi^* + q_1 .$$

Lastly, the model in Figure 4.47d) takes the stiffness of both the drive shaft and the connecting rod into consideration. Then the equations are

$$J\ddot{q}_1 + b_1\dot{q}_1 + c_1 q_1 = \left[\Pi(\phi^* + q_1)\right](b_2\dot{q}_2 + c_2 q_2), \qquad [4.31]$$

$$m\ddot{q}_2 + b_2\dot{q}_2 + c_2 q_2 = -F - m\ddot{s}$$

or

$$J\ddot{q}_1 + b_1\dot{q}_1 + c_1 q_1 = e\frac{z_1}{z_2}\left[\sin(\phi^* + q_1)\frac{z_1}{z_2}\right](b_2\dot{q}_2 + c_2 q_2), \qquad [4.32]$$

$$m\ddot{q}_2 + b_2\dot{q}_2 + c_2 q_2 = -F - m\ddot{s} .$$

For cases c) and d), s does not equal s^* because, at the input of the mechanism, the rotation of the drive shaft equals $\phi = \phi^* + q_1$ and the driven mass moves in accordance with $y = s + q_2$ where $s = \Pi(\phi^* + q_1)$.

From Equation (4.3) we have

$$\ddot{s}^* = \Pi''(\phi^*)\dot{\phi}^{*2}$$

and

$$\ddot{s} = \left[\Pi''(\phi^* + q_1)\right](\dot{\phi}^* + \dot{q}_1)^2 + \left[\Pi'(\phi^* + q_1)\right]\ddot{q}_1 . \qquad [4.33]$$

By substituting Equation (4.33) into expressions (4.32), (4.31), and (4.30), we obtain a system of equations that can be solved with respect to q_1 or q_2. These additional motions q_1 and q_2 create the dynamic errors or deviations. Equations (4.29), (4.30), (4.31), and (4.32) become linear when $\Pi'(\phi)$ is constant; otherwise, we must deal with nonlinear equations.

An example of such a linear situation is shown in Figure 4.48, in which mass m is driven by a gear transmission engaged with a toothed rack. The diameter of wheel z_3 equals D, and we have

$$s = \Pi(\phi) = \frac{z_1 D}{2z_2}\phi \qquad [4.34]$$

and

$$\Pi'(\phi) = \frac{z_1 D}{2z_2} = \text{const.} \qquad [4.35]$$

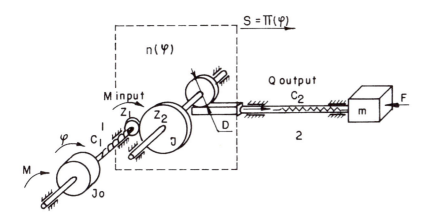

FIGURE 4.48   Example of a mechanism described by linear equations.

When designing, the engineer (if it is important) must use these approaches (choosing the assumptions as required) and numerically solve the appropriate equations. It must be mentioned that dynamic errors (q_1, q_2, etc.) are often serious obstacles in the effort to increase the efficiency, accuracy, quality, and/or productivity of newly designed equipment.

Let us make a short digression and consider an example relating to the recent history of typewriters and the essence of their dynamics. The classical structure of the typewriter included a carriage that holds the paper and moves it along the typed line, providing the correct intervals between the characters. The typebars are fastened onto specially shaped levers that are actuated manually or electrically. The wider the paper sheet or the more copies being typed simultaneously, the larger must the mass of the carriage and the dynamic effort of the mechanism be. To compensate, it was necessary to limit the typing speed and the dimensions of the parts. These limitations were overcome by the introduction of the IBM concept, where the carriage does not move along the lines and thus no inertial forces occur. The line of characters is typed by a small, moving "golfball" element. It is made of light plastic, and therefore its mass is much less than that of the carriage in the old concept. The dynamic efforts are thus considerably reduced and do not depend on the paper width and the number of copies. The speed of manual typing is not limited by this concept. However, problems can (and certainly do) appear when this kind of typewriter is attached to a computer, which can type much faster.

Going back to our subject, let us now consider an example of dynamic disturbances in an industrial machine. Figure 4.49 shows an indexing table drive (we will deal with these devices in more detail in Chapter 6). This drive consists of a one-revolution mechanism like the one discussed earlier and shown in Figure 4.21. In this case, one-revolution mechanism 1 (when actuated) drives spatial cam 2 (into which the one-revolution mechanism is built): the cam is engaged with a row of rollers 3 fastened around the perimeter of rotating table 4. The mechanism is driven by motor 5 and transmission 6. The wheel 11 is the permanently rotating part of the mechanism. Key 7 is kept in the disengagement position by "teeth" 10 on cylinder 8. The latter is actuated by electromagnet 9 whose movement rotates cylinder 8, disconnecting "teeth" 10 from

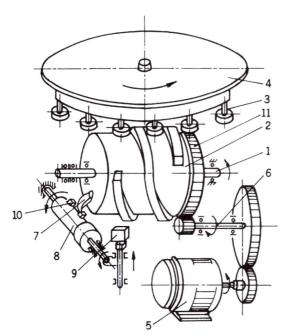

FIGURE 4.49 Layout of an indexing table as an example of a source of dynamic errors.

key 7, thus engaging cam 2 with one-revolution mechanism 1 and thus causing rotation of the cam. During this revolution the magnet must be switched off and "teeth" 10 are put in the way of the rotating cam with key 7 attached to it, leading to disconnection of cam 2 from mechanism 1. During this one revolution of cam 2 it causes rollers 3 to move, rotating table 4 for a corresponding angle (one pitch between the rollers).

The profile of the cam is designed in such a way as to optimize the dynamic behavior of the table during its rotation. The main aim is to accelerate the process of indexing; that is, to rotate for one pitch as quickly as possible, and to shorten transient processes such as parasitic oscillations as much as possible. The shorter the indexing time, the better the device. The best devices existing on the market complete this process in 0.25–0.3 seconds. By experimentally analyzing the mechanism in Figure 4.49 and measuring the dynamics of its behavior, we obtained the graphs shown in Figure 4.50a) and b), for the rotational speed and acceleration, respectively, of the cam when actuated by the one-revolution mechanism. These graphs imply that, instead of uniform rotation of the cam (for which it was calculated and designed), the rotation is essentially nonuniform, especially at the beginning of the revolution. How can we predict such behavior before the mechanism is built? What happens at the beginning of the engagement between the cam and the drive of the one-revolution mechanism?

When "teeth" 10 (see Figure 4.49) free the key 7, it collides with the surface of the half-circular slot on the permanently rotating body of the mechanism (see Figure 4.21). As a result of this collision, the cam with the key and the rotating body rebound. From this moment these parts move independently until a new collision takes place. Thus, two modes of operation occur at the time of engagement. In the first, the two rotating bodies move together, connected to each other through the elastic key (Figure 4.51a)); and in the second, after the rebound, when the bodies rotate independently (Figure 4.51b)).

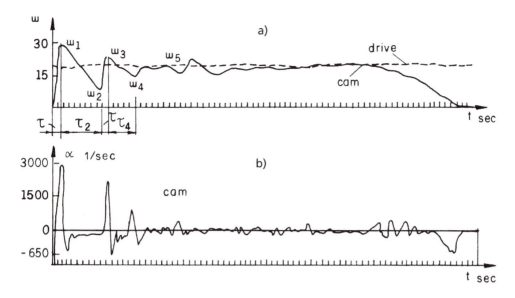

FIGURE 4.50   Records of a) speed and b) acceleration of the cam in the mechanism shown
in Figure 4.49. Here, τ = intervals of common drive-cam motion; τ_2, τ_4 = intervals of
independent drive-cam motion.

Here we use the simplest model for the inertial behavior of bodies possessing
moments of inertia, respectively J_c and J_d. The equations for the first mode of operation are

$$\begin{cases} J_d\ddot{\phi}_d + c(\phi_d - \phi_c) = 0, \\ J_c\ddot{\phi}_c + c(\phi_c - \phi_d) = 0. \end{cases} \qquad [4.36]$$

Here,

J_d = moment of inertia of the drive,
J_c = moment of inertia of the cam,
c = stiffness of the key,
ϕ_d, ϕ_c = rotational angles of the drive and cam, respectively.

The initial conditions for $t = 0$ are $\phi_{di} = \phi_{ci}$, $\dot{\phi}_{di} = \omega_{di}$, and $\dot{\phi}_{ci} = \omega_{ci}$, where $i = 0, 2, 4, ..., 2n$
for $n = 0, 1, 2$, the number of time intervals (here the collision intervals are considered).
For the second mode of motion we have the following equations:

$$J_c\ddot{\phi}_c + b_c\dot{\phi}_c = 0,$$
$$J_d\ddot{\phi}_d + b_d\dot{\phi}_d = 0. \qquad [4.37]$$

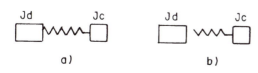

FIGURE 4.51   Models of two modes of movement of the parts of a one-revolution mechanism: a) The rotating masses function as
one system; b) The rotating masses move
independently.

The initial conditions for $t = 0$ are $\phi_{dj} = \phi_{cj}$, $\dot{\phi}_{dj} = \omega_{dj}$ and $\dot{\phi}_{cj} = \omega_{cj}$ where $j = 1, 3, 5, ..., 2n - 1$, for $n = 1, 2, 3, ...$. Here b_c, b_d are the lumped coefficients of viscous friction of the cam and drive, respectively.

Omitting transformations, we obtain the following solutions for the system of Equations (4.36):

$$\dot{\phi}_{d(i+1)} = u_{dt}\,p\cos pt + A_i ,$$
$$\dot{\phi}_{c(i+1)} = u_{ci}\,p\cos pt + A_i ,$$

[4.38]

where

$$p = \sqrt{\frac{c(J_d + J_c)}{J_d J_c}} ,$$

[4.39]

$$u_{dt} = \frac{-(\omega_{di} - \omega_{ci})J_d}{p(J_c + J_d)} ,$$

$$u_{ci} = \frac{-(\omega_{di} - \omega_{ci})J_c}{p(J_c + J_d)} ,$$

[4.40]

and

$$A_i = \frac{J_d \omega_{di} + J_c \omega_{ci}}{J_d + J_c} .$$

[4.41]

We obtain the following formula for the durations τ:

$$\tau = \frac{\pi}{p} .$$

[4.42]

The stiffness c is assumed to be the stiffness of the key 7 supported as shown in Figure 4.52. Thus

$$c = \frac{3E}{l^2{}_2\left(\dfrac{l_1}{J_1} + \dfrac{l_2}{J_2}\right)} .$$

[4.43]

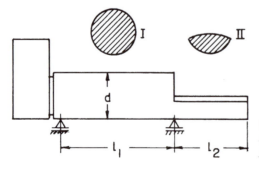

FIGURE 4.52   The key of the mechanism as an elastic beam and its cross sections. A model for stiffness calculation.

Here E is Young's modulus; l_1 and l_2 are clear from Figure 4.52, and J_1 and J_2 are inertial moments of the cross-sections I and II, respectively.

The solution of the independent motion of the cam from Equation (4.37) is as follows:

$$\dot{\phi}_{c(j+1)} = \omega_{cj}e^{-nt}, \tag{4.44}$$

where $n = b_c/J_c$.

Equations (4.38) and (4.44) are presented graphically in Figure 4.53. Here it is clear how the proposed model causes the changes in rotational speed and results in motion disturbances or errors. The intervals τ of common motion and τ_2, τ_4, and τ_1 are determined here according to the modes of independent motion of the drive and cam. Figure 4.54 presents a comparison of the calculated and measured processes. The solid line represents the measured cam speed during its engagement with the drive, while the dashed line represents the process calculated according to Expressions (4.38) and (4.44). Considering the roughness of some assumptions made here, the curves do not look bad.

4.5   Damping of Harmful Vibrations

The dynamic errors or disturbances we discussed in the previous section are obviously not useful. They cause noise, fatigue, wear of the links, and inaccuracy in their movements, and result in decreased quality and productivity of the whole mechanism. The most obvious way to prevent such vibrations is to get rid of the forces that cause them. However, this is usually impossible or inconvenient. It may also be possible to move away from resonance conditions, by changing the mass (or the moment of inertia) of the vibrating element or the elasticity of a spring. There are several approaches to this, which we will discuss later. However, even the method of moving away from resonance conditions does not always work.

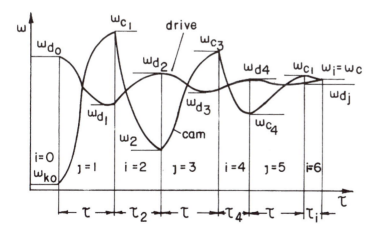

FIGURE 4.53   Calculated speeds of the drive and cam during acceleration of the indexing table.

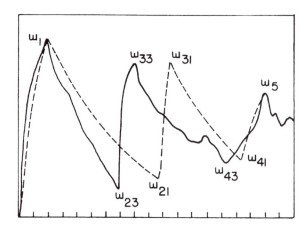

FIGURE 4.54 Comparison between the experimental (solid line) and the calculated (dashed line) speed of the cam of the one-revolution mechanism during operation.

The third possibility is to try to use a dynamic vibration damper or dynamic vibration absorber such as that shown schematically in Figure 4.55. This design is based on the finding that when the natural frequency

$$\omega_a = \sqrt{k_2 / m}$$

equals the frequency ω of the excitation force $P = P_0 \sin \omega t$, the main mass M will not oscillate, because of the vibrations of the added absorber's mass m. Thus, the force $k_2(x_2 - x_1)$ of the stretched spring will balance the force P acting on the main mass M. The equations describing the behavior of this system (assuming that friction is negligible) are as follows:

$$\begin{cases} M\ddot{x}_1 + (k_1 + k_2)x_1 - k_2 x_2 = P_0 \sin \omega t, \\ m\ddot{x}_2 + k_2(x_2 - x_1) = 0. \end{cases}$$ [4.45]

The solutions of this system are

$$x_1 = a_1 \sin \omega t,$$
$$x_2 = a_2 \sin \omega t,$$ [4.46]

where for the amplitudes a_1 and a_2 we obtain

$$a_1 = \frac{P_0 / k_1 \left[1 - (\omega / \omega_a)^2 \right]}{\left[1 - (\omega / \omega_a)^2 \right]\left[1 + k_2 / k_1 - (\omega / \Omega)^2 \right] + k_2 / k_1},$$ [4.47]

$$a_2 = \frac{P_0 / k_1}{\left[1 - (\omega / \omega_a)^2 \right]\left[1 + k_2 / k_1 - (\omega / \Omega)^2 \right] + k_2 / k_1}.$$ [4.48]

Here $\Omega = \sqrt{k_1 / M}$.

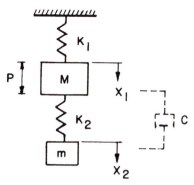

FIGURE 4.55 Model of a dynamic damper.

Equation (4.47) proves what was stated earlier, i.e., that when $\omega_a = \omega$ the numerator of this expression, and obviously also the amplitude a_1, equals 0.

Let us consider a model of free vibration. The model shown in Figure 4.55 is free to oscillate when no excitation force is applied, i.e., $P = 0$. For this case the equations for the mass movements are

$$\begin{cases} M\ddot{x}_1 + k_1 x_1 + k_2(x_1 - x_2) = 0, \\ m\ddot{x}_2 + k_2(x_2 - x_1) = 0. \end{cases} \quad [4.49]$$

The solutions are in the form

$$x_1 = a_1 \sin \omega t, \quad [4.50]$$
$$x_2 = a_2 \sin \omega t.$$

To find the natural frequencies we must solve the following biquadratic equation:

$$\omega^4 - \omega^2 \left\{ \frac{k_2}{m} + \frac{k_1 + k_2}{M} \right\} + \frac{k_1 k_2}{Mm} = 0. \quad [4.51]$$

The amplitudes are related as follows:

$$a_2 = a_1 \frac{m\omega^2 - k_2}{-k_2}. \quad [4.52]$$

From the latter expression we obtain the condition for the minimum value of a_2 in the form

$$\omega^2 = \frac{k_2}{m}. \quad [4.53]$$

Obviously, if the models discussed here represent rotational vibrations, the mass characteristic must be replaced by the moments of inertia and the springs must be described by their angular stiffness. Rotational vibrations have very important effects on indexing tables (see Chapter 7), which require some time to come to a complete rest after every step. An example is shown in Figure 4.56 of a pneumatically driven indexing table. In case a) the table, which has moment of inertia $J_1 + J_2$ when stopped, comes to rest as illustrated by the acceleration recording shown below. This process

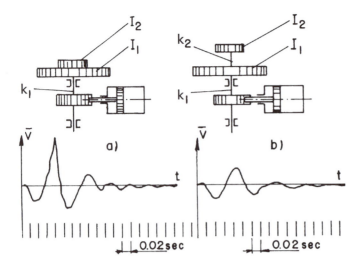

FIGURE 4.56 Indexing table (pneumatic drive) and its
acceleration at the end of each rotation: a) Undamped;
b) Dynamically damped.

takes about 200 milliseconds. In case b) the masses are separated by a rod with an
angular stiffness k_2. The acceleration recording shows considerably lower acceleration
(and thus also lower impact forces) and a shorter resting time.

Now we must consider a situation where, in the layout given in Figure 4.55, an
energy absorber is installed parallel to spring k_2. Then the equations describing the
motion of the masses will be as follows (c-damping coefficient):

$$\begin{cases} M\ddot{x}_1 + k_1 x_1 + k_2(x_1 - x_2) + c(\dot{x}_1 - \dot{x}_2) = P, \\ m\ddot{x}_2 + k_2(x_2 - x_1) + c(\dot{x}_2 - \dot{x}_1) = 0. \end{cases} \qquad [4.54]$$

Using the method of complex numbers we obtain, for mass M (which is the point of
interest), the following expression:

$$x_1 = P_0 \frac{(k_2 - m\omega^2) + j\omega c}{\left[(-M\omega^2 + k_1)(-m\omega^2 + k_2) - m\omega^2 k_2\right] + j\omega c(-M\omega^2 + k_1 - m\omega^2)}, \qquad [4.55]$$

where $j = \sqrt{-1}$.

To put this answer in the form of real numbers, we must express the complex
numbers in terms of their absolute values. Thus, for the amplitude of vibrations of
mass M we derive, from Equation (4.55),

$$x^2_1 = P^2_0 \frac{(k_2 - m\omega^2)^2 + \omega^2 c^2}{\left[(-M\omega^2 + k_1)(-m\omega^2 + k_2) - m\omega^2 k_2\right] + \omega^2\omega^2 c^2(-M\omega^2 + k_1 - m\omega^2)^2}. \qquad [4.56]$$

By substituting $c = 0$ into Equation (4.56) we obtain the result shown earlier in (4.47).
For the opposite case, i.e., when $c \to \infty$, the model under consideration becomes a
system with one degree of freedom where the oscillating mass is $M + m$. In both cases,
when $c = 0$ and $c \to \infty$, no energy is consumed in the absorber. This brings us to the

conclusion that there must be some value of c at which a maximum amount of energy is absorbed during the relative movement of the vibrating masses, and at which the vibrational amplitude of mass M is thus minimized. Denoting $\omega_a/\Omega = f$, it can be proved that the minimum vibrational amplitude of the main mass M is reached when

$$f = \frac{1}{1+m/M}.$$ [4.57]

For this condition the vibrational amplitude a_1 of the main mass M can be estimated from the expression

$$a_1 = \frac{P_0}{k_1}\sqrt{1+\frac{2M}{m}}.$$ [4.58]

The last specific case applied to the model shown in Figure 4.55 that we will consider here describes the situation where $c \neq 0$, and $k_2 = 0$; that is, when masses M and m are connected only by means of the absorber. Such an absorber is called a Lanchester damper; it is based on viscous friction. The implementation of such a damper for industrial purposes can have various forms. An example is shown in Figure 4.57. This design consists of two discs 1 freely rotating on bearings 2. The latter are mounted on bushing 3 fastened onto the vibrating shaft 9. This bushing has disc 4 with friction gaskets 5. Springs 6 can be tuned by bolt 7 and nut 8 to develop the required frictional torque.

The optimum amplitude (the minimum vibrational amplitude) in this case is estimated by the Formula (4.58). (Obviously, for rotational vibration the masses and stiffnesses must be replaced by the appropriate concepts: masses by moments of inertia,

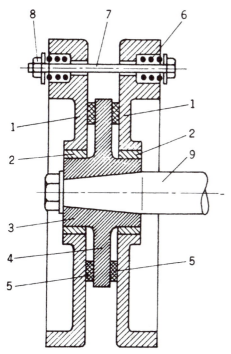

FIGURE 4.57 Lanchester damper—cross section.

stiffnesses by angular stiffnesses, linear displacements by angles.) The optimum ratio m/M is 0.08–0.1.

It must be mentioned that absorbers are less effective as vibration dampers than are dynamic dampers. The extinction of vibrations by dynamic means is more effective. However, as follows from the above discussion, damping can be achieved when the parameters of the system are tuned accurately for all methods of damping. Unfortunately, accurate tuning is often almost impossible to achieve, for various reasons. For instance, the moments of inertia can change during the action of the manipulator, and the stiffnesses can also change because of changes in the effective lengths of the shaft when a sliding wheel moves along it. Vibration dampers that rely on friction are especially inaccurate. This brings us to the idea of using automatically controlled vibration damping. The next section is devoted to this question and describes some ideas on adaptive systems that can automatically tune themselves so as to minimize harmful vibrations. More detailed information on this subject can be found in the excellent book *Mechanical Vibrations* by J. P. DenHartog (McGraw-Hill, New York, 1956).

4.6   Automatic Vibration Damping

We consider here some ideas for dynamic damping that can be useful for automatic vibration control under changing conditions. As a first example we discuss a rapidly rotating shaft. As is known, a shaft with a concentrated mass on it, as in Figure 4.58, vibrates according to the solid curve shown in Figure 4.59. The amplitudes A are described by the following formula

$$A = \frac{m\omega^2 e}{c - m\omega^2}.$$ [4.59]

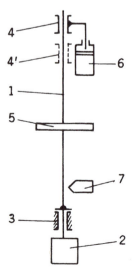

FIGURE 4.58 Dynamically damped rapidly rotating shaft.

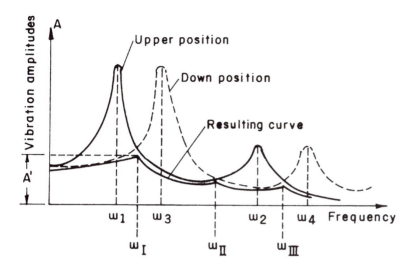

FIGURE 4.59 Behavior of the automatically damped shaft of Figure 4.58.

Here,

m = the concentrated mass,
ω = rotation speed,
c = stiffness of the shaft,
e = eccentricity of the mass on the shaft.

From Expression (4.59) it follows that at some speed ω_0 the amplitudes increase indefinitely. This speed equals

$$\omega_0 = \sqrt{c/m}\,.$$ [4.60]

The stiffness c of a shaft (assuming it is supported at two points and loaded symmetrically) is

$$c = 48\frac{EI}{l^3}\,.$$ [4.61]

Here,

E = Young's modulus,
I = cross-sectional inertia moment,
l = distance between the supports.

Thus, by automatically changing the length l of the shaft, one can influence the speed ω_0 at which the natural frequency ω_0 occurs. Therefore, we can propose a solution for a relatively silent (restricted values of vibration amplitudes), rapidly rotating shaft as presented in Figure 4.58. The system consists of shaft 1 driven by drive 2 and located in bearings 3 and 4. The shaft rotates massive disc 5 which causes the vibrations we are dealing with. Upper bearing 4 is movable by, say, pneumocylinder 6, and can be set in two positions, 4 and 4′. Speed pick-up 7 creates the feedback to make the system work automatically. Obviously, this plan is an approximation of various rotating shafts with wheels, turbines, rotors, barrels, etc. Briefly, this automatic vibration-restricting system operates as follows (see also Figure 4.59). When the shaft begins to accelerate from zero speed, the upper bearing is in position 4' and the vibrational amplitudes develop according to the dashed curve. After speed ω_1 is reached, the pick-up gives a command to raise the bearing into position 4. Thus, the behavior of the system is "switched" from the dashed curve to the solid curve, the resonance that would occur at ω_3 (dashed curve) is avoided, and the amplitudes continue to develop along the solid curve until the speed reaches ω_{II}. At this moment, the pick-up commands the bearing to return to its initial position 4', thus avoiding the approach of the next critical situation at speed ω_2 while further acceleration develops along the broken curve until speed ω_{III}, and so on. Decelerating the pick-up involves a similar series of commands. Thus, the vibrational amplitudes never exceed the value A' (see Figure 4.59) and are characterized by the resulting curve. In Figure 4.60 we show an example of measurements made on a real device. The upper recording was made without automatic damping, and the lower recording with automatic control of vibrational amplitudes. The shaft's speed of rotation was about 12,000 rpm.

Our second example uses a cam mechanism (Figure 4.61). Here, the problem is to decrease the dynamic errors q of the motion of the follower (see Equations 4.31 and 4.32), which is the difference between the real follower displacement s and the desired follower motion s^*. To solve this problem, we divide the follower in our scheme into two parts so that the dynamic model of the mechanism is close to that shown in Figure 4.55. Thus, changes in stiffness k_2 are enabled by moving auxiliary mass 4 along special rods 5, which also serve as springs. In this design cam 1 drives follower 2 which in turn drives mass 3 and auxiliary mass 4. The latter is connected to mass 3 by means of two rods 5 (with stiffness k_2). Mass 4 can be moved along rods 5 by cylinder 6 and fork 7, which are controlled by cam speed sensor 8 and position sensor 9. Control unit 10 processes the information necessary to move the piston of cylinder 6, and thus fork 7,

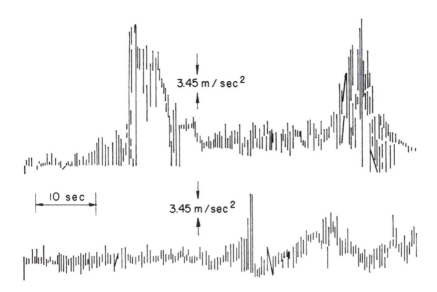

FIGURE 4.60 Experimental comparison of the acceleration of a shaft
a) without and (b) with damping.

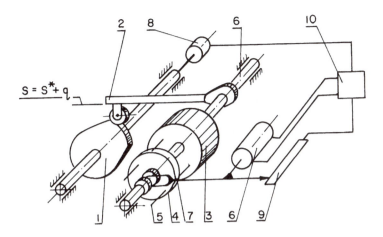

FIGURE 4.61   Dynamically damped cam mechanism.

so that mass 4 on rods 5 will have the required parameters as a dynamic damper. The disturbances (or errors) q change their frequency at different cam speeds; therefore, the damper must be tuned accordingly. Figure 4.62a) shows the usual (without damping) acceleration of the follower at two cam speeds, and Figure 4.62b) shows automatically damped acceleration of the follower, which clearly indicates that the damper considerably reduces the dynamic errors in the follower's motion.

The third example illustrates dynamic damping in a multimass system, shown in Figure 4.63. The device consists of base 1 and four beams 2 on which masses 3 and 4

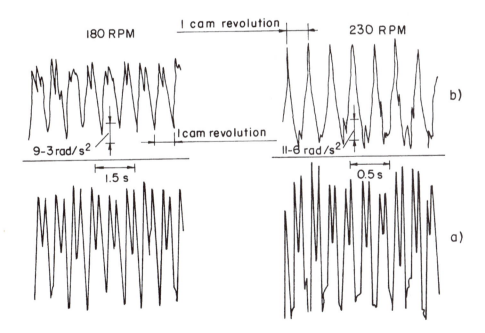

FIGURE 4.62   Experimental recordings of acceleration of follower shown in Figure 4.61: a) Not damped; b) Automatically damped.

are mounted. Masses 3 can be moved along the beams automatically by motors 5 and cables 6, the latter supported by pulleys 7. Mass 4 can be fastened manually at any point on its beam. This mass is provided with a drive to rotate unbalanced rotors 8 at different speeds, thus exciting the system with alternating forces of two different frequencies. The beams are connected by ties 9 (changeable rods or springs of linear or nonlinear nature). The vibrations of the beams are measured by strain gauges 10 glued to the beams close to the fastening points. The positions of masses 3 are measured by potentiometers 11. The information about the deflections of the beams and the positions of masses 3 is processed by a computer so as to move masses 3 into the proper positions for minimizing vibrations of beam A (or any other beam). Several algorithms were tried, for instance, to produce the minimum vibrational amplitudes in beam A while the rotation speed of masses 8 slowly changed. Figure 4.64 shows the results of two independent experiments. The upper record in both cases shows the vibration amplitude (in volts) of beam A before damping was attempted (indicating the frequencies in the excitation force). The lower record shows the damped vibrations. To the left of each recording, the relative locations of masses 3 before damping and after optimization was reached are schematically shown. The algorithm used for this example was based on a random-search strategy.

The fourth technique for reducing dynamic errors that is considered here is based on continuous tuning of the natural frequency of a damper by changing the stiffness of springs by means of electromagnetic means.

4.7   Electrically Controlled Vibration Dampers

We have developed a family of dynamic dampers (DDs) whose natural frequency is controlled by low-level direct current (Israel patent #95233 of 30/7/90 by R. Mozniker

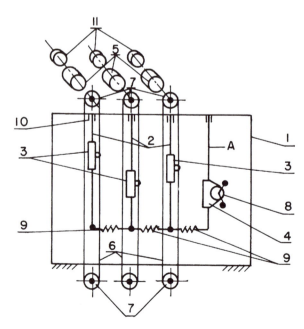

FIGURE 4.63   Layout of four-mass vibrating system.

& B-Z. Sandler). These DDs are suitable for damping vibrations with frequencies of about 15–100 Hz or more.

As a first case, we consider the DD layout presented in Figure 4.55. Here, two masses M and m connected by means of springs k_1 and k_2 and energy absorber c (in this case, friction sources are proportional to the relative speed of the vibrating bodies) are shown. A force P acts on the mass M. Figure 4.65 illustrates the "amplitude-frequency" characteristic of such a device for ideal conditions, i.e., zero friction.

It is clearly seen here that at about $k_2/m = 14$ the amplitudes of oscillations a1 = 0.

Figure 4. 66 shows that this damping occurs in a very narrow band of the ratio k_2/m (in this example, $k_2/m \approx 20$). To reach such a control level in real time is possible, if at all, only by electronic means.

The DD designed in accordance with the patent mentioned above is shown in Figure 4.67 and consists of a base 1, a spring system 2, a mass 3, a magnet core 4, and a coil 5.

By changing the voltage (about 15–20 millivolts) in the coil 5, we control the restoring force developed by the flat springs 2, or, in other words, the stiffness of the springs, and, as a result, the natural frequency of the DD. In Figure 4.68 we show a photograph of an experimental device.

When the frequency is essentially lower, the device becomes too "soft" and less practical. Therefore, the concept of active damping (AD) must be introduced. For instance, we measured the transient vibrations of a robot's arm in several orthogonal directions. The average frequency of vibrations for different arm lengths and masses kept in the gripper was about 1.5–3 Hz (10–20 rad/sec). The natural frequency of this device is in the neighborhood of 15 rad/sec, which brings us to very low spring stiffness and makes the device less practical for industrial use.

The convenience in using this device lies in the fact that no mechanical displacement of any kind is needed to tune it. Tuning is done by purely electrical means, which simplifies the interaction between the damper and other automatization systems— such as electronic circuits, microprocessors, or even computers. The disadvantage of this damper is that it is essentially nonlinear and therefore when the vibrational amplitude of the vibrating base changes, the natural frequency of the damper must be retuned. An analytical approximation of the nonlinear stiffness $\bar{k}$ of this damper is:

$$\bar{k} = k - 2P\frac{\delta^3(\delta^2 + 3x^2)}{(\delta^2 - x^2)^3}. \qquad [4.62]$$

Here (see Figure 4.67),

> k = the constant stiffness of the mechanical spring or elastic element of the damper;
> P = the force developed by the electromagnet, which is a function of the DC current in the coils;
> δ = the air gap between the damper's mass and the magnet; and
> x = the displacement of the damper's mass during vibration.

Another approach to this problem is to apply an active force to the vibrating mass, thus creating an Active Damper (AD). The AD device generates a variable force P applied to the oscillating mass M, as is shown in Figure 4.69. This force changes as the acceleration of the mass changes and is opposed to it.

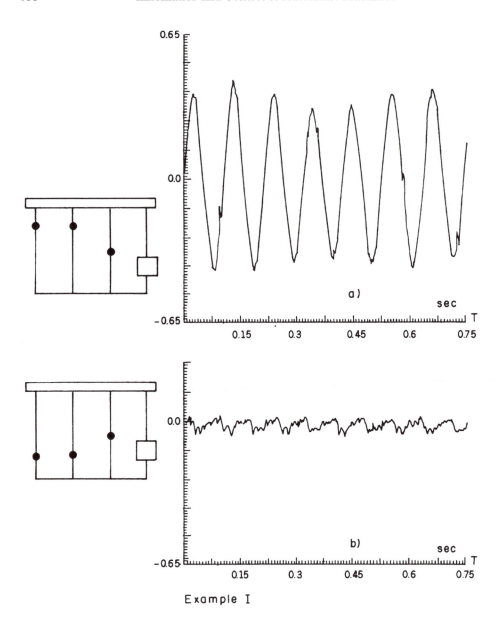

FIGURE 4.64 Two examples (I and II) of the behavior of beam A: a) Without damping; b) Automatic (computer-aided) damping.

In MATHEMATICA language the process then is described by the following expression:

g1 = Integrate[Cos[w y+Pi] Exp[-2 (t-y)] Sin[20 (t-y)],{y,0,t}]

In Figure 4.70 we show the resulting behavior of the mass M in this case.

A plot of surface amplitude versus frequency w of the force $P(t)$ and versus time shows a clear "valley" where the oscillations are almost damped out completely.

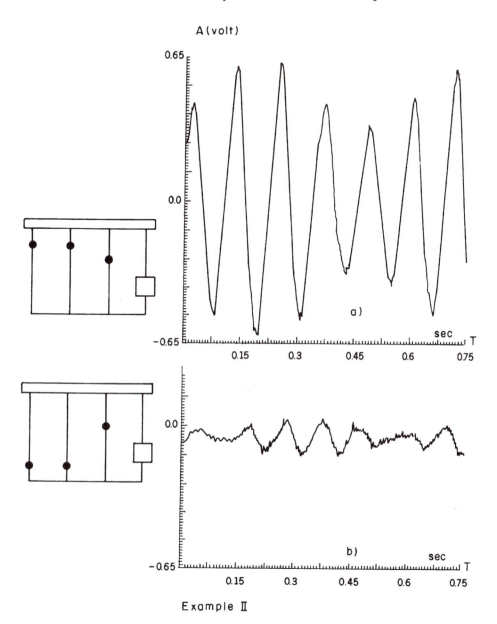

Example II

It can be seen in Figure 4.71 that the cross sections of the surface given in Figure 4.70 produced at different times have a minimal value of oscillation amplitudes at a certain value of the frequency of the applied force $P(t)$.

The force generator proposed in this case is an electromagnet fastened to the vibrating mass M (say, the arm of the robot). The magnet consists of a core 2, a coil 3, and an armature 4. An elastic layer (not shown in Figure 4.71) is placed between 3 and 4. When energized, the magnet develops a force P, pulling the armature. As a result, a

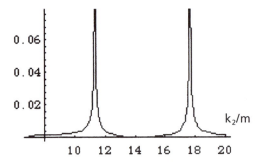

FIGURE 4.65 Oscillation amplitudes a1 of the mass M versus ratio k_2/m. (Example).

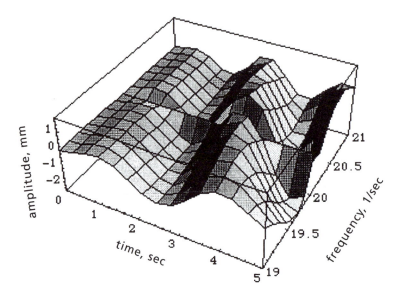

FIGURE 4.66 Vibration amplitudes of mass M versus ratio k_2/m and time during 5 seconds of the process (Example).

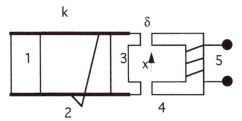

FIGURE 4.67 Layout of a DD device.

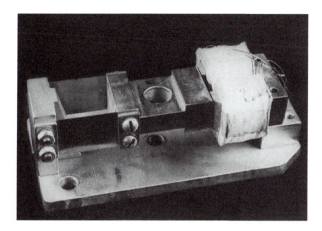

FIGURE 4.68   Photograph of one of the DDs used in our experiments.

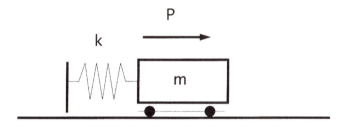

FIGURE 4.69   Layout of an active damper. Force P changes depending upon the free vibrations of the mass m.

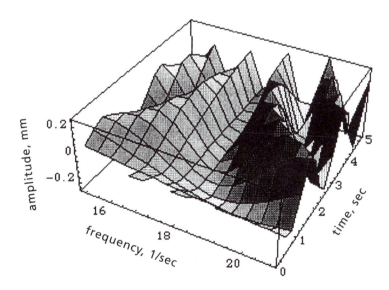

FIGURE 4.70   Oscillation amplitude versus frequency of P and time during the first 5 seconds. A "valley" of almost zero amplitudes at frequency about $\omega = 18$ 1/sec is clearly seen.

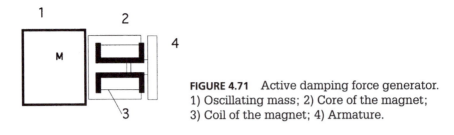

FIGURE 4.71   Active damping force generator.
1) Oscillating mass; 2) Core of the magnet;
3) Coil of the magnet; 4) Armature.

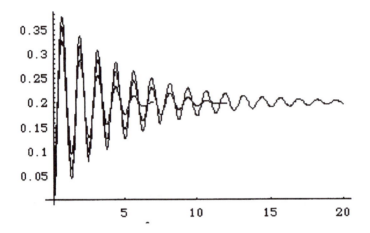

FIGURE 4.72 Comparison of the free oscillation of mass M
(computation) without (damping takes about 20 sec) and
with actuation of the AD (damping takes about 10 sec).

force is applied to the mass M. Obviously, the bigger the mass of the armature 4, the bigger the force.

The core 4 is fastened to the arm of the manipulator (or any other object).

An example of a comparison of the vibrations damping processes is shown in Figure 4.72. One process, taking about 20 seconds, is calculated for a usual system, without any artificial damping means, while the other, taking about 11 seconds, is the result of AD use.

A special control system that carries out all signal transformations must be used for this method. Its general layout for one control channel is shown in Figure 4.73. The accelerometer and the active damper are placed on the end of a robot's arm. The signal

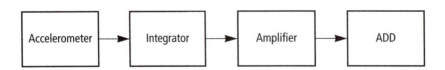

FIGURE 4.73   Layout of the proposed AD system.

from each accelerometer is doubly integrated and amplified. Thereafter, the obtained power signal enters the active damper where it generates the force $P(t)$ required for damping.

This latter idea of electrically controlled damping being designed to suit different and various mechanical systems (including manipulators) seems to be a very fruitful means for increasing accuracy of automatic manufacturing machines. The main advantage of this idea is the possibility to interact between the mechanics and the control electronics or computer. This kind of interaction recently has been given the name *mechatronics*.

Exercise 4E-1

For the mechanisms shown in Figure 4E-1 a) and b), write the motion functions $y = \Pi(x)$ and $y' = \Pi'(x)$, respectively.

For case a) calculate the speed $\dot{y}$ and the acceleration $\ddot{y}$ of link 2 when $x = 0.05$ m, $\dot{x} = 0.1$ m/sec, $\ddot{x} = 0$, and $L = 0.15$ m, and the force acting on link 1 to overcome force $F = 5$N acting on link 2.

For case b) calculate the speed $\dot{y}$ and the acceleration $\ddot{y}$ of link 3 when $\phi = 30°$, $\dot{\phi} = 5$ rad/sec, $\ddot{\phi} = 0$, $AO = 0.2$ m and $AC/AB = 2$.

Exercise 4E-2

A cam mechanism is shown in Figure 4E-2. The radius of the initial dwelling circle is $r_0 = 0.08$ m. The follower moves along a line passing through the camshaft center O (i.e., $e = 0$). The law of motion of the follower $y(\phi)$ is given by:

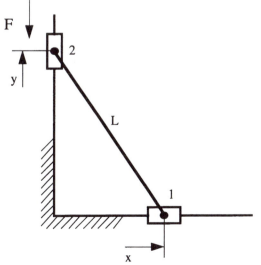

FIGURE 4E-1a)

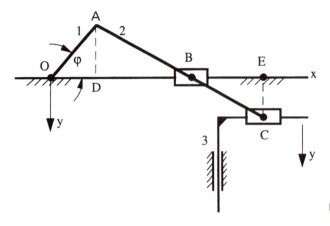

FIGURE 4E-1b)

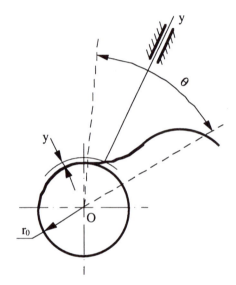

FIGURE 4E-2

$$y = 0.5h\left[1 - \cos\left\{\phi\pi/\theta\right\}\right].$$

During rotation for $\theta = 45°$, the cam's profile completes the displacement of the follower for a distance h. Calculate the maximum allowed value h which provides the condition where the pressure angle α does not exceed the permitted value $\alpha_{max} = 20°$; calculate the profile angle ϕ^* at which the pressure angle becomes worse.

5

Feedback Sensors

Referring to Figure 1.5 we see that, beginning from level 8, feedbacks are introduced into the design of an automatic machine or robot. These serve to control the machine or process, assuring automatic correction response of the system when conditions change. Sensors are the principal elements of a feedback system. This chapter presents a brief review of the most important feedback domains and sensors appropriate to them. The sensors can be divided into two main groups: analog and digital. To the first group belong those sensors that respond to changes in the measured value by changing some other physical value in their output, say, voltage, resistance, pressure, etc. In contrast, digital sensors transform the measured value into a sequence of electrical pulses. Information is carried encoded as the amount of pulses (say, the higher the number of pulses, the larger the measured dimension), as the frequency of pulses, or as some other pulse-duration parameter. The amplitude of the pulses usually has no importance in information transmission.

5.1   Linear and Angular Displacement Sensors

The most common task of a feedback is to gather information about the real locations of robot or machine links using, for example, sensors that respond to displacement or changes in location. There are several kinds of these sensors, some of which will be considered here.

Electrical sensors

The simplest displacement sensor is a potentiometer: a variable electrical resistor in which the slide arm is mechanically connected to the moving link. Thus, the resistance changes in accordance with the displacement. The electrical displacement or location sensors are usually a part of an electrical bridge, the layout of which is shown in Figure 5.1a). When a constant voltage V_0 is introduced, the off-balance voltage ΔV can be expressed as follows:

$$\Delta V = V_0 \frac{R_1 R_4 - R_3 R_2}{(R_1 + R_2)(R_3 + R_4)}. \qquad [5.1]$$

There are several methods to use these bridges. For instance, keeping the resistances R_1, R_2, and R_4 constant so that $R_1 = R_2 = R_4 = R$ and using the resistance R_3 as a sensor, i.e., a variable resistor responding to changes in the measured value, we can rewrite Expression (5.1) as

$$\Delta V = V_0 \frac{R - R_3}{2(R_3 + R)}. \qquad [5.2]$$

Substituting here $R_3 = R \pm \Delta R$, where ΔR is a small change of the resistance, so as $\Delta R \ll R$ we obtain, from (5.2),

$$\Delta V \cong V_0 \frac{\pm \Delta R}{4R}. \qquad [5.3]$$

In the simplest case, the displacement (or the measurement of some dimension) is transformed directly into displacement of the slide arm of the resistor. Thus, as follows from Relation (5.3), the change in the output voltage ΔV across the bridge's diagonally opposite pair of terminals a-a is directly proportional to the displacement (for small displacements). However, it is possible to increase the sensitivity of the bridge by using a so-called differential layout, as shown in Figure 5.1 b. For this case, by substituting the following in Expression (5.1),

$$R_1 = R_2 = R \quad \text{and} \quad R_3 = R + \Delta R, \quad R_4 = R - \Delta R,$$

we obtain

$$\Delta V = \frac{V_0 \Delta R}{2R}. \qquad [5.4]$$

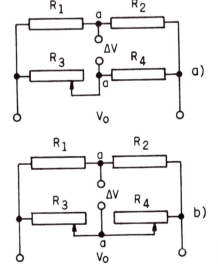

FIGURE 5.1 Layout of an electrical measurement bridge: a) Common circuit; b) Differential circuit.

This concept of a bridge feedback can be realized in a design such as that shown in Figure 5.2. This layout is called a *compensating bridge*. Here resistors R_1 and R_3 are variable. The slide arm of resistor R_1 indicates the location of cutter support 1 driven by motor 2 via screw drive 3. The slide arm of resistor R_3 is connected to feeler 4 which traces the program template 5 (master cam) fastened onto carrier 6, driven by motor 7 via screw drive 8. Thus, when resistance R_3 changes its value due to the template's displacement, the balance of the bridge is disturbed and voltage ΔV occurs on the output of the circuit. This voltage is amplified by amplifier 9 and actuates motor 2, which moves the cutter so as to change the value of resistance R_1 until the imbalance of the bridge vanishes. Thus, motor 2 compensates for the disturbances in the circuit caused by motor 7. From Expression (5.1), by substituting $R_1 = R + \Delta R$ and $R_3 = R - \Delta R$ while $R_2 = R_4 = R$, we obtain

$$\Delta V = V_0 \frac{2R\Delta R}{4R^2 - \Delta R^2}.$$ [5.5]

Assuming $\Delta R << R$ this can be rewritten as

$$\Delta V \cong V_0 \frac{\Delta R}{2R}.$$ [5.6]

The accuracy of such sensors is not high, about 0.5%, and absolute values of about 0.25 mm can be measured. When the resistors have a circular form, angular displacements can be measured.

Sometimes a sensor that gives a functional dependence between the rotation and output voltage is required. Figure 5.3 gives an example. Here, bases 1 are wound with high resistance wire 2 so that subsequent winds touch one another. Arm 3 is able to rotate around center 0. The function this device provides is

$$V_{\text{out 1}} = V_0 \sin\alpha,$$
$$V_{\text{out 2}} = V_0 \cos\alpha.$$ [5.7]

Figure 5.4 shows a rotating resistance sensor that produces a trapezoidal relation between the angle and the output voltage. Here 1 is a resistor, 2 is a conductor, and 3 is a slide

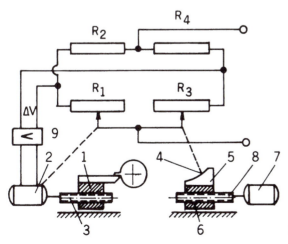

FIGURE 5.2   Electrical bridge used for feedback in tracking machine.

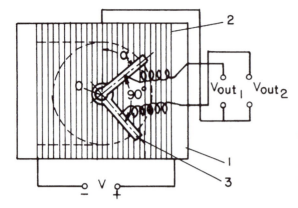

FIGURE 5.3 Resistance sensor for measuring angular displacements with a harmonic relation between the measured angle and the output voltage.

arm. The resistance wire must be wound uniformly to provide linearity during the appropriate rotation intervals. The angles $2\alpha_0$ are made of high-conductivity material.

Much higher sensitivity can be achieved by using variable-induction sensors (also called variable-reluctance pick-ups). The layout of the simplest of this kind of sensor is shown in Figure 5.5. It consists of a core 1, coils 2, and armature 3. The coils are fed by alternating current with a constant frequency ω. The alternating-current resistance Z in this case can be expressed in the form

$$Z = \sqrt{R^2 + X^2_L},$$ [5.8]

where R = ohmic resistance, and X_L = inductive reactance. The latter is described as

$$X_L = \omega L,$$ [5.9]

where L = inductance of the system. For the layout in Figure 5.5 this parameter is described by the following formula:

$$L = \frac{\mu Q W^2}{2\delta},$$ [5.10]

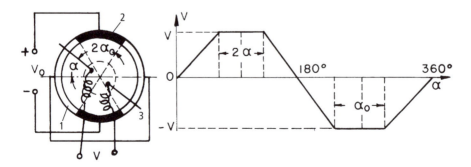

FIGURE 5.4 Resistance sensor for measuring angular displacements with a trapezoidal relation between the measured angle and the output voltage.

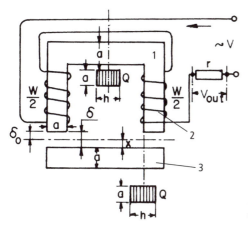

FIGURE 5.5 Layout of an induction displacement sensor.

where

    μ = magnetic permeability,
    Q = cross-sectional area of the core ($Q = a \cdot h$),
  a, h = the dimensions of the cross section of the magnetic circuit,
    W = the number of winds,
    δ = the width of the gap.

We assume here (to make the formula simple) that the cross-sectional areas of the core and armature are equal, as are the materials of which they are made. Obviously, the gap can be represented as the following sum:

$$\delta = \delta_0 + x, \qquad\qquad [5.11]$$

where δ_0 = initial gap and x = the measured displacement.

    Substituting (5.11) into (5.10) and the latter into (5.9), we see that (5.8) is a function of x.

    A more complicated design for an induction sensor is shown in Figure 5.6. This device consists of housing 1, made of ferromagnetic material with a high magnetic permeability, which constitutes the core of the sensor. Two coils 2 and 3 generate the

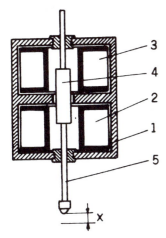

FIGURE 5.6 Differential induction sensor for displacement measurements. Cross-sectional view.

magnetic flux. Armature 4 is mounted on rod 5, which serves as a pick-up for the displacement x. (The rod is made of a nonmagnetic material.) The magnetic flux is divided into two loops going through the coils and armature 4. The length of the armature's sections in each loop determines the inductive reactance of each coil. Thus, these coils, which are a part of a bridge, change its balance (as in the case presented in Figure 5.1b)). Induction sensors are usually limited to measuring ranges not larger than, say, 15–20 mm. However, the accuracy is on the order of 10^{-3}–10^{-4} mm.

Another useful modification of an induction position sensor is shown in Figure 5.7. Here a lead screw 1 with a certain pitch (large enough to suit the design) and profile interacts with an induction pick-up 2. The alternating current resistance of its coil 3 depends on the relative position (see the above explanations) of the thread and the poles of the magnetic core. Thus, fractions of the screw's revolution can be measured. This design is thus made very effective.

The next kind of sensor we consider is the variable-capacitance pick-up. The bridge layout of such a sensor is shown in Figure 5.8. The capacitances C of gaps A and B are described by the following expressions, respectively:

$$C_A = \frac{\varepsilon S}{\delta - x} \quad \text{and} \quad C_B = \frac{\varepsilon S}{\delta + x}, \quad [5.12]$$

where,

   ε = dielectric permittivity,
   S = area of the capacitor's plates,
   δ = initial gap between the plates, and
   x = measured displacement.

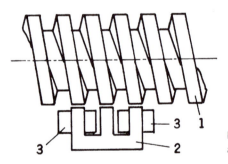

FIGURE 5.7 Induction position sensor based on a lead screw.

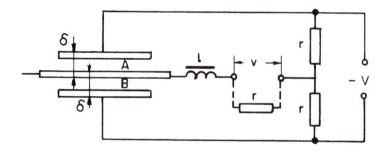

FIGURE 5.8 Layout of variable-capacitance pickup.

Thus, a circuit without a choke ($X_L = 0$) has an alternating current resistance Z:

$$Z = \sqrt{\left(\frac{R}{2}\right)^2 + \left(\frac{1}{2\omega C}\right)^2}, \qquad [5.13]$$

where the capacitance C is calculated from (5.12) for each gap, and ω = the frequency of the alternating current.

The sensitivity of this layout and sensor is high and can be estimated about 10^{-4} mm. However, the measuring range is small.

Specific optical effects can be used as the basis of a very powerful displacement measurement method. We will briefly describe the principle of a Michelson interferometer that can be applied for accurate displacement determination in industrial systems where machine elements must move with high precision. Interference results from the algebraic addition of the individual components of two or more light beams. If two of the light beams are of the same frequency, the extent of their interference will depend on the phase shift between them. In Figure 5.9 we show the layout of an interferometer for precision measurement of the location of some machine element 1. This device consists of a laser light source 2 and two mirrors 3 and 4, which are fastened to the moving element 1 and the base, respectively. There are also a beam splitter 5, a transparent plate 6, and a signal detector 7. The beam splitter 5 is usually a plane-parallel transparent plate of appreciable thickness, bearing a partially reflecting film 8 on one surface, which divides the light from source 2 into two beams. One beam traverses the splitter and strikes mirror 3, placed normal to the beam, and then returns to the splitter where part of it is reflected and enters detector 7. The other beam is reflected by mirror 4 and part of it is transmitted by the splitter to the detector. This latter beam serves as a reference to which the beam reflected from moving mirror 3 is compared (mirror 4 is strictly immobile). Because of the interference due to the phase shift occurring between these two beams, the detector obtains (and processes) information about the movement of mirror 3 (and element 1). It is easy to see that the beam striking mirror 3 traverses the thickness of the splitter three times before entering the detector, whereas the beam reflected from mirror 4 traverses it only once. Although this plate does not alter the direction of a ray passing through it, it shifts it laterally and introduces additional path length. To correct for this, a second plate 6, identical to 5 except that it bears no partially reflecting film, is placed in the path to mirror 4 and parallel with 5.

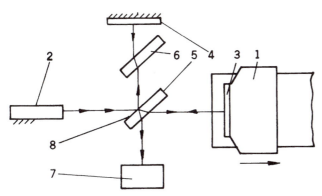

FIGURE 5.9 Layout of a Michelson interferometer for precise positioning.

This plate is called the compensating plate, and it is easy to see that the paths of the two beams are then identical as regards their passage through refracting plates.

The device described above is named for its inventor, Michelson. Its accuracy is very high—about 0.0001 mm. The detector obtains information as a sequence of bright and dark fringes. Thus, the system works in a digital mode, by counting the fringes. This kind of feedback measuring device, because of its high accuracy, is practically the only solution for automatic robotic machines in manufacturing integrated circuits.

Another optical sensor for displacement measurement is based on photosensitive elements, for example, as shown in Figure 5.10a). The element, the location of which is to be measured, is provided with a transparent scale 1, beaming a grating of transparent and opaque stripes 2. The scale is illuminated with a parallel light beam obtained via condensor lens 3 from source 4. The shadow of the scale is projected onto reticle 5, which has an identical grating 6. Obviously, the amount of light going through reticle 5 at any instant is an almost linear function of the position of the scale. This light is detected by photocells 7 and transformed into digital electric signals, which are counted and translated into distances.

A problem arises when the direction of movement must be distinguished. For this purpose an auxiliary grating is placed on the same scale. This idea is illustrated in Figure 5.10b). Line 1 is the main grating while line 2 is the auxiliary one. When the scale moves rightward, the two gratings produce a sequence of pulses in which an auxiliary pulse comes at $T - \tau$ after every pulse from the main grating. Conversely, when the scale moves leftward, this time interval equals τ. Thus, for $T - \tau \neq \tau$, the system can distinguish the displacement direction. The sensitivity for the described system is about 0.01 mm.

The principle of the device shown in Figure 5.10 can easily be transformed for measurement of rotation. Such rotation encoders are widely used in machine tool and manipulator designs.

Photoelectric cells also permit creating analog-type displacement sensors. One possible example is shown in Figure 5.11, the so-called optical wedge. This device is a pho-

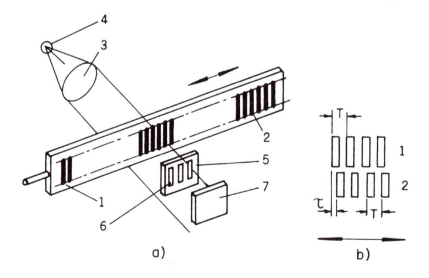

FIGURE 5.10 Digital optical displacement sensor: a) Layout of the sensor; b) Layout of the grating used for determining direction.

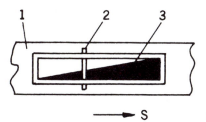

FIGURE 5.11 Analog optical displacement sensor.

toelectric element that has a variable response to illumination along its surface. When a diaphragm 1 with a narrow slit 2 moves along optical wedge 3, the latter's response corresponds to the relative position of these elements. It is not a highly sensitive device. However, there are situations where it is appropriate.

Pneumatic sensors

We now consider pneumatic sensors. The basic model we will consider is shown in Figure 5. 12. Its main elements are the nozzles in sections I and II. Let us consider the continuity of flow through these nozzles, which is described in the following form:

$$\alpha_1 f_1 \rho V_1 = \alpha_2 f_2 \rho V_2 .$$ [5.14]

Here,

α_1, α_2 = coefficient of flow rates in sections I and II, respectively,
f_1, f_2 = cross-sectional areas of the nozzles I and II,
ρ = density of the gas, assumed to be constant,
V_1, V_2 = velocities of gas flow within I and II,
H = working pressure before nozzle I,
h = working pressure before nozzle II.

Now we make some assumptions: first, that the gas density in the two sections is practically equal; second:

$$\alpha_1 / \alpha_2 = 1.$$ [5.15]

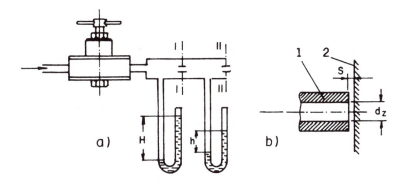

FIGURE 5.12 Layout of pneumatic position sensor.

Thus, we can express the gas velocity within each nozzle as it is accepted for dropping liquids:

$$V_1 = \sqrt{\frac{2}{\rho}(H-h)} \quad \text{and} \quad V_2 = \sqrt{\frac{2}{\rho}h} \, . \tag{5.16}$$

Substituting Expressions (5.16) into Equation (5.14), we obtain

$$\alpha_1 f_1 \sqrt{H-h} = \alpha_2 f_2 \sqrt{h} \, . \tag{5.17}$$

From Equation (5.17) and the second assumption it follows that

$$h = \frac{H}{1+(f_2/f_1)^2} \, . \tag{5.18}$$

The area f_1 is obviously

$$f_1 = \frac{\pi d^2{}_1}{4} \, . \tag{5.19}$$

We assume that the area f_2 can be calculated for the model given in Figure 5.12b) in a relatively simple way. In this scheme 1 is an enlarged diagram of nozzle II (see Figure 5.12a)) and 2 is the surface of an element, the position of which is to be measured. The distance between this surface and the face of the nozzle is s. Experiments show that, if it is true that

$$\frac{\pi d_2^2}{4} > \pi d_2 s \tag{5.20}$$

or

$$s < \frac{d_2}{4} \, ,$$

then for f_2 we can use the formula for the area of the side of a cylinder, namely:

$$f_2 = \pi d_2 s \, . \tag{5.21}$$

Substituting (5.21) and (5.19) into (5.18) we obtain

$$h = \frac{H}{1+16(d_2 s)^2 / d_1^4} \, . \tag{5.22}$$

The latter formula shows the dependence of the pressure h on the distance s. Below are shown some examples of the use of pneumatic measurements of distances and linear dimensions. The main advantages of this kind of sensor are:

1. The possibility of carrying out the measurements without direct mechanical contact between the sensor and the surface of the checked element, if necessary.
2. The relatively high sensitivity of this method, which is about 0.001 mm or even better.

These advantages permit, for instance, carrying out the checking of dimensions during rotation of the measured part, saving time and money.

Figure 5.13 shows a design for automatic continuous measurement of the thickness of a metal strip. In this layout, air from a compressor passes through pressure stabilizer 1, inlet nozzle 2 of electrocontact transducer 3, and then to pneumatic gauge 4. The transducer is fitted with mercury and has two tubes 5 and 6, the lower ends of which are immersed in the mercury. The tubes contain contacts located at different levels. When the mercury level in tube 5 reaches its contact, the coil of relay R_2 is energized and the contacts of this relay are actuated. The normally open contacts become closed and the normally closed, open. Thus, lamp 11 is lit. When the pressure increases further and the mercury level reaches the contact in tube 6 (note that the contact in tube 5 is also closed then), the coil of relay R_2 is energized and lamp 9 is lit. When no contact is closed, lamp 8 is switched on. In addition to the signal lamps, this layout actuates relays 7 which control devices to correct the thickness of the strip. To make the measurement reliable a cleaning device 10 is installed.

Figure 5.14 shows a layout for controlling the dimensions of holes. This device consists of an air pressure stabilizer 1, filter 2, first nozzle 3, pressure transducer 4, and measuring head 5. In both cases considered here the measurements do not require mechanical contact with the measured object. In addition, in both cases, the sum of two air gaps is measured. Thus, no high-precision tuning of the device is needed. The

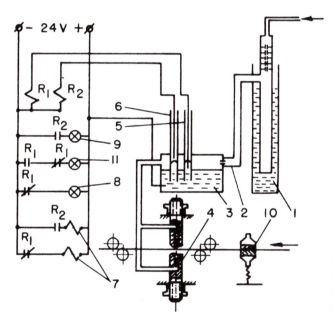

FIGURE 5.13 Layout for automatic, continuous measurement of the thickness of a metal strip.

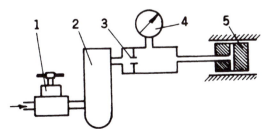

FIGURE 5.14 Pneumatic device for hole-diameter control.

device measures the difference between the initial distance (between the nozzles) and the distance as affected by either the thickness of the strip or the diameter of the holes.

This method is useful for controlling the center distance between two openings. Figure 5.15 presents a measuring head for estimating the dimension B. By reversing the directions in which the pick-ups face, we can measure the value A. Obviously, the center distance L is

$$L = (A + B)/2.$$ [5.23]

The device in Figure 5.16 permits measuring L directly. Four measuring heads, 6, 7, 8, and 9, can be mounted on rod 5 in accordance with values A and B. These heads are connected to pressure transducer 3 (here this is replaced by a liquid barometer, on whose scale 4 the result can be read out). Deviations in the diameters of the holes do not affect the measurement because this deviation actuates both sides of the transducer. Nozzles 1 and 2 must be strictly identical to obtain accurate results.

Figure 5.17 shows a device for checking conical shapes. (This device requires partial contact with the item being measured.) Here 1 is the measured conical shape, 2 is the

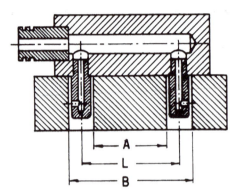

FIGURE 5.15   Pneumatic device for center distance control.

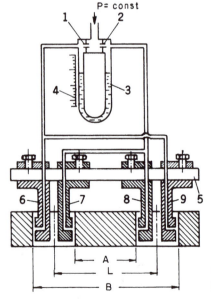

FIGURE 5.16   Center distance control that allows for deviations in the opening diameters.

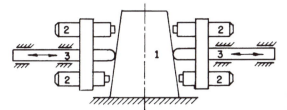

FIGURE 5.17 Pneumatic control for conical surfaces.

measuring nozzles, and 3 is the mechanical supports. The operating strategy is clear from the figure. The same concept is useful for measuring conical holes in solid materials. Figure 5.18 presents such a device. In this device the measurement is made so as to obtain the average deflection from the required conical shape. Nozzles 3 and 4 are connected to pressure connector 1, and nozzles 5 and 6 to the connector 2.

Another example, given in Figure 5.19, shows a pneumatic method for monitoring spherical surfaces, for instance, optical lenses. The lens 1 to be checked is installed on support ring 2. The gap between the lens and the measuring nozzle 3 depends on the

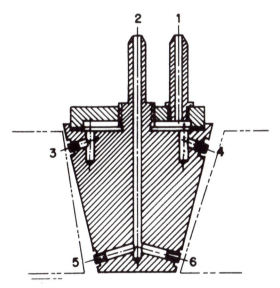

FIGURE 5.18 Pneumatic control for conical opening.

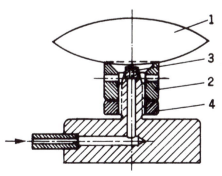

FIGURE 5.19 Pneumatic control of spherical bodies.

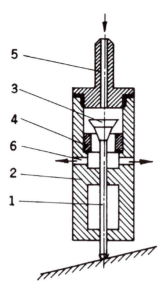

5
3
4
6
2
1

FIGURE 5.20 Pneumomechanical gauge for tracking or measuring.

curvature of the lens. The curvature for several positions can be examined by moving the lens. This device is tuned for different lenses by adjusting support 2 and locking it with nut 4. This device is suitable for both convex and concave items or lenses.

We complete our listing with an example of a mechanically actuated pneumatic pick-up, shown in Figure 5.20. Gauge 1 is directed by guides in housing 2. The gauge has a conically shaped head 3 which, together with bushing 4, creates a valve. Union 5 provides both the air flow through the device (note the outlet openings 6) and the connection to the pressure transducer.

Length of continuous materials (mechanical sensor)

When some continuous material must be measured (wires, strips, ribbons, etc.), an effective means consists of a roller engaged with the running material, e.g., by friction. By counting the revolutions of the roller (knowing its circumference), we can estimate the length of material consumed. This kind of device is very common, for instance, in textile production.

5.2 Speed and Flow-Rate Sensors

The main situations where measurements of speed are needed are

1. When uniform speed must be maintained, and
2. When a transient process in movement is to be controlled.

The first case is important for maintaining uniform processing, thus assuring consistency in the product's quality during, for instance, the processes of drawing (wires, pipes, etc.), rolling (profiles from metals, strips, etc.), extruding, plating, coloring, and coating.

The second case is typical for point-to-point movements of tools and elements of machine manipulators. For accuracy it is important to approach the desired points at minimal speeds to avoid dynamic effects such as vibration and overshoot. In addition, some special speed-control situations may also require speed sensors.

The simplest way (conceptually, not technically) to solve the problem of speed measurement is to obtain the derivative of the displacement of appropriate elements versus time. It is easy to imagine that a variable voltage obtained from the output of a displacement-measuring bridge could be differentiated by means of some microprocessor or even some more simple device. For a digital displacement system, speed estimation V requires a timer and a pulse counter. The timer turns the counter on and stops it at the end of a certain time interval Δt. Thus, the counter indicates the number of pulses n per time interval.

$$V = \frac{n}{\Delta t}.$$ [5.24]

For example, the device shown in Figure 5.10 can be and is used also for speed measurement, especially when a digital readout is desired.

An analogous kind of speed sensor, of an electrical nature, is shown in Figure 5.21. This device consists of a permanent magnet 1 fastened to the moving element and an immovable coil 2. When relative movement occurs between these two elements, an electromotive force (EMF) appears in the coil. This EMF can easily be transformed into voltage, as is shown in the figure. The value of the EMF is proportional to the first derivative of the magnetic flux Φ:

$$E = -a\frac{d\Phi}{dt},$$ [5.25]

where a = constant and t = time.

As is obvious, we can also calculate speed by measuring and integrating the acceleration. Accelerometers will be considered in the next section.

For measuring rotational speed, small electricity generators are used. These devices are called tachogenerators and can generate direct or alternating current. The output voltage or frequency is proportional to the speed of rotation being measured.

Another widely used principle for the same purpose is shown in Figure 5.22. Here the measured rotation speed is transmitted to permanent magnet 1, whose rotation creates a rotating magnetic field. As a result, alternating current is induced in metal disc 2. The interaction between the field and the current creates a rotational torque

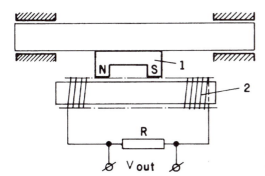

FIGURE 5.21 Electromagnetic sensor for speed measuring.

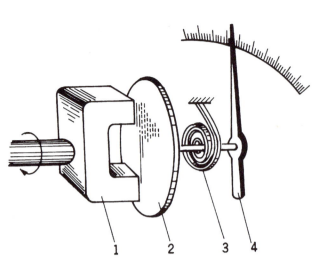

FIGURE 5.22 Electromagnetic device for measuring speed of rotation (car speedometer).

proportional to the speed of rotation. Therefore, the torque angle of spring 3 and the rotational angle of indicator 4 are also proportional to the speed of rotation.

The method of length measurement mentioned at the end of the previous section (5.1) is easily transformed into speed.

Flow-rate sensors are also a kind of speed-measuring device which provide information about the flow rates of gases or liquids. Here we consider some ideas used for this purpose. Figure 5.23 shows an example of a design for an impeller-type sensor. Impeller 1 (a miniature turbine) rotates in agate bearings 2, installed in supports 3 inside a pipe. Axial displacement is limited by agate thrust bearings 4. The pipe section and the impeller are made from a nonmagnetic material. A small piece of a magnetic material 5 is set into one side of the impeller shaft. Outside the pipe a permanent magnet 6 with coil 7 wound around it is installed. During rotation the impeller's shaft turns piece 5, and thus the magnetic flux changes depending upon its position relative to the permanent magnet. This induces an alternate electromotive force of a frequency double that of the impeller's speed of rotation. The rotation speed of the

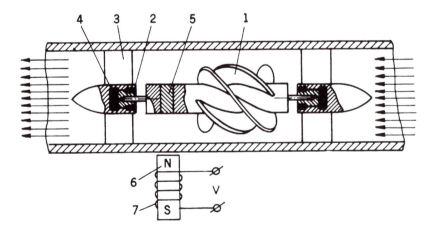

FIGURE 5.23 Gas or liquid flowmeter with an impeller.

impeller (with small losses) is proportional to the specific kinetic energy W of the flow, which is described by the following expression:

$$W = \frac{\rho V^2}{2},$$ [5.26]

where ρ = density of the liquid and V = speed of the liquid.

The liquid's speed multiplied by the cross-sectional area of the pipe gives the flow rate or consumption. The accuracy of these devices is about 1–0.3%.

It is conceivably possible to use the well-known Venturi or Pitot tubes for the same (flow-rate measuring) purpose. However, these sensors require differential pressure pick-ups which, for low flow rates, may be too coarse. This follows from Expression (5.26). (Note that the dimensions of the specific energy Nm/m^3 and pressure drop N/m^2 are equal.) The chain of information transfer in the turbine-type device is shorter than in Pitot or Venturi devices. Therefore, in the latter, sensitivity and accuracy get lost to some extent on the way.

To prevent the negative effect caused by slip of the impeller and to increase the precision of measurement, devices without mechanically moving parts can be introduced. In Figure 5.24 a thermal flow-rate sensor is presented. The pipe section serves as a housing for the device and is provided with a heater 1. Before reaching the heater (say, from the left side), the flowing liquid or gas has a temperature t_0, and after it passes the heater its temperature rises to the value $(t_0 + \Delta t)$. These temperatures are registered by thermoresistors (or other temperature sensors) 2 and 3, respectively, which together with constant resistors R create a bridge. Warming the gas by a certain temperature increment (here Δt) requires different quantities of heat energy introduced by the heater for different mass flow rates of the gas. The voltage ΔV that appears when the bridge is out of balance is amplified by amplifier 4, and controller 5 changes the heating current that feeds the heater. The value of the current I gives an indication of the flow

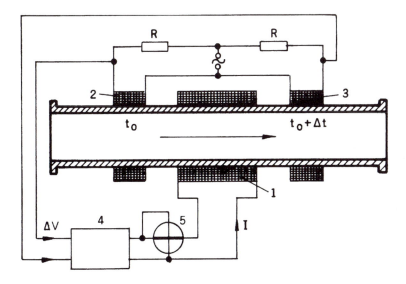

FIGURE 5.24 Thermal flow-rate sensor.

rate. These devices are suitable for a wide range of flow rates, from very small to thousands of cubic meters per hour. The accuracy is 1–2%.

To increase the sensitivity of the device and make it suitable for rapidly changing flow rates, the thermoresistors must have very small masses to reduce their thermal inertia. Figure 5.25 schematically shows such a thermoresistor. On a base 1, in the form of a truncated cone made of a material with low heat conductivity, a thin (1-mm) film 2 of platinum (or other metal) is applied. The other dimensions of this film are about 0.2 by 1 mm. The film is provided with leads 3. The plane of the film is oriented according to the flow's speed vector. Such sensors can respond to fast flow changes of up to 100 kHz.

Positive displacement devices constitute a special class of flowmeters. One possible type is a wobbling-disc flowmeter like that schematically shown in Figure 5.26. Here a specially shaped housing 1 contains a disc 2 able to wobble relative to the center of spherical support 3. The disc has a slot that embraces a partition wall 4. The inlet and outlet ports 5 and 6 are located on either side of this partition. Spherical support 3 has a pin 7 which traces a cone as the disc wobbles. This motion can be mechanically or electrically (by photo or inductive sensors) monitored to determine flow rates. A portion of liquid (or gas) flowing into the device through inlet 5 is restrained by partition wall 4, housing 1, and inclined wobbling disc 2. To proceed, the liquid (or gas) pushes the disc until the initial portion reaches outlet 6. Thus, each turn of the wobbling disc frees a certain volume of liquid (or gas). This volume can be easily calculated from the device's geometry. Of course, in reality there is some slip between the liquid (or gas) volume and the wobbling disc due to leakage between the disc and the housing.

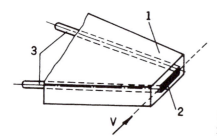

FIGURE 5.25   High-sensitivity thermoresistor for flow-rate sensors.

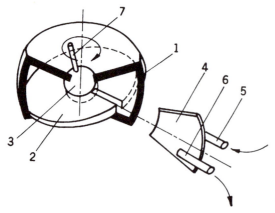

FIGURE 5.26   Wobbling-disc flowmeter with positive displacement.

5.3 Force Sensors

Considering force measurements, we will discuss pure force sensors, as well as acceleration and pressure pick-ups. Obviously, acceleration can be derived from measured speed by differentiation. However, as is known, this entails loss of accuracy (the opposite, integration of acceleration to obtain speed and distance values, is more accurate). This is one of the reasons for using force measurements based on Newton's law for acceleration transducers.

In Figure 5.27 we show a transducer that can detect the displacement, speed, and acceleration of a vibrating element of a machine or robot. Here, mass 3, in the form of a permanent magnet with a circular gap, is installed on base 1 by means of lever 2. This lever is kept in balance with soft spring 4. Thus, the device has a low natural frequency of several Hz, and the sensor is inductive. A coil 5 is fastened on base 1 and is located within the gap. The electromotive force appearing in the coil is proportional to the speed of the vibrating magnet. This signal can be amplified, and when integrated the displacement x is obtained, and when differentiated the acceleration $\ddot{x}$ is obtained. Vibrations of 20 Hz to 500 Hz can be measured.

Often the piezoelectric effect is used for measuring acceleration. In Figure 5.28 such a piezoelectric transducer is represented. Housing 1 is threaded to connect it, by means of its thread 2, to the object being measured. A piezoelement 3 is glued to the bottom of the housing. Usually it is made of quartz, a zirconium and barium compound, titanium ceramics, etc. Inertial mass 4 is fastened above the element. Tungsten alloys are used here since their density is about $18\mathrm{g/cm^3}$, approximately 2 or 3 times that of steel. The natural frequency of this device is about 50 to 100 kHz. During acceleration mass 4 develops a force $F = ma$ (m = mass, a = acceleration). The force acts on piezoelement 3 to induce an electromotive force proportional to the mechanical force. The signal is led out by cable conductor 5, which is connected to metal mass 4, and coaxial cable 6, whose braiding is connected to the housing.

Accelerometers for measurement of constant or slowly changing accelerations can be based on mechanical oscillation systems. However, here very soft springs must be used (to obtain a low natural frequency). This entails small restitution forces which in turn require special means to reduce the frictional forces between the inertial mass and its guides. The device shown in Figure 5.29 is one way of solving this problem. (Its

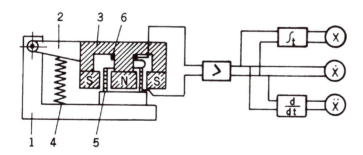

FIGURE 5.27 Transducer able to transmit displacement x, speed $\dot{x}$ and acceleration $\ddot{x}$.

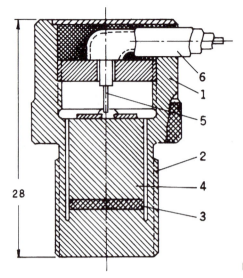

FIGURE 5.28   Piezoelectric acceleration sensor.

theory will be explained in Chapter 9.) Here inertial mass 1 is suspended on spiral springs 2 and 5 and can move along guides 3 and 4. These guides are kept permanently rotating by means of transmission 8 and motor 9. The transducer of the information is a variable resistor 7 and a brush 6 fastened through an insulator to mass 1. The design of the guides ensures that the force of friction is proportional to the speed (in other words, dry friction is replaced by viscous friction). The accelerations measured by this kind of device are in the range 5–200 m/sec^2, and its accuracy is about 1–1.5%.

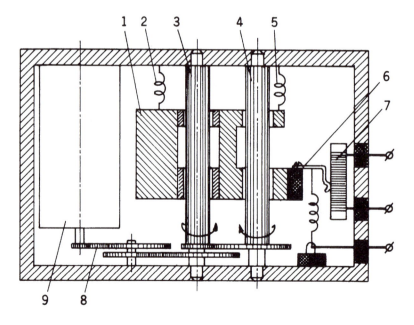

FIGURE 5.29   Oscillating system for sensing slowly changing acceleration.

The devices described so far transform the acceleration into force, and the latter is in some way transformed into an electrical signal. All these devices involve contact with the item being measured. A noncontact method for acceleration measurement during vibrations is based on the Doppler effect that appears when a laser beam reflects from the vibrating surface. This method is suitable for research measurements under laboratory conditions rather than for use in industrial automatic machines and robots, and will not be described here. The method is an effective but still very expensive means.

When the purpose is to measure pure force, strain-measuring sensors can be used. Figure 5.30 shows the plan of resistance-type strain gauges used for tension and compression stress measurements (case a)) and for twisting stress measurements (case b)). Resistance-type strain gauges are widely used and are made in a variety of shapes, sizes, and materials. They operate on the principle that the electrical resistance of some conducting (or semiconducting) materials changes when the wire they are made of is stretched or compressed. Strains as small as 10^{-6} can be measured. In both cases (a) and b)), strain gauge 1 is glued on the surface of the part under load. The direction of the device is chosen so as to be parallel to the direction of maximum stress. To compensate for the influence of temperatures on the readout, another strain gauge 2 of the same type and material is glued perpendicular to the first. By connecting them in opposite arms of the measurement bridge, only those changes in resistance that are due to stresses will be detected. These devices are characterized by a sensitivity coefficient k, which is defined in the following way:

$$k \cong \frac{\Delta R}{\Delta l} \frac{l}{R}.$$

[5.27]

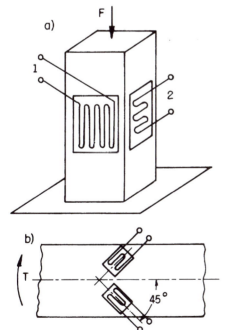

FIGURE 5.30   Strain-gauge sensors: a) Tension or compression measurements; b) Twisting or torque measurements.

Here,

 $\Delta l/l$ = strain value,
 ΔR = resistance increment due to strain, and
 R = initial resistance of the gauge.

Modern devices have k values of several hundred.

Some load sensors are designed as shown schematically in Figure 5.31. In case a) strain gauge 2 is glued on rod 1 to protect it from outside disturbances. The rod is placed in housing 3, and the load is applied via threaded elements 4. This device is suited for large loads. Case b) is similar; however, strain gauges 2 are glued on both sides of hollow cylinder 1 placed inside housing 3. The load (for instance, compression) is applied through the bottom 4 of the device and element 5. Because of the cylindrical shape, the strain is greater and thus the device is more sensitive.

Simpler devices are based on potentiometer sensors. An example is shown in Figure 5.32. The measured load is applied to flanges 1 and 2, which have threaded couplings, and is transmitted through bars 3 and 4 connected by flat springs 5. As a result, bars 3 and 4 can shift in proportion to the applied load (within the limits of elasticity of the springs 5). Thus, pin 6 presses lever 7, which has a slide contact that moves along variable resistor 8. Spiral spring 9 serves to bring lever 7 back. The accuracy is about ±3%. This device is suitable for slow force variations up to about 3 kN.

A torque-measuring device as shown in Figure 5.33, for example, will be considered now. An insert consisting of elastic shaft 1 with two toothed wheels 2 and 3 on its ends is installed between drive 6 and load 7 (a machine or mechanism). Opposite the teeth are located magnetic sensors 4 and 5. One is shown on an enlarged scale in Figure 5.34. The magnetic circuit is designed so that when one pole is located opposite a tooth the other pole is opposite a space between teeth. Thus, during rotation of the shaft, the magnetic flux crosses the magnet coils alternately (see Figure 5.34). The electromotive force appearing there changes with a frequency proportional to the speed of

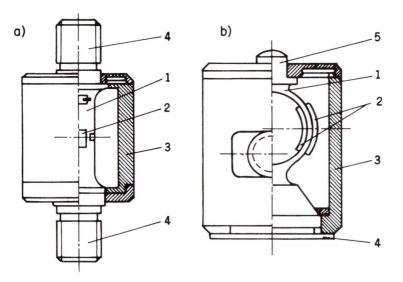

FIGURE 5.31 Load sensors: a) For tension load; b) For compression load.

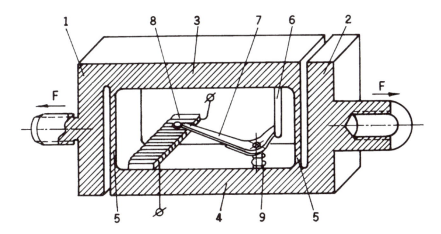

FIGURE 5.32   Potentiometer load sensor.

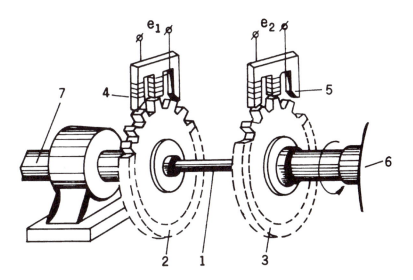

FIGURE 5.33   Inductive sensor for measuring torque.

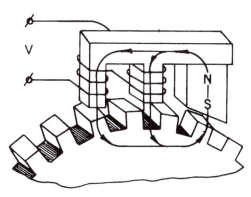

FIGURE 5.34   Magnetic flux in the torque sensor shown in Figure 5.33.

rotation. When no torque is applied to shaft 1, the teeth pass the magnets simultaneously. However, when the shaft is loaded, a rotational shift between the two wheels
occurs. As a result, there is a phase shift between the electromotive forces, which is
proportional to the torque. The phase shift does not depend on the speed of rotation.
The accuracy of the device depends on the properties of shaft 1's material and the stability, and is about 0.1–0.2% at constant temperatures.

Another application of force measurements is in pressure sensors, which are used
either for direct pressure indication or for obtaining the pressure as secondary information, for instance, in pneumatic devices for dimension measurements, or for flow-
rate measurements with Venturi or Pitot devices. Figure 5.35 shows a typical pressure
sensor. The pressure p affects membrane 1 which presses against flat spring 2 due to
the pusher 4. The latter has strain gauges 3 glued to it. The action of this device is
obvious. The accuracy is about ±1.5%, and the range is 0.1–0.6 MN/m². Other elastic
pressure-sensitive elements can also be used in combination with appropriately glued
strain-gauge sets, e.g., bellows, Bourdon tubes, etc. In cases where the elastic element
undergoes relatively large displacements in response to the pressure, appropriate
sensors without amplifiers can be used.

For lower (less than 0.1 MN/m²) and rapidly changing pressures, inductive and
capacitance sensors are widely used. Figure 5.36 shows one kind of inductive sensor.

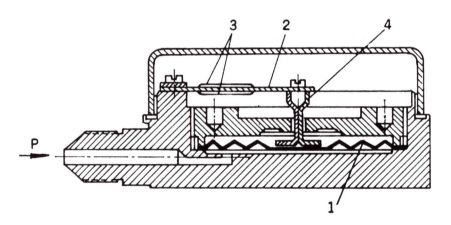

FIGURE 5.35 Pressure-sensitive pickup.

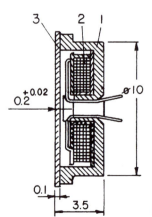

FIGURE 5.36 Inductive pressure sensor.

This sensor consists of magnetic circuit 1, coil 2, and membrane 3 made of ferromagnetic material and influenced by the measured pressure. The initial air gap between 3 and 2 is about 0.2–0.5 mm.

Figure 5.37 shows a capacitance sensor consisting of membrane 1, made as one piece with the housing, and an immobile electrode 2 insulated from the housing by bushing 3. The air gap between 1 and 2 must be as small as possible. The accuracy of these devices is about 2%.

To improve the accuracy of pressure sensors, a different approach than that described above must be used. In devices like those shown in Figures 5.35–5.37, the force affects an elastic element and is balanced by the elasticity of the system via either a special spring or the membrane itself. The higher the measured forces, pressures, or acceleration, the less accurate are the measured values. However, for small forces a more effective approach is based on balancing the measured force with an artificially created force. An example is presented in Figure 5.38. Here, the measured force *P* or pressure *p* through rod 1 actuates lever 2, on one end of which armature 3 is fastened. This armature works in concert with an inductive displacement sensor 7, the signal of which (after transformation by circuit 4) is transmitted into coil 5 (mounted on the other end of lever 2) of electromagnetic transformer 6. The larger the deflection of lever 2 (and of armature 3), the higher the value of the force developed by transformer 6. This force tends to return the lever to its initial position. Current I is thus used both for pressure compensation and as the output of the device.

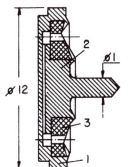

FIGURE 5.37   Capacitance pressure sensor.

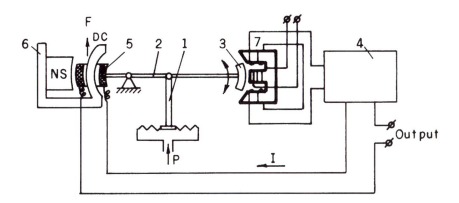

FIGURE 5.38   Counterbalancing pressure sensor.

5.4   Temperature Sensors

We consider three groups of phenomena used in the design of temperature sensors:

1. Electrical phenomena,
2. Thermal expansion,
3. Optical phenomena.

1. Electrical sensors are based mainly on thermoresistors and thermocouples. In addition, an electroacoustic method, based on the dependence of the speed of sound upon gas temperature, and a thermonoise method are in use.

Figure 5.39 shows a bridge layout including ratiometers R_0 and $R_0{}^1$. Permanent resistors R_1, R_2, and R_3 form three bridge arms, while thermoresistor R_t serves as the fourth arm. Resistors R_4 are used for precise tuning of the circuit. Resistor r_t compensates for the effects of temperature changes on the circuit (except for in the thermoresistor). This kind of device is very accurate: for instance, at about 1000°C the error is less than 0.01°C for a platinum thermoresistor.

Figure 5.40 shows a possible layout for a thermocouple, with compensation for its "cold arm." In a thermocouple, two wires made of two different metals or alloys, whose ends are welded together, are used. When placed in environments with different temperatures the weld points create an electromotive force E, which is described well enough by the expression

$$E = \alpha(T_1 - T_2).$$ [5.28]

Here, T_1 and T_2 are temperatures of the environments at the two ends of the thermocouple, and α is a coefficient that depends on the specific properties of the materials the device is made of. In Figure 5.40, point 1 is the hot end of the thermocouple (temperature T_1) and point 2 is the "cold" one. As implied by Expression (5.28), the EMF depends on the difference between the temperatures. However, automatization usually requires measuring the absolute value of T_1. Therefore, a compensation element is inserted between points a-a in the design. This compensator is a bridge in which one arm is thermoresistor R_t. The circuit is tuned so as to create voltage ΔV, which is added to or subtracted from that created by the thermocouple. The accuracy is about 0.04% at around 10°C. The need for a constant voltage source is a disadvantage.

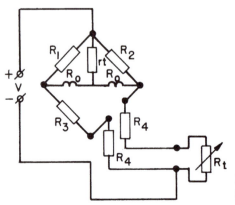

FIGURE 5.39   Resistance-type temperature sensor.

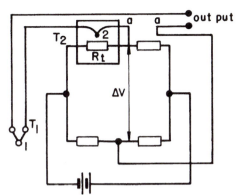

FIGURE 5.40 Thermocouple-based temperature sensor.

2. The phenomenon of thermal expansion of solids and liquids can be utilized in temperature sensors. For example, a bimetallic strip, as shown schematically in Figure 5.41, is an effective on-off temperature sensor. The bimetallic strip 4 consists of two different metallic strips 1 and 2 welded together. When heated or cooled this strip bends because the thermal expansion coefficients of each layer are different. (The strip is flat only at one specific temperature value.) As a result, contact 4 touches immovable contact 3, thus closing gap Δ.

In Figure 5.42 an expansion thermosensor based on a liquid is schematically shown. The liquid is contained in capsule 1, which is located on or attached to the measured element. When heated or cooled the liquid changes its pressure because of its change

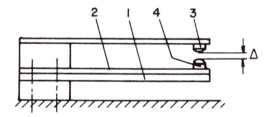

FIGURE 5.41 Bimetallic "on-off" temperature sensor.

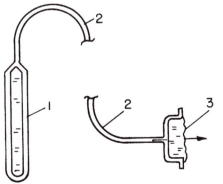

FIGURE 5.42 Temperature sensor based on thermal expansion of liquid.

in volume. This pressure is transmitted through thin pipe 2 to pressure sensor 3 of, say, a membrane type. The deformation of this membrane is transformed by a displacement sensor into a useful signal.

3. Temperature measurements can also be made by a noncontact optical (pyrometer) method based on the fact that, for an absolute black body, the radiant energy E depends on its temperature, as follows:

$$E = \sigma T^4, \tag{5.29}$$

where σ is constant.

Measuring this energy allows estimation of the temperature of the radiating body. Figure 5.43 shows a design for a pyrometer consisting of tube l in which thermoresistor 2 is placed at the focal point of spherical mirror 3. To prevent the influence of reflected light, the thermoresistor is protected by screen 4 and the internal surface of the tube is covered with black ribs 5. Protector 6 shields the device from dirt. The thermoresistor is connected in a bridge circuit. Such a device can be used both for low temperatures, 20°–100°C, and for high temperatures, 100°C–2,500°C. The radiating surface 7 is placed at a certain distance from the pyrometer. The intensity of radiation is inversely proportional to the square of the distance and influences the accuracy of measurement, which is usually about ±2.5%.

5.5   Item Presence Sensors

As we have seen in earlier chapters and sections, a checking position is often needed for automatic industrial machines, to find out at certain manufacturing stages whether the hardware and the process are still all right. The importance of checking can be demonstrated (with reference to an assembly machine) as follows: let us consider the mass assembly of the product shown in Figure 5.44, which is made of three parts. It is

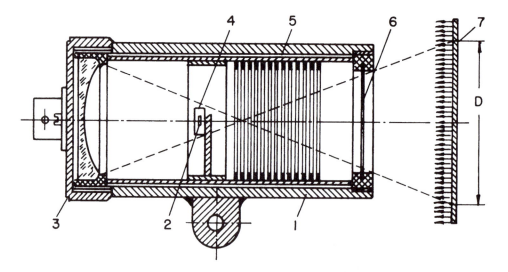

FIGURE 5.43   Pyrometer sensor actuated by light radiation from a heated surface.

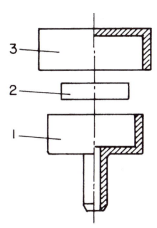

FIGURE 5.44   Example of an assembly consisting of three parts.

natural to begin its assembly with part 1. Let us imagine that, because of some technical defect, this part did not appear in its position, and the second part (part 2) is put in its place by the manipulator without checking. As a result, this part (as well as part 3 because it will also be blindly put in place) will simply be thrown out and lost for production. Let us now imagine another situation. Part 1 successfully reached its position; however, technical defects occurred with part 2, which thus did not appear. Then part 3 is added, which closes the assembly. The end result looks on the surface like a completely finished product and might be mistakenly sold as such. To examine every item after assembly to exclude such defective items may be overly expensive. Therefore, checking the situation during the manufacturing process, at every appropriate processing stage, may increase the cost of the machine somewhat but it will prevent the problems associated with selling defective products (even when the chance of producing and selling such incomplete items is relatively low).

The simplest way to solve this problem is to use a photosensor. Figure 5.45 shows a concept of how to use one for the assembly presented in Figure 5.44. Light source 1a and detector 2a check for the presence of part 1 in the socket of indexing table 3. In addition, light source 1b and detector 2b check for part 2 in the assembly. When either or both parts are not in the right positions the sensors will actuate some system to stop the machine, prevent the assembly process from continuing to the next step, or alert

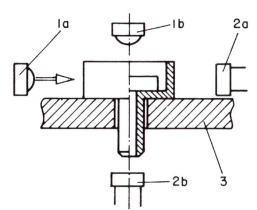

FIGURE 5.45   Optical method for checking the presence of parts in their proper places.

a human worker. Here, interruption of the light beam attests to the presence of a part on its place. Sometimes it is necessary to check for the absence of any kind of contaminant in certain places—a piece of a chip, a dropped bolt, screw, or nut, etc. This method also permits checking whether the correct part is placed in a certain socket. For this purpose elements of the geometrical shape of the part—holes, slots, planes—would serve as markers.

Pneumatic sensors also can be used for checking part presence. As explained in Section 5.1 (2), pneumatic sensors can work without contacting the measured surface. Thus, in Figure 5.12b), when there is no element 2 near nozzle 1, the pressure in the device would drop. In this case high accuracy is not required, and the device is used in the "on-off" regime.

Very often an electromechanical sensor for part presence or absence is used. Figure 5.46 shows the design of a possible solution. Here gauge 1 checks the presence of part 2 located in a socket of transporting device 3 (see Chapter 6). Gauge 1 is held in housing 4 and is pressed towards the part under examination by means of spring 5. The housing is moved in guide 6 by means of a pneumocylinder, magnet, or mechanical system (not shown in the figure). When rod 7 on which housing 4 is mounted goes down in accordance with the machine's timing diagram, it closes contact 8. If part 2 is not present in the socket, the pair of contacts 9 also closes (they are always closed when the gauge is not pressed) and the relay coil R is energized and actuates a signal, stops the machine, or performs some other action. Thus, the relay is energized (both contacts 8 and 9 closed) only when there is no part in the socket.

In Figure 5.47 we show another kind of electromechanical device for checking the presence of wire. The device consists of gauge 1 in the form of a fork freely suspended on its axis 2. The latter is joined to lever 3 which is brought into oscillation by crank 4 and connecting rod 5. When the fork, during its movement leftward, meets wire 6 (or a thread, filament, or rod), it lifts the right, wedge-shaped end 7 of gauge 1. In addition, there is a lever 8 suspended on an immovable axis 9 and a push button 10 that actuates contacts 11. These contacts are normally closed, due to spring 12. During operation this device checks the presence of wire 6 in the following manner. As was mentioned earlier, the wire lifts end 7 of lever 1, and thus its oscillations do not have any effect. However, when the wire is absent, the fork does not meet resistance; thus end 7 of

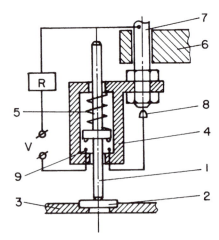

FIGURE 5.46 Electromechanical gauge for checking the presence of a part in its position.

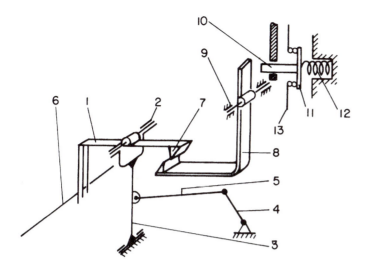

FIGURE 5.47   Electromechanical device for checking presence
of wire, rod, strip, thread, etc.

lever 1 stays low and engages with the correspondingly formed end of lever 8. This
causes the latter to rotate so that its upper end actuates push button 10, disconnecting
contacts 11 and interrupting circuit 13. This signal can be used in any convenient way:
to stop the machine, signal the personnel about some defect, supply the wire, etc. This
device is actually a very old invention used in looms for checking wire presence.

6

Transporting Devices

6.1 General Considerations

We have already touched on the problem of the general configuration of industrial automatic machines and robot systems. In Chapter 1 it was shown that there exist two main options: the linear approach, schematically presented in Figure 1.23, and the circular one as shown in Figures 1.20 and 1.22. Both involve a set of functional mechanisms, tools, manipulators, and a transporting device which conveys the part or product being processed along a straight line or around a circle. In the linear arrangement the tools and manipulators are placed along the conveyor, on one side (Figure 1.23)—(this gives easier access to the blanks and instruments), or on both sides of the transporting device (this saves room and allows the whole machine to be smaller). Analogously, in the circular layout the tools are located around the rotating conveying device either on its outside (Figure 1.20) or with some on the inside (Figure 1.22).

We also have to distinguish between periodically working transporters and continuous transportation. This chapter deals with some specific designs used for these purposes and their main features and properties.

6.2 Linear Transportation

First, we consider continuous transportation, for instance, as shown in Figure 1.24, which represents the layout of a rotary printing machine. This kind of manufacturing process, as was already noted, allows continuous processing through continuous transportation. Typical problems that arise in this kind of automatic machine (among others) are:

- Maintaining constant tension (in the paper, fabric, wire, threads, etc., that are handled);
- Maintaining constant speed of the running material;
- Handling the problem of stretching of some materials during their processing (thermal treatment, humidity and drying, plastic deformation, etc.);
- Maintaining a constant rate of consumption of the transported material. (This is close to, although not exactly the same as, keeping the speed constant.)

Figure 6.1 shows a plan for a tension-sensitive device. The wire (thread, ribbon, etc.) 1 runs over two rolling guides 2, passing under roller 3, to which is fastened lever 4. The tension force T causes the appearance of a resultant force R, which is balanced by spring 5. Thus, the angle of inclination of lever 4 responds to the tension in the wire. By using a position sensor for measuring angle (see Chapter 5, Section 5.2), we can control speed V_1 at the input of the system so as to keep the tension of the wire within the desired limits.

The problem of maintaining a constant transportation speed can be solved as in the example shown in Figure 6.2. Here running material 1 (fabric, plastic, metal sheet, etc.) passes guide rollers 2 and drive rollers 3. The speed of the drive rollers is determined by the corresponding motors and transmissions 4, while guides 2 are driven by the material itself. Thus, their rotation speed (measured by, for instance, tachogenerators 5) indicates the material's speed or rate of consumption in the machine.

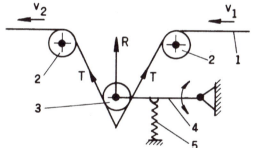

FIGURE 6.1   Design of wire or thread tension-regulating device.

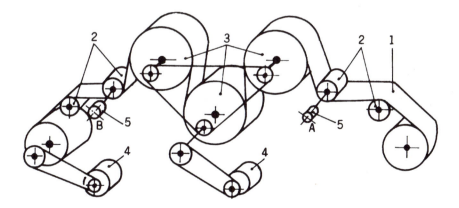

FIGURE 6.2   Design of a constant transportation speed system.

The same arrangement can deal with the problem of stretching of the transported material. For this purpose, the readouts of two tachometers driven by guides A and B are compared, to yield information about stretching (or shortening) of the material. How to use this information depends on the type of material and on the nature of the product treatment process.

Another way to handle the problem of material stretching is to use a length compensator. A possible design is presented in Figure 6.3. Transported material 1 passes guide rollers 2. The loop created by the excess material passes an additional guide roller 3. The shaft of this roller is placed in slide guide 4 and is connected to chain 5 supported by sprockets 6. Load 7 partly balances the weight of roller 3. The ratio between the inlet and outlet speeds V_1/V_2 determines the length x of the loop. The higher this ratio, the larger is x, and vice versa. Two limit switches 8 respond when the value of x is outside the permitted limits. Thus, the following condition exists:

$$L_1 < x < L_2 .$$ [6.1]

Usually adjustments are made by changing the value V_1 (the material inlet speed). Reducing V_1 shortens the loop whereas increasing V_1 causes elongation of the loop. The same effect can obviously be achieved by changing speed V_2. Alternatively, instead of using limit switches, the angle can be measured and the condition

$$\phi_1 < \phi < \phi_2$$ [6.2]

permits controlling the speed ratio in the same manner.

Next we consider processes where continuous transportation is applied to separate parts or items. This kind of transportation is typical of galvanizing, painting or dyeing, thermal treatment, chemical treatment, and similar processes. Figure 6.4 shows a layout of such a transport device, whose purpose is to submerge items 1 into, say,

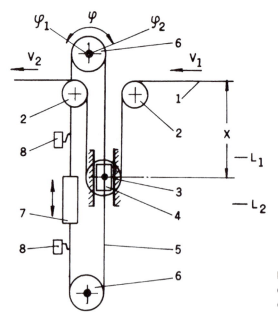

FIGURE 6.3 Design of a length compensator for an automatic continuous processing machine.

liquid 2 (or gas, vapor, environment, etc.), which is contained in tank 3. Conveying chain 4 is guided by rollers 5 and driven by sprocket 6. To compensate for possible length changes in the chain, compensating roller 7, which can slide and is pulled by load 8, is used. At position A the parts are loaded onto the chain (separately or in baskets) while at position B they are unloaded. The length of the tanks and the speed of the chain determine the time of treatment and the output of the device. Often, a process requires several dippings (or other treatments). Then the conveying system consists of more rollers 5 for lowering and lifting parts 1 during passage from one tank (chamber, area zone, etc.) to the next and, of course, the chain is much longer. Attention must be paid to the fact that in this case a certain number of parts are always travelling in the gaps between tanks. This entails:

- More parts loaded on the machine, thus greater masses to be moved;
- Greater time interval when the parts are not being treated and are in contact with air.

In cases where these circumstances must be avoided, the arrangement shown in Figure 6.5 can help. Here, a line of tanks 1 is served by a special conveyor system consisting of horizontal transporting chain 2 and several vertical conveyors 3. The speed of horizontal transportation V_1 is much slower than the speeds V_2 of the vertical conveyors. When the parts (baskets, blanks, etc.) are suspended on horizontal chain 2 in region A they begin to move leftward at speed V_1 until they reach the first vertical chain 3, which catches hangers 4 and transfers them with speed V_2 (i.e., rapidly) into the first tank, and returns the hangers to the horizontal conveyor. It is worthwhile to explain

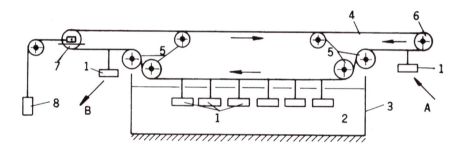

FIGURE 6.4 Design of a continuous transportation device.

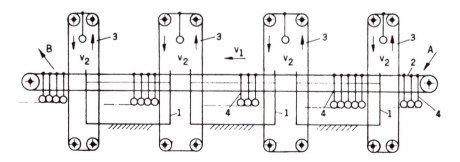

FIGURE 6.5 Design of a time-saving continuous transportation device.

this process in more detail. Figure 6.6 shows the meeting point of the horizontal and vertical conveyors on a larger scale. Hanger 4 holds basket 5 in which the parts being processed are put. The hanger is suspended from rod 6, which is carried by chain 2 and slowly (speed V_1) moves towards vertical chain 3. The latter has hooks 7 which move vertically with speed V_2. When rod 6 reaches chain 3 it is caught by the nearest hook 7 that is moving upwards, and begins its test travel to the next tank. On the other side of the vertical conveyor where the hanger comes down, rod 6 is transferred to chain 2 while chain 3 continues to move down, carrying away hook 7. Thus, rod 6, together with hanger 4 and basket 5, begins its slow travel inside the next tank while the basket is submerged in the liquid contained there.

Now we will consider periodic transportation devices. This discussion must be carried out in terms of time intervals. If T is the duration of a period, we can state that this value is composed of smaller time intervals:

$$T = T_1 + T_2 . \qquad [6.3]$$

Here, T_1 is the time the transporting device rests (regardless of whether it is linear or circular), and T_2 is the time the transporting device moves.

Usually, the aim is to design the transportation device so as to minimize T_2. The restriction that must be taken into account is the value of the acceleration occurring when the motion begins (as well as the deceleration as the motion stops). The smaller T_2 is, the higher are the accelerations that appear in the system, the greater the inertial forces, the higher the power needed for the drive, and the more expensive the whole machine. Keeping this in mind, we pass on to describe some of the technical ways of providing periodic movement of parts under processing or other technological treatment.

First, we consider a chain-like conveyor consisting of specially shaped links. During each period, the chain drive moves forward so that one link is replaced by the next.

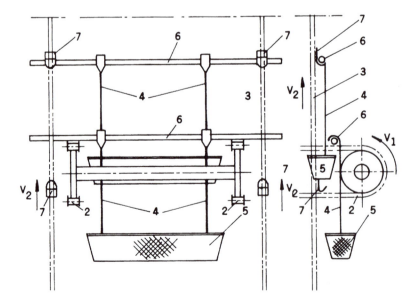

FIGURE 6.6 Cross section of the layout shown in Figure 6.5. The meeting of the horizontal and vertical conveyors.

Thus, the pitch of the movement equals the pitch of the chain or the length of one link. In some cases the pitch of the movement equals some whole number of chain pitches. This is illustrated schematically in Figure 6.7. To drive chain 1, drum 2 must be driven. The simplest way to do so is to use an appropriate kind of Geneva mechanism (see Figure 4. 1). A strict ratio T_2/T_1 is typical for these mechanisms. Indeed, let us suppose the driven link of the mechanism has z slots. Then, as it follows from the sketch shown in Figure 6.8 (this is an instantaneous look of a Geneva mechanism at the moment of engagement between a slot and driving pin), we can write

$$\phi = \pi / z \tag{6.4}$$

and

$$\psi = \pi\left(\frac{1}{2} - \frac{1}{z}\right) = \pi\frac{T_2}{T_1 + T_2} = \pi\frac{T_2}{T}. \tag{6.5}$$

From Expression (6.5) we obtain

$$T_2 = T\left(\frac{1}{2} - \frac{1}{z}\right), \tag{6.6}$$

and, taking into account (6.3), from (6.6) we have

$$\frac{T_2}{T_1} = \frac{T(1/2 - 1/z)}{T - T(1/2 - 1/z)} = \frac{z-2}{z+2}. \tag{6.7}$$

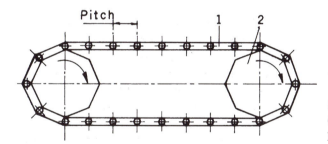

FIGURE 6.7   Design of a chain-type conveyor for periodic automatic processing.

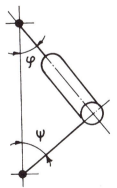

FIGURE 6.8   One slot of the Geneva mechanism engaged with the driving link of the mechanism.

In Table 6.1 we see the ratios for several numbers of slots in the driven links.

Any other mechanism that provides interrupted rotation can obviously also be used for driving the drum. We will consider some possibilities in the next section of this chapter while discussing the drives of circular transporting devices.

Attention must be paid to the fact that conveyor-type transporting devices consume a great deal of power just to move the mass of the chain or belt, etc. Obviously, this power is in essence useless, and one must reduce such power losses. One way would be to find a chainless solution, where the additional masses moving together with the part or body being treated are minimal. Figure 6.9 represents one such solution. This device consists of two systems of mechanical pushers, one immovable 1 and the second 2 moving. The pushers are provided with a set of pawls 3 actuated by springs 4. The moving pusher slides in guides 5 and is driven by, say, a pneumocylinder 6. Blanks 7 are located on the machine table 8. The mechanism operates as follows. Cylinders 6 (or any other drive—electromagnetic, mechanical, etc.) pull pusher 2 rightward to the position shown in the figure. Then pawls 3 of this pusher are retracted into the body of the pusher due to their inclined back face (the blanks press against the pawls and overcome the force of spring 4). At the same time pawls 3 of immovable pushers 1 stay extended and keep the blanks in their places (the straight front faces of the pawls prevent "retreat" of the blanks). The second stage of operation begins with the reverse movement of cylinder 6. Then, pawls 3 of pusher 2 push the blanks leftward, while pawls 3 of immovable pushers 1 now (because of the inclined faces of the pawls) do not prevent displacement of the blanks. Obviously, as a result of this reciprocating movement of element 2, all blanks on table 8 are moved by one pitch of the system. The resting time of the blanks is not restricted, and the control mechanism can be tuned as necessary. The time in motion depends on the dynamic parameters of the system and can be estimated during design.

Another approach for an analogous task is presented in Figure 6.10. This mechanism also consists of two systems. One keeps blanks 1 in their positions during processing. This system consists of shaft 2 rotating in bearings 3 provided with clamps 4 and lever 5 driven by pneumocylinder 6 (or some other drive). Thus, shaft 2 is only able to rotate in its bearings. The other system must move the blanks and consists of shaft 7 that can both rotate and slide in guides 8, and which has pawls 9. Shaft 7 is rotated by cylinder 10 through lever 11 and key 12. The reciprocating motion of this shaft is provided by cylinder 13 through cross-piece 14. The connection between the

TABLE 6.1 Time components ratios on the output of the Geneva mechanism for different numbers of slots on driven link.

z	T_2/T	T_2/T_1
3	0.167	0.200
4	0.250	0.333
5	0.300	0.429
6	0.333	0.500
7	0.357	0.555
8	0.375	0.600

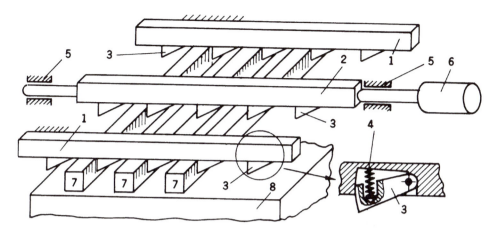

FIGURE 6.9   Design of a chainless transportation device with one degree of freedom.

cross-piece and the shaft is made so as to allow rotation of the shaft. During operation, this mechanism goes through the following sequence of motions. First, pawls 9
must be brought into the down-right position (next to the blanks on their right side),
and clamps 4 are down to hold the blanks. The events are shown also in the timing
diagram (Figure 6.10a). (The description begins at $t = 2$.) Transportation begins with
lifting clamps (cylinder 6); afterwards cylinder 13 begins its motion leftward, moving

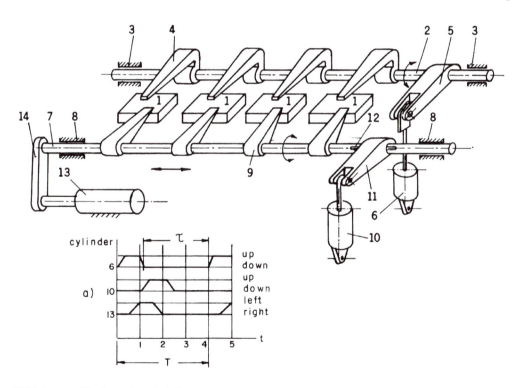

FIGURE 6.10   Design of a chainless transportation device with two degrees of freedom.
a) Timing diagram for this device.

the blanks. Now clamps 4 go down to hold the blanks. At this moment, processing (drilling, assembly, cutting, etc.) may begin and lasts for time interval τ. Sometime during this interval the system must be returned to its initial state. The pawls must be lifted (cylinder 10) and moved rightward (cylinder 13) where they are lowered (again, cylinder 10). This completes the cycle of the transporting device, whose duration cycle equals T.

We may make the following note here. The transportation device discussed above consists of two manipulators working in concert. One has one degree of freedom (clamping mechanisms); the second has two degrees of freedom. Of course, both of them can be controlled by systems of different levels of flexibility. The controls can be stiff (carried out by a cam mechanism tuned once forever), a computerized system of steppers allowing flexible programming, or any other level in between.

Before finishing this section, we discuss here one more example of a linear periodic transporting device with a variable T_1/T_2 ratio. This device is useful for processes where the material, parts, or products being processed are submerged in several tanks, pools, chambers, etc., in sequence. The plan of this device is given in Figure 6.11 and consists of frame 1 installed above a row of tanks 2. On the frame a system of levers 3 is mounted so that all of them rotate simultaneously and in phase. These levers are driven by worm gears 4 and worms 5. The worms are rotated by shaft 6, which is driven

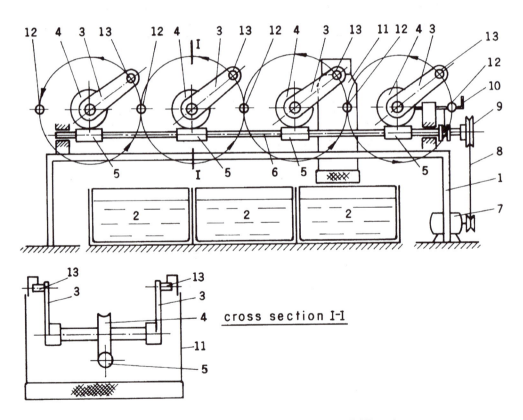

FIGURE 6.11 Design of a transportation device with variable T_1/T_2 ratio.

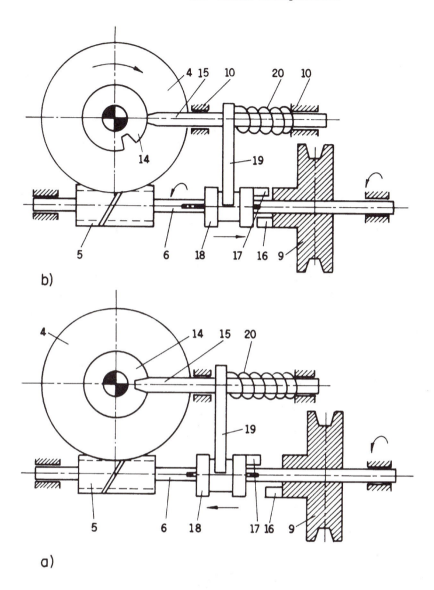

FIGURE 6.12 Design of one-revolution mechanism used in the layout shown in Figure 6.11: a) Locked state; b) Driving state.

by motor 7 and belt 8. Pulley 9 always rotates, but due to the one-revolution mechanism shown in Figure 6.12 (discussed below), its rotation is transferred to shaft 6 only intermittently. Commands to actuate the one-revolution mechanism are given by a controller (not shown) through, say, lever 10. In the resting state the levers 3 are in the horizontal position, and the transporting baskets or fixtures 11 carrying the parts being processed rest on supports 12. During movement, pins 13 on levers 3 catch the fixtures and transfer them on a circular path to the next supports, leaving the fixtures to rest on these supports 12. In the meantime levers 3 finish their revolution and stop, ready for the next cycle.

We use this opportunity to illustrate another one-revolution mechanism (the previous one was presented in Figure 4.21). This mechanism is shown in its resting state in Figure 6.12a). Worm gear 4 (in Figure 6.11 the same numeration) and worm 5 engaged with it are not moving. In addition, holder 15 locks worm gear 4 by engaging with slot 14 in its hub. Pulley 9 always rotates freely on shaft 6. The pulley has pin 16 which is disconnected in this situation from the other pin 17 of bushing 18. The command to begin the movement acts to pull out holder 15 from slot 14 of the worm gear hub. This also pulls fork 19, which is engaged with bushing 18, rightward and brings the pins 16 and 17 into engagement. Thus, pulley 9 begins to drive bushing 18 and, through a key, it drives shaft 6 and worm 5 (here we show only one worm transmission). Worm wheel 4 begins to rotate, carrying around with it the levers (in Figure 6.11, levers 3). Now holder 15 may be freed and spring 20 will press it against the hub, although it cannot disengage pulley 9 from bushing 18. Only when slot 14 reaches its initial position (the revolution is finished) does the holder fall into the slot and disconnect pins 16 and 17. The system is thus returned to the state shown in Figure 6.12a).

The above transporting devices are widely used; however, there are specific cases for which these devices are not suitable. Then special means must be undertaken or special devices designed. An illustration will be worthwhile. In Figure 6.13 a layout of an automatic sorting machine is shown. Here the transportation and measurement processes are closely related. Transportation is carried out by pushers 1 which go up and down. The measured part 2 rolls down to gauge 4 and stays there if the gap between this gauge 4 and the partition wall 3 is smaller than the part. In this case pusher 1 lifts the part to the upper lever, where the part rolls to the next gauge. This process comes to an end when at last the part is smaller than the gap and falls through into channel 5. Thus, gravity and the pushers transport the parts, while gauges 4, after being appropriately adjusted, do the sorting.

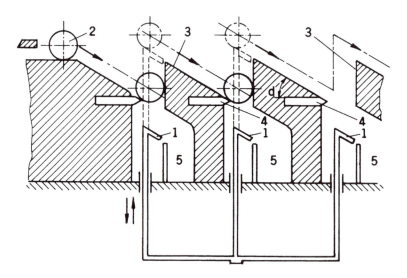

FIGURE 6.13 Design of an automatic sorting machine, as an example of a special type of transportation.

6.3   **Rotational Transportation**

Rotational transportation devices are often also called indexing tables. In this section we consider the drives used most often to achieve interrupted rotation, where, as before, T_1 is the resting time and T_2 is the time the transporting device moves.

One mechanism—the Geneva drive—has already been discussed (Figures 4.1, 4.4, 4.8). In addition we can mention that combining these drives with gear transmissions allows variation of the difference between the number of slots in the mechanism and the number of positions on the indexing table. For instance, a four-slotted Geneva mechanism transmitting the rotation of the driven link through a gear transmission having a 1:2 ratio would drive an eight-position indexing table. (The ratio T_2/T_1 stays the same as for a four-slotted Geneva mechanism.) To get around the rigid dependence of Geneva mechanisms on the T_2/T_1 ratio, cam mechanisms of the sort shown in Figure 6.14a) can be used. Here spatial cam 1, provided with scroll-like profile 2, is engaged with a rotating follower 3 by means of pins 4 (often having rollers to reduce friction). The follower drives indexing table 5. This mechanism is sometimes modified so that pins 2 are mounted on the underside of indexing table 1 which slides or rolls on ball bearings in circular guide 3 (Figure 6.14b)).

The design of the cam's profile is somewhat flexible, and thus the resting time of the indexing table and the time of its rotation are not rigidly related. However, restrictions due to the allowed pressure angle value do not leave too much freedom for this flexibility. One solution to this problem lies in the use of interrupted rotation of the cam. For this purpose the cam is combined with a one-revolution mechanism. This combination was described in Section 4.4 of Chapter 4, as was the corresponding mechanism, and is represented schematically in Figure 4.49. Indexing times in the range 0.2–0.5 second are usually obtained. As was shown in Chapter 4, the mechanism (of course, it depends on the mass of the table) works even for a time 0.4 second, under dynamically intense conditions. The combination of a cam mechanism with a one-revolution coupling gives the possibility of actuating the indexing table at every instant and allows much higher flexibility for the whole machine.

Now we consider a pneumatic drive for an indexing table. A design for such a drive is illustrated in Figure 6.15. The indexing table (not shown) is fixed on shaft 1. On the same shaft are mounted ratchet wheel 2 and lock wheel 3. Lever 4 is free to rotate around shaft 1 and carries pawl 5 fastened on axis 6. Another pin 7 connects lever 4 with piston rod 8 of cylinder 9, which can oscillate around pin 10. Another cylinder 11 is used to drive stop 12. This mechanism operates as follows. When the indexing table

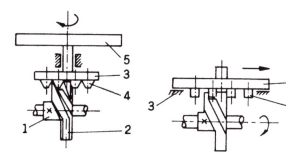

FIGURE 6.14   Spatial cam drives for a circular transporting device: a) Follower separated from the rotating table; b) Follower part of the table.

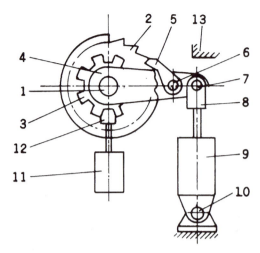

FIGURE 6.15 Design of a pneumatically driven indexing table.

must be driven to the next position, stop 12 is pulled out from the slot of lock wheel 3. Next, cylinder 9 is energized and, by means of pawl 5, drives ratchet wheel 2 and shaft 1. Piston rod 8 moves until stop 13 is reached. The latter can be adjusted for precise indexing. Before the rotation of the table is finished, stop 12 is actuated by cylinder 11 and begins its return toward the next slot of lock wheel 3. Thus, the indexing cycle is completed.

It is easy to imagine hydraulic cylinders or electromagnets instead of the pneumocylinders. The choice must be made according to the job and according to the masses the table must move. These mechanisms are flexible with respect to timing so that the ratio T_2/T_1 depends almost entirely on the controlling device.

Another possibility we consider here is based on the use of an electromotor drive in combination with a gear transmission. A layout of such an approach is presented in Figure 6.16. To optimize the dynamics of this drive, a feedback, for instance, a speed sensor, must be included in the design. Thus, the speed of rotation of the indexing table is controlled so as to obtain smooth changes in the accelerations and movements.

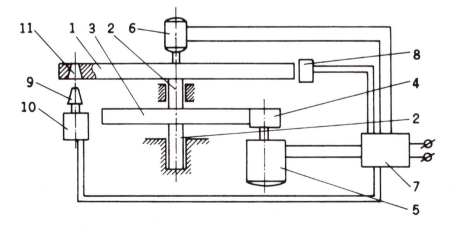

FIGURE 6.16 Design of an indexing table driven by an electric motor.

Table 1 is mounted on shaft 2, which also bears gearwheel 3. The latter is engaged with pinion 4 driven by motor 5. Through control unit 7, tachogenerator 6 creates feedback to control the speed of motor 5. Another sensor 8 measures the angle of rotation of indexing table 1 and gives commands to the driving motor 5, through unit 7. To ensure a precise stopping of the table, a suitable brake must be employed. The further it is located from the axis of rotation, the higher the accuracy of indexing that can be achieved. This brake can be designed, for example, as shown in Figure 6.16. Here, conical brake 9, which is driven by electromagnet 10, provides the needed immobilization of the table when it is inserted into the corresponding conical opening 11. This electromagnet also is controlled by unit 7. The motor 5 chosen for the indexing table drive can be a DC or stepping motor and unit 7 must be designed accordingly. The indexing mechanism considered here belongs to the class of polar manipulators with one degree of freedom.

Often indexing tables or similar mechanisms carry out two motions, that is, the movement has two degrees of freedom. The additional degree is usually movement along the rotating shaft. We have already considered an example of this kind. The layout of an automatic membrane tin-plating (Figure 2.10) machine includes an indexing mechanism of this kind. Here the mechanism has suction cups. Alternatively, these may be electromagnets energized at appropriate times to grip the part being processed, or mechanical grippers actuated by any means.

Figure 6.17 shows a design for an indexing mechanism with two degrees of freedom. Here, indexing table 1 is fixed on shaft 2 provided with a key 3. Cross 5 of the Geneva mechanism is fastened onto bushing 4. Due to key 3, torque is transferred to shaft 2 without hindering the shaft from moving along its length. Cross 5 is driven by roller 6 on carrier 7. The carrier is rotated by shaft 8. Roller 6 engages the slots in cross 5. For

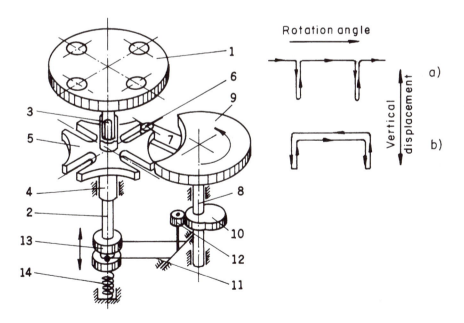

FIGURE 6.17   Design of an indexing mechanism with two degrees of freedom.
a) and b): Two variations of the indexing mechanism movement.

holding the cross during its rest, the curved parts of the cross are locked by disc 9. Shaft 8 also drives cam 10 which is used for raising and lowering the indexing table with the aid of fork-like follower 11, which is connected to roller 12. Fork 11 is engaged with bushing 13. The indexing table is balanced by spring 14. This structure allows the double motion of the indexing mechanism to be carried out. This mechanism can be used, for example, to take parts (by means of suction cups or other grippers) from their positions, transfer them to the next positions, and lower them into their places. The movement of the suction cups is graphed in Figure 6.17a).

The indexing mechanism can also be designed so as to carry out recurrent angular motion in combination with reciprocating vertical motion. A diagram of this motion is shown in Figure 6.17b). The advantage of this kind of motion is that flexible means of communication can be used, e.g., electric wires, hoses, pipes, whereas in the previous case special methods are required to connect suction cups to a central vacuum pump to avoid leakage of air into the system and prevent the hoses from twisting.

A simplified transporting device for this kind of motion (when the number of positions is small) is shown in Figure 6.18. It consists of arm 1 fastened onto rod 2. With two degrees of freedom, as indicated in the figure by arrows, the arm can transfer items from position I to position II. A series of such devices operating in synchrony can be used when more than two positions must be served.

Finally we discuss a special type of rotating transporting mechanism, an example of which (a sorting machine) is presented in Figure 6.19. This machine must take rollers of a roller chain from feeding device I and sort them into four groups, according to size, into positions, II, III, IV and V (Patent #213542, 1962, USSR, Janson A. F.). Transportation is carried out by arm 1 which has elastic gripper 2 that holds the part by means of elastic and frictional forces. The center of the gripper moves so as to transcribe hypocycloids in a plane, moving in sequence from I to II to III to IV to V and at last back to position I. To achieve this kind of trajectory, arm 1 is fastened onto shaft 3 which is driven by gear wheel 4. Shaft 3 rotates on bearings placed in lever 5 which is driven by central shaft 6. When lever 5 rotates, wheel 4 rolls over immovable ring gear 7. This design, although correct, kinematically is too heavy to use. A more elegant and practical kinematic solution can be proposed (Figure 6.20). Here, arm 1, which carries gripper 2, is driven by pinion 3. The latter rotates in bearings located at the end of lever 4, which is driven by shaft 5. The sun wheel 6 remains immobile. As a result of lever 4's rotation, planet gear 7 revolves around sun wheel 6 and drives pinion 3 together with arm 1. If the ratios of the wheels are chosen correctly, the gripper will move along the same trajectory as in the case given in Figure 6.19.

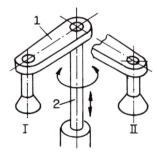

FIGURE 6.18 Simplified transportation device— automatic arm with two degrees of freedom.

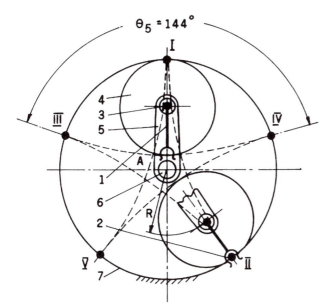

FIGURE 6.19   Design of a rotating machine for sorting rollers.

How to make the right choice of ratios? We will demonstrate on the basis of this specific example of a five-position machine. We denote:

ω_3 = angular speed of pinion 3,
Ω = angular speed of lever 4,
z_3 = number of teeth on pinion 3,
z_6 = number of teeth on sun gear 6.

Then we can write

$$\frac{z_6}{z_3} = \frac{\omega_3 - \Omega}{-\Omega}.$$

[6.8]

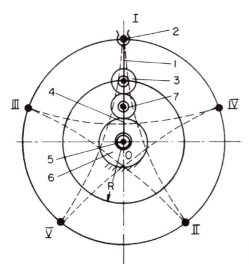

FIGURE 6.20   Another kinematic solution of the mechanism shown in Figure 6.19.

This mechanism must be equivalent to that shown in Figure 6. 19, which means that rotation angle θ_5 of lever 5 and rotation angle θ_4 of wheel 4 (or the rotation angle of arm 1) are related by some definite ratio. It is easy to understand that, between any two positions, $\theta_5 = 144° = (2/5)360°$. With regard to angle θ_4, attention must be paid to the fact that the length of arc III–IV (as well as those of arcs I–II, II–III, and IV–V, dashed lines in Figure 6.19) equals the circumference of wheel 4. From this fact it follows that angle θ_4 may be expressed as

$$\theta_4 = 360° - \theta_5 = 216°. \tag{6.9}$$

The consequence of this statement is

$$\frac{\Omega}{\omega_3} = \frac{144°}{216°} = 0.666. \tag{6.10}$$

Thus, Expression (6.8) can be rewritten as

$$\frac{z_6}{z_3} = \frac{\omega_3(1-0.666\ldots)}{-\omega_3 0.666} - 0.5. \tag{6.11}$$

The other condition we must provide is the radius R. This is made by choosing the proper dimensions for planet gear 7.

In Figure 6.21a) we show a cross section 1-0 of the schema in Figure 6.20. This cross section shows the feeding position of the sorting machine. Magazine 8 is loaded with rollers 9 (the elements in the roller chain), which, by gravity, fall down to base 10. Gripper 2 then catches roller 9 and arm 1 brings it out from feeding position 1. The next roller falls down, awaiting its turn after lever 4 completes a revolution. At the other four positions, the rollers are measured. Figure 6.21b) shows the design of a measuring position. Arm 1 brings roller 9 (held in gripper 2) into the measuring position, which consists of catch 11 and adjustment screw 12. What happens now depends on the dimension of the roller. Two outcomes are possible here: one, the gap between base 10 and tip 13 of the catch is wider than the height of the roller so that the roller remains in the gripper and continues to travel to the next position, where the gap is

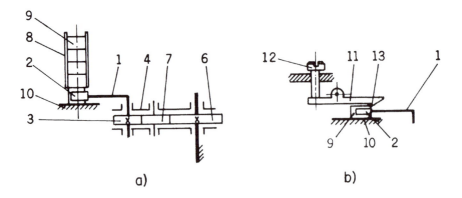

FIGURE 6.21 a) Cross section of the drive of the mechanism in Figure 6.20; b) Cross section of the measuring device of the same machine.

smaller; or two, the gap between base 10 and tip 13 of the catch is less than the height of the roller so that the roller is pulled out from the gripper and remains in this position. (When the next roller is caught here it will push this roller out into a collector.)

This mechanism provides unstressed catching and loading of blanks (details, parts) in appropriate positions. The speed of the gripper at catching and loading moments is zero. It is also possible to use several arms that work simultaneously. The speed and acceleration of the moving elements change smoothly, while some of the moving masses rotate permanently. Thus, the dynamics of this mechanism are much better than with the Geneva mechanism, for instance. This results in high productivity, because of the possibility of using much higher speeds than allowed in the other mechanisms discussed earlier.

As a disadvantage we must mention the fact that the parts do not absolutely stop at any point. The gripper centers (and hence the centers of the gripped details) actually move about 0. 1–0.15 mm during the rotation of the driving lever by about 5°–10° around the "stop" points of positions I, II, III, IV and V.

From the point of view of dynamics, this mechanism is nearly a permanently rotating device. This reminds us of the continuously acting automatic rotational machines mentioned in Chapter 1 (see Figure 1.25). It is worthwhile to mention that it is possible to combine these permanently rotating rotors with the hypocyclic transporting device. To do so, the speed of the gripper at the top of the hypocycloid (points A in Figure 6.19) must be equal to that of the rotor.

6.4   Vibrational Transportation

The layout of a vibrotransporting device, with a single-mass transporting tray with an electromagnetic drive, is shown in Figure 6.22. Tray 1 is fastened by springs 2 to the base. These springs oscillate with a constant amplitude, with respect to a certain angle of

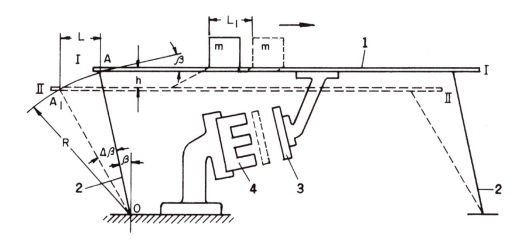

FIGURE 6.22   Diagram of a vibrating transportation tray.

inclination β. Armature 3 of the magnet is fastened to the tray, while magnet 4 is mounted on the base of the device. Now we consider the operation of one spring. Its lower end is fastened to the base at point O and its upper end A is connected to the tray. When the coil of electromagnet 4 is energized, the armature 3 is pulled towards the magnet. Tray 1 moves together with the armature, and the springs deflect by a corresponding angle. As a result, the initial inclination angle β increases by $\Delta\beta$ while point A moves to point A_1 along an arc with radius R. Thus every point of the tray is displaced horizontally by distance L and vertically by length h, and the whole tray passes from position I-I to position II-II. When the voltage feeding the coil of the electromagnet is switched off, the elastic forces of springs 2 return the tray to its previous position, thus completing one cycle of the device's operation. With the standard alternating current network, one can obtain either 100 cycles per second or 50 cycles per second (with a rectifier).

Let us consider the behavior of a body possessing mass m and located on the tray. This body exerts a downward pressure on the tray by its weight, which obviously equals mg. A frictional force P_F appears in the horizontal direction. This force depends on the frictional coefficient μ (characteristic of the tray and body materials) and on the vertical force N that the body exerts on the tray during the tray's motion. During this motion, varying horizontal and vertical acceleration components appear. The vertical

FIGURE 6.22a General view of a vibroconveyer driven by an electromotor. This device works according to the diagram shown in Figure 6.22. It was designed and built in the Engineering Institute of Ben-Gurion University by Dr. R. Mozniker for investigative purposes, and is a miniature copy of industrial vibroconveyors. Since it is driven by a motor and crankshaft, it maintains a constant vibrational amplitude.

acceleration component A_v determines the vertical pressure that the body exerts. Obviously, when A_v is positive (directed upwards) the pressure P_v can be expressed as

$$P_v = m(g + A_v), \qquad [6.12]$$

and when the vertical component is negative, we have

$$P_v = m(g - A_v). \qquad [6.13]$$

The horizontal component A_h also becomes positive (rightward) and negative during the cycle of motion. This component engenders horizontal inertial forces P_h which equal

$$P_h = \pm mA_h. \qquad [6.14]$$

These forces can be smaller or larger than the frictional force P_F. We can now express the frictional force, through Expressions (6.12) and (6.13), as follows:

$$P_F = \mu m(g \pm A_v). \qquad [6.15]$$

Obviously, horizontal displacement of the body relative to the tray will take place when

$$A_h > \mu(g \pm A_v). \qquad [6.16]$$

Analysis of these expressions shows that there are several different possibilities for the bodies' behavior on the tray. These possibilities can be described qualitatively as follows:

1. No motion occurs between the body and the tray. This happens when the value mA_h is always smaller than the frictional force.
2. Motion along the tray occurs, because

$$A_h > \mu(g - A_v). \qquad [6.17]$$

   However, the body does not rebound. It is always in contact with the tray because

$$A_v < g. \qquad [6.18]$$

3. Motion along the tray occurs because of both a) condition (6.17) and b) rebounds during the intervals when $A_v \geqslant g$ and there is no contact between the tray and the body. Therefore, relative motion of some sort takes place.
4. Relative motion between the tray and the body occurs but the body does not proceed in any definite direction because the values of the frictional coefficient are very low. (Balls or rollers on the tray.)

In practice, case 2 is preferable. In this case the body proceeds smoothly along the tray in the direction shown by the arrow.

Vibrating transporting trays are used because of their simplicity, high reliability, high transporting speed, simple ways to control this speed, and simple means that are adequate for stopping the transported bodies (simple mechanical stops are used).

In Chapter 7 we will speak about vibrofeeders and consider the properties of vibro-conveying in greater detail.

Exercise 6E-1

The vibrotransporting tray shown in Figure 6E-1 carries a mass m. The flat springs are inclined at an angle $\alpha = 10°$ to the vertical. The coefficient of friction between the tray and the mass is $\mu = 0.2$. Calculate the minimum amplitude of vibrations of the tray that will cause movement of the mass m if the vibration frequency is 50 Hz or 314 rad/sec; calculate the minimal frequency of vibrations if the vibrational amplitude a is about $a = 0.01$ mm that will cause movement of the mass m. Assume the vibrations are harmonic.

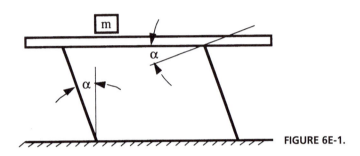

FIGURE 6E-1.

7

Feeding and Orientation Devices

7.1   Introduction

As we have seen in the previous chapters, every automatic manufacturing machine is provided with at least one feeding position. In this chapter we discuss aspects of feeding for automatically acting equipment. These automatic feeding devices or systems can be classified according to the form of the fed materials, which can be:

- Liquids of different viscosities;
- Powders or other granular materials;
- Wires, strips, or ribbons, etc.;
- Rods of various profiles; or
- Individual parts, blanks, or details.

In addition, the specific chemical and physical properties of the materials must be considered. These properties may or may not be exploitable for automatic feeding.

Automatic feeding devices must usually provide the following actions and conditions:

- Dosing of fluid or continuous materials;
- Keeping discrete items in a definite arrangement or orientation;
- Carrying out the action at the right moment, at the required place, and as quickly as possible.

Sometimes feeding coincides with some other process. For example, several feeding devices can work in parallel and bring materials or parts together during feeding. Screws and washers can be assembled during feeding and can be transported together to the next operation, which would logically consist of inserting the screw into a part.

7.2   Feeding of Liquid and Granular Materials

We begin the discussion with automatic feeding of liquids, which includes, for example:

- Automatic filling of bottles, cans, and other containers with milk, beer, oil, dyes, lubricants, etc.;
- Automatic distribution of fuel, dye, glue, etc., to definite positions and elements of an automatic machine;
- Automatic lubrication of machine joints, guides, shafts, etc.

Here, two kinds of feeding exist—continuous and dosewise.

Flowmeters of every kind provide automatic control for continuous feeding of liquids. Such flowmeters were discussed and illustrated in Chapter 5. They are included in the control layout and create feedbacks ensuring the desired level of consumption accuracy. These flowmeters are useful for providing uniformity of dye consumption in automatic dyeing machines.

Industrial painting systems can serve as a clear example for the strategy of liquid feeding during processing, including a method for preventing losses of dye and for providing high efficiency, i.e., uniform coloring of the parts, and good penetration of the dye into crevasses. The system shown in Figure 7.1 consists of a dye sprayer 1, a chain transporting device 2 provided with hooks 3 on which metal parts 4 to be colored are hung. An electrostatic field is created in the chamber in which this system is installed by connecting the chain to the positive and the sprayer to the negative poles. Thus, the negatively charged dye fog is attracted towards the parts (while the chain is protected by screen 5).

Let us next consider an automatic device for dosewise filling of bottles or cans. Figure 7.2 shows three states of an element involved in the process of filling bottles. The mechanism consists of transporting device 2 that moves bottles 1 rightward, dosing cylinder 3, and nozzle-moving cylinder 4. The latter first moves nozzle 5 down into the bottle, and then pulls it up relatively slowly, while the bottle is simultaneously filled with the liquid. To provide this movement, piston 6 is mounted on the nozzle, which also functions as a piston rod. Valve 7 controls the motion of this piston inside cylinder 4. By changing the position of the valve, the system connects the appropriate end

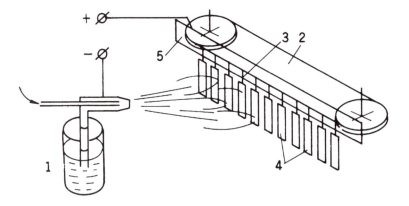

FIGURE 7.1   Design of an automatic dyeing machine with electrostatic dye application.

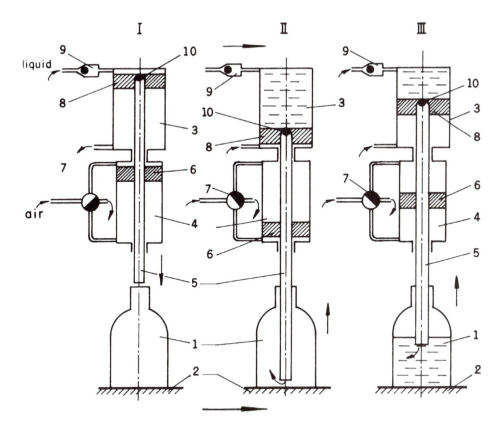

FIGURE 7.2 Design of automatic device for filling bottles with liquid.

of cylinder 4 to the air pressure. The upper end of the nozzle is provided with another piston 8, which serves as a pump. During the downstroke of this piston the liquid is sucked into the upper volume of the doser, and during the upstroke the liquid is transferred to the bottle. This sequence of liquid displacements is due to two one-way valves 9 and 10. Thus, filling of the bottle occurs as the nozzle is slowly pulled out of the bottle. This action sequence prevents bubbling, foaming, and dripping of the liquid. The lifting speed of the nozzle is kept equal to the rate at which the liquid level rises, so that its tip stays below the liquid during filling. It follows from this description that the volume of the dosing cylinder must equal the volume of the bottle. The first state of the mechanism shown in the figure (I) is the situation at the moment when the bottle is brought into position under the filling mechanism and the nozzle begins its movement downward. In state II the nozzle has reached the lowest point and dosing cylinder 3 is filled with the liquid. The bottle is still empty. In state III of the filling process the nozzle is about halfway out of the bottle, the bottle about half full, and the dosing cylinder about half empty. The bottle-filling process may be carried out while both the bottles and the dosing devices are in continuous motion.

Now we consider an example of feeding granular materials in portions. This situation is typical, for instance, of casting, molding, or pressing from powders or granular material. A plan of this sort of device is shown in Figure 7.3. Rotor 2 rotates around immobile axle 1. The rotor consists of a system of automatic scales that include levers 3, force sensors 4, and pockets 5 in which bowls 6 are located. Hopper 7 is placed at

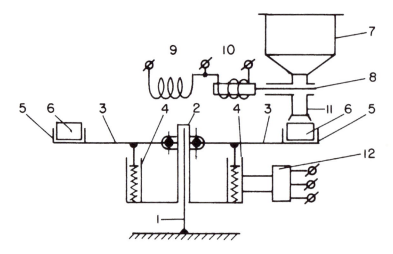

FIGURE 7.3 Plan for automatic weighing machine for granular material.

one position above the rotor. This hopper has gate 8 controlled by two electromagnets 9 and 10, which receive commands from control unit 12 connected to force sensors 4. An empty pocket 5 with bowl 6 stops under sleeve 11. At this moment, force sensor 4 produces a signal through control unit 12 which actuates electromagnet 9 to open gate 8. When the weight of the material reaches the value the scale is set for, sensor 4 produces another command to energize electromagnet 10 and close the gate. At this moment the rotor rotates for one pitch, putting the next empty pocket under the hopper. The filled pockets may then be handled and used for specific purposes.

We have just considered an interrupted feeding process. Belt conveyors, which are useful for a wide range of capacities, are often used for continuous feeding of granulated matter. An effective feeding tool is the vibrating conveyer described in Chapter 6. By changing the vibrational amplitudes or frequency, the feeding speed can be tuned very accurately.

The last mechanism we consider for feeding this kind of material is the auger or screw conveyor, a design for which is presented in Figure 7.4. Screw 1 rotates on its

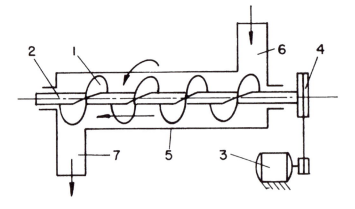

FIGURE 7.4 Screw conveyor for feeding granular material.

shaft 2 which is driven by motor 3 via transmission 4 (here a belt transmission is shown). The screw is located inside tubular housing 5, which has inlet and outlet sleeves 6 and 7, respectively. The material is poured into sleeve 6 and due to rotation of the screw, is led to sleeve 7 where it exits for subsequent use or distribution. Obviously, the speed of the screw's rotation defines the rate of consumption of the material.

7.3   Feeding of Strips, Rods, Wires, Ribbons, Etc.

Linear materials are often used in manufacturing. Their advantage is that they are intrinsically oriented. (We will discuss orientation problems later.) Thus, the feeding operation requires relatively simple manipulations. Indeed, in unwinding wire from the coil it is supplied on, only one point on this wire needs to be determined to completely define its position. Thus, an effective technical solution for feeding this kind of material is two rollers gripping the wire (strip, rod, etc.), from two sides and pulling or pushing it by means of the frictional forces developed between them and the material. We have already used this approach in examples considered in Chapter 2 (for example, Figures 2.2 and 2.4). Continuous rotation of the rollers provides, of course, continuous feeding of the material, which is effective for continuous manufacturing processes. However, for a periodical manufacturing process, feeding must be interrupted. One way to do this is based on the use of a separate drive controlled by the main controller of the machine. Such an example was discussed in Chapter 2. When the feeding time is a small fraction of the whole period, this solution is preferable.

When the feeding time is close to the period time, the solution presented in Figure 7.5 may be proposed. Here, lower roller 1 is always driven, and upper roller 2 is pressed against roller 1 by force F to produce the friction required to pull material 3. The force F can be produced by a spring or weight. (The latter needs more room but does not depend on time and maintains a constant force.) Roller 1 has a disc-like cam 4, which protrudes from the roller's surface for a definite angle ϕ. Thus, during part of the rotation of the driving roller 1, i.e., that corresponding to angle ϕ, upper roller 2 will be disconnected from the wire (rod, strip, etc.) 3, and the mechanism will therefore stop

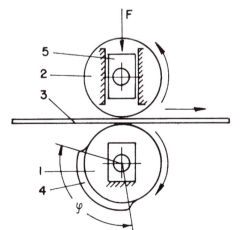

FIGURE 7.5 Frictional roller device for continuous feeding of wires.

pulling or feeding the material. Obviously, other means to disconnect the roller are available; for instance, a mechanism to lift slider 5.

Another sort of device for interrupted feeding of materials is also based on creating frictional forces; however, feeding is done by pure pulling and pushing of the materials. Let us consider the scheme in Figure 7.6. Here, lever 1 is pressed by force Q against strip 3 by means of spring 2. Strip 3 is clamped between the lever and surface 4. Due to this pressure, frictional forces F occur at points A and A' (we assume that the net forces acting on the surfaces can be considered at these points). Quantitative relations between the forces are derived from the following equilibrium equations written with respect to lever 1:

$$\begin{cases} 1. \ F - F_0 = 0, \\ 2. \ N - Q - N_0 = 0, \\ 3. \ Fa - Nb - Qc = 0, \\ 4. \ F = \mu N. \end{cases} \qquad [7.1]$$

Here μ = frictional coefficient between the materials of the strip and of the lever at point A. We assume that the same condition exists at point A'. The four Equations (7.1) contain four unknown quantities: N, N_0, F, and F_0. By substituting Equation 4 into Equation 3 we obtain

$$N = \frac{Qc}{b - \mu a} \quad \text{and} \quad F = \frac{\mu Qc}{b - \mu a}. \qquad [7.2]$$

By substituting Equation (7.2), into the first equation, we obtain

$$F_0 = \frac{\mu Qc}{b - \mu a}. \qquad [7.3]$$

From Equations (2) and (4) it follows that

$$N_0 = \frac{Qc}{b - \mu a} - Q. \qquad [7.4]$$

The derived results reveal a very important fact: when

$$\mu = \frac{b}{a}, \qquad [7.5]$$

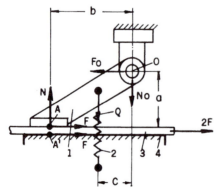

FIGURE 7.6 Frictional clamping device (lever type).

no spring (no force Q) is needed—the system is self-locking. The harder we try to pull the strip, the stronger it will be clamped. The force the device applies to the strip equals $2F$ because there are two contact points A and A′ where the strip is caught, and frictional forces F affect the strip from both sides.

The structure shown in Figure 7.7 works analogously. Here, strip 1 is clamped between surface 2 and roller 3. To produce clamping forces, the roller is pushed by force N_c (due to a spring not shown in the figure). The equilibrium equations with respect to the immobile rollers 3 have the following forms:

$$
\begin{aligned}
&1.\ N_A \sin\alpha - F_A \cos\alpha - F_B - N_C = 0, \\
&2.\ N_B - F_A \sin\alpha - N_A \cos\alpha = 0, \\
&3.\ F_A \le \mu N_A, \\
&4.\ F_B \le \mu N_B.
\end{aligned}
\qquad \text{[7.6]}
$$

Pay attention to inequalities 3 and 4 in the latter system of equations. The friction force at a point "B" is determined by the pulling force developed by the device, while the friction force at a point "A" fits the equilibrium of all the components of the force.

We assume that the frictional coefficients at points A, B, and C are identical. The unknown forces here are F_A, N_A, F_B, and N_B. Substituting Equations 3 and 4 into Equations 1 and 2, we obtain

$$
\begin{aligned}
&N_A \sin\alpha - \mu N_A \cos\alpha - \mu N_B = N_C, \\
&N_A \cos\alpha + \mu N_A \sin\alpha - N_B = 0.
\end{aligned}
\qquad \text{[7.7]}
$$

From this it follows that

$$
N_B = N_A(\cos\alpha + \mu \sin\alpha)
\qquad \text{[7.8]}
$$

and

$$
N_A = \frac{N_C}{(1 - \mu^2)\sin\alpha - 2\mu\cos\alpha}.
\qquad \text{[7.9]}
$$

Finally, we have

$$
N_B = \frac{N_C(\cos\alpha + \mu \sin\alpha)}{(1 - \mu^2)\sin\alpha - 2\mu\cos a}.
\qquad \text{[7.10]}
$$

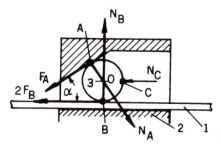

FIGURE 7.7 Frictional clamping device (roller type).

Obviously, when

$$\tan\alpha = \frac{2\mu}{1-\mu^2},$$ [7.11]

self-locking occurs, and no N_C force (no spring) is needed to lock the strip, wire. etc.

The devices in Figures 7.6 and 7.7 must be designed so that they do not reach the self-locking state, to ensure easy release of the material when the direction of the applied force is changed. Thus, the relations usually should be

$$\tan\alpha > \frac{2\mu}{1-\mu^2} \quad \text{and} \quad \frac{b}{a} > \mu.$$ [7.12]

The principles described above allow an effective feeder to be designed. A possible layout is shown in Figure 7.8. Here, two identical units I and II work in concert so that one (say, I) is immobile and the other carries out reciprocating movement, with the length L of a stroke equal to the length L of the fed section of the strip, etc. Each unit consists of housing 1, two rollers 2 pressed against inclined surfaces inside the housing, and spring 3 exerting force N_C. The housings have holes through which the strip, ribbon, etc., passes. How does this device act? First, unit II moves to the right. Then the material is clamped in it due to the direction of the frictional force acting on the rollers, while in unit I the material (for the same reason) stays unlocked and its movement is not restricted. As a result, the material is pulled through unit I while clamped by unit II. Afterwards, unit II moves backward the same distance. This time, the frictional forces are directed so that unit I clamps the material and resists its movement to the left. Unit II is now unlocked and slides along the strip as it moves. At the end of the leftward stroke, the device is ready for the next cycle. In the cross section A–A in Figure 7.8 another version of the clamps is shown. Here, instead of two rollers (which are convenient for gripping flat materials), three balls in a cylindrical housing are shown. This solution is used when materials with a circular cross section (wires, rods, etc.) are fed.

Finally, we show another strip-feeding device which is suitable when the time τ during which the material is stopped is relatively short in comparison to the period T; that is, $T \gg \tau$. The mechanism is shown in Figure 7.9a) and consists of a linkage and

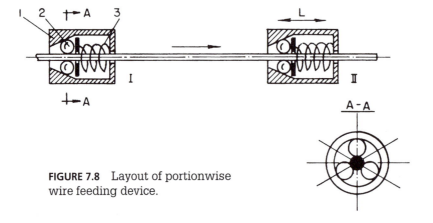

FIGURE 7.8 Layout of portionwise wire feeding device.

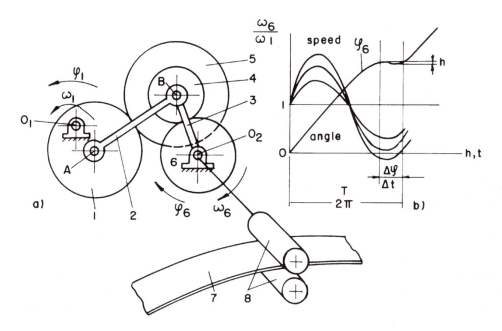

FIGURE 7.9   a) Geared linkage as a drive for roller friction feeder for interrupted feeding; b) Speed and angle changes versus time, with this device.

gears. Crank 1 is a geared wheel, rotating around immobile center O_1, whose geometrical center A serves as a joint for connecting rod 2. The latter drives lever 3. A block of gear wheels 4 and 5 is assembled on joint B. Wheel 5 is engaged with driven wheel 6. The sum of the links' and wheels' rotation speeds (when the tooth numbers are chosen properly) allows this mechanism to have a variable ratio ω_6/ω_1, which is shown graphically in Figure 7.9b). During rotation interval Δt, wheel 6 is almost immobile (the backlash that always exists in gear engagement makes this stop practically absolute). Imagine now strip 7 fed by rollers 8 driven by wheel 6, and you have an interrupted feeding, although driving link 1 is always rotating. Because of the smooth speed and displacement curves, the dynamics of this mechanism are rather good.

7.4   Feeding of Oriented Parts from Magazines

There are essentially two approaches to the parts-feeding problem: first, feeding of previously oriented parts; second, feeding from a bulk supply.

We begin with the first: feeding of the previously oriented parts. For this purpose some classical solutions and several subapproaches exist. They will be discussed here on the basis of some practical examples.

Example 1

Electronic elements such as resistors, capacitors, and some types of diodes are shaped as shown in Figure 7.10a). To make the feeding of these parts effective, they are

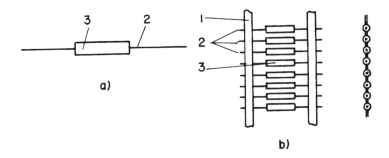

FIGURE 7.10   Separate parts arranged for automatic feeding in a band-like form, by means of tapes.

assembled into a band by means of tapes or plastic ribbons 1 (Figure 7.10b). The leads 2 of the resistors 3 are glued between two tapes, making a band convenient for storage (wound on a coil), for transportation to the working position of an automatic machine, and for automatic feeding. Obviously, additional orientation of the resistors is unimportant. It is relatively easy to bring them to the appropriate position accurately enough so that a gripper or other tool can handle them.

Example 2

Very often in mass production, parts are stamped out from metal or plastic strips or ribbons. To make them convenient for further processing, the following method can be used. Let us consider a detail made of a thin metal strip, as shown in Figure 7.11a). It can also be handled in a band form; however, in this case the procedure is simpler because this form can be made directly by stamping a strip (without additional effort).

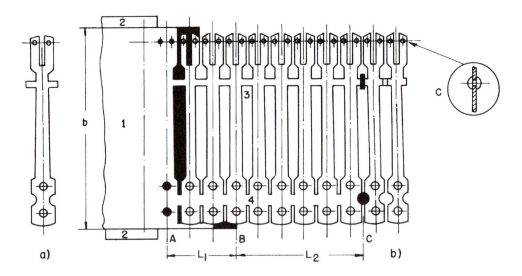

FIGURE 7.11   Stamping sequence to make a product convenient for automatic handling. a) Final product—a contact bar of an electromagnetic relay; b) Intermediate processing stages; c) Cross section of the contact rivets.

Figure 7.11b) shows how this can be done for a contact bar of an electromagnetic relay. Platinum-iridium contacts are riveted in the two small openings in the split end of the bar (see cross section in Figure 7.11c)). This riveting is much more convenient to do while the bars are together in a band-like structure, as in the illustration. Strip 1 is introduced into the stamp. It has a certain width b and is guided into the tool by supports 2. At line A the openings (blackened in the illustration) are cut. In the next step the split end of the bar is shaped and next the lower end is completed. Thus, section L_1 is needed to produce the bar. From line B the band-like semiproduct is ready. However, the bars are kept connected by two cross-pieces 3 and 4. The contact is riveted in section L_2, either on the same or another machine. An example of this process is explained in Chapter 8. Obviously, in either case no special efforts are needed to bring the bar oriented to the riveting position. When the contact is in its place the bars must be separated. This happens at line C by means of two punches which cut the remaining cross-pieces (blackened spots in the illustration).

The above examples (Figures 7.10 and 7.11) are typical high-productivity automatic processes, where automatic feeding of parts must be as rapid as possible. Therefore, the contrivances described above are justified. However, often the processing time is relatively long and the automatic operation does not suffer much if feeding is simplified. This brings us to the idea of hoppers or magazines. The classical means of automating industrial processes use a wide range of different kinds of hoppers, some of which are discussed below.

Tray hoppers are manually loaded with parts which then slide or roll under the influence of gravity, as shown in Figure 7. 12. A shut-off device is installed at the end of the tray to remove only a single part from the flow of parts on the tray. The design of these devices depends, of course, on the shape of the part they must handle. The rough estimation of the moving time along the inclined tray was considered in Chapter 2, Section 2.1.

A phenomenon which must always be taken into account in designing tray hoppers is *seizure,* which is schematically illustrated in Figure 7.13. To ensure reliable movement of the part along the tray, one must keep the seizure angle γ as large as possible. This angle depends on the ratio L/D (the length L of the part to its diameter or width

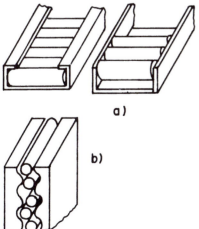

a)

b)

FIGURE 7.12   Tray hoppers: a) Usual type; b) Tortuous slot shape for a hopper.

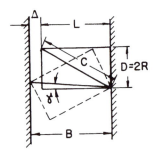

FIGURE 7.13 Graphical interpretation of seizure
of parts in a tray.

D), and values of $L/D < 3$ are good enough. In practice the clearance Δ must be chosen
correctly to prevent seizure. From Figure 7.13 it follows that

$$\cos\gamma = \frac{L+\Delta}{c},$$ [7.13]

which, by substituting

$$c = \sqrt{L^2 + D^2},$$

yields

$$\Delta = \sqrt{L^2 + D^2}\,\cos\gamma - L.$$ [7.14]

To avoid seizure in the design shown in the figure, the seizure angle γ must be larger
than the friction angle ρ, which means

$$\tan\gamma > \tan\rho = \mu.$$ [7.15]

Here μ is the frictional coefficient between the tray sides and the part. Expressing $\cos\gamma$
through $tg\,\gamma$, we obtain the clearance from Equation (7.14) in the following form:

$$\Delta = \frac{\sqrt{D^2 + L^2}}{\sqrt{1 + \mu^2}} - L.$$ [7.16]

Contrary to case a), case b) in Figure 7.12 is suitable for parts with $L/D > 3$ because,
due to the tortuous slot shape, the part cannot fall sideways and achieve dangerous
values of angle γ. This design is useful for many other applications in machinery where
seizure can take place.

   The length of the tray depends, obviously, on the processing time and must provide
a reasonable amount of parts without frequent human interference. To elongate the
tray and increase the number of parts stored in it, zigzag or spiral trays are used (see
Figures 7.14a) and b)). The zigzag hopper, in addition, limits the falling speed of parts,
which is sometimes important, for instance, when they are made of glass.

   Tray hoppers are sometimes modified into a vertical sleeve or channel, as shown
in Figure 7.15. In case a), hollow cylindrical parts are fed, and in case b), flat parts. Here
we see the shut-off mechanisms: a cylindrical pusher in a) and a flat slider in b), which
carry out reciprocating motion. The pace of motion is dictated by the control system;
however, it must allow the free fall of the parts in the hopper. It may be possible to

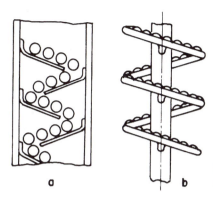

FIGURE 7.14   High-volume a) zigzag and b) spiral hoppers.

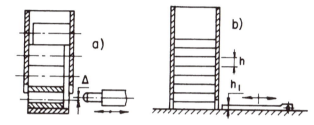

FIGURE 7.15   Examples of vertical sleeve, tube, or channel hopper.

drive the parts in the hopper pneumatically or with a spring. The latter is generally used in automatic firearms. To be reliable, cut-off of the fed parts requires a certain degree of accuracy in the mechanism. Thus, the gap Δ is restricted to a value of about 0.05 to 0.1 mm, the value $h_1 \approx h - (0.05$ to 0.1 mm$)$, and $h \geq 0.5$ mm.

Vertical box hoppers are more compact. Figure 7.16 illustrates several such hoppers. Case a) consists of box 1 in which the blanks are loaded in several layers, tray 2, and shut-off pusher 3 which takes the blanks out of the hopper by pushing along their axis. View b) shows the cross section of this hopper, and here agitator mechanism 4 is shown. The purpose of this mechanism is to prevent creation of a bridge of blanks which disturbs their free movement towards the outlet. Case c) shows a similar hopper where

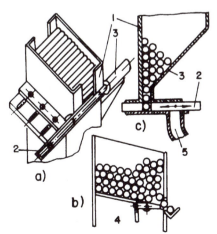

FIGURE 7.16   Vertical box hopper.

shut-off mechanism 2 pushes the blanks sideways, bringing them from the bottom of the box to channel 5.

For flat details or blanks, *horizontal box hoppers* are used. Two examples are illustrated in Figure 7.17. The height of these details may not be more than 50–70% of their width or diameter. Case a) consists of inclined tray 1 provided with edges 2 and agitator 3. The parts move by gravity. The oscillations of the agitator destroy any bridges that might impede movement of the parts. In case b) the hopper consists of a horizontal circular box with rotating bottom 1, circular wall 2, and agitator 3. Friction between the bottom and the blanks advances them to outlet 4. The danger of seizure appears here, also. The layout shown in Figure 7.18 explains the geometry of this phenomenon, which happens when the angle α approaches the friction angle, i.e.,

$$\tan\alpha = \frac{h}{B} = \tan\rho = \mu.\qquad[7.17]$$

Here, ρ is the friction angle, and μ is the coefficient of friction.

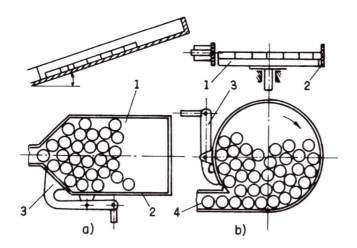

FIGURE 7.17 Horizontal box hoppers: a) Gravity drive; b) Friction drive.

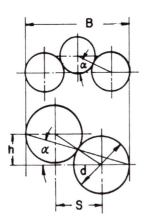

FIGURE 7.18 Graphical interpretation of parts seizure.

Obviously,

$$h = \sqrt{d^2 - S^2}$$ [7.18]

and

$$B = S + d.$$ [7.19]

Thus, by substituting Equations (7.18) and (7.19) into Equation (7.17), we obtain

$$(S + d)\mu = \sqrt{d^2 - S^2},$$

and from here,

$$S = d\frac{1 - \mu^2}{1 + \mu^2}$$ [7.20]

and

$$B = \frac{2d}{1 + \mu^2}.$$ [7.21]

This formula defines the width of the tray at which two parts cause seizure. For n parts in a row, we analogously derive

$$B = (n - 1)S + d$$

and

$$B = d\frac{n - (n - 2)\mu^2}{1 + \mu^2}.$$ [7.22]

Finally, we consider a hopper used for feeding parts in an automatic machine for welding aneroids (an example is described in Chapter 2). The hopper is shown in Figure 7.19a), and consists of cylindrical housing 1 having spring 2 for lifting membranes 3 previously fastened pairwise at, say, three points by point welding. At the top of the hopper a shut-off device is installed. This device consists of two forks 4 and 5, each of which has two prongs 41 and 42, and 51 and 52, and rotates around pins 6 and 7, respectively. Prongs 41 and 51 are connected by spring 8. (In Figure 7.19b) the forks are shown separately to facilitate understanding.) The prongs are seen in cross section at the upper part of the hopper. Note that the prongs are located diagonally, i.e., the upper right and lower left belong to fork 5, and the upper left and lower right to fork 4. When situated as in Figure 7.19 view I, prongs 41 and 51 hold the upper aneroid by its flange while spring 2 lifts the column of blanks. Magnetic gripper 9 in the meantime approaches the uppermost blank. At this moment force F is applied simultaneously to forks 4 and 5, moving them as arrows a and b show (Figure 7.19b)). This brings the shut-off device to the position shown in view II. Prongs 41 and 51 move apart while prongs 42 and 52 are pushed together, holding the flange of the penultimate aneroid and leaving the uppermost aneroid free to be taken by the magnetic gripper. We showed in Chapter 2 that welding one aneroid takes about 30 seconds. Keeping about 120 blanks in the hopper will allow 1 hour of automatic work without human intervention. The thickness of one

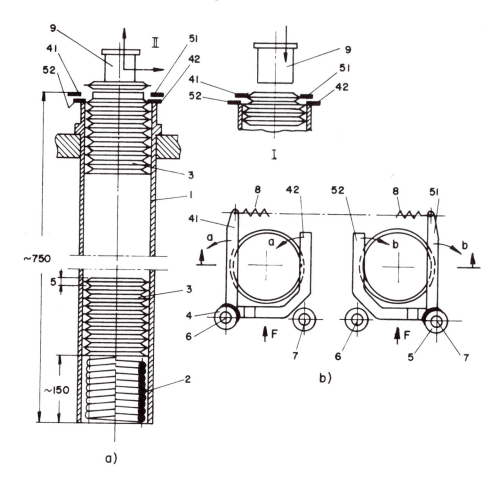

a)

b)

FIGURE 7.19 Tube-like hopper for an automatic machine for welding aneroids. a) General view of the device; b) Plan view of the shut-off mechanism.

aneroid is about 5 mm: therefore, the height of the column of blanks is about 600 mm. Together with the compressed spring, the hopper is about 750 mm long.

7.5   Feeding of Parts from Bins

In the feeding devices discussed in this section, the parts are fed from bulk supplies. The device must issue the parts in the required amount per unit time and, what is most important, in a definite orientation. Feeding bins can issue the parts by the piece, by portions of parts, or as a continuous flow of parts. We illustrate each approach here.

First, the pocket hopper will be considered. A typical feeder of this kind is shown in Figure 7.20. This device consists of rotating disc 1 placed at the bottom of housing 2. The whole device is tilted, and outlet channel 3 is located at the upper point of the bottom. Disc 1 is driven by, say, worm transmission 4. The disc is provided with pockets of a shape appropriate to the parts the device handles. Figure 7.20 shows three ways of locating these pockets. The point is that, depending on the l/d ratio, the parts find

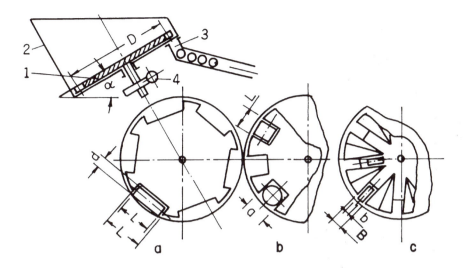

FIGURE 7.20   Pocket hopper: a) Pockets for elongated details; b) Pockets for short details; c) Radially oriented pockets.

their preferred orientation so as to minimize the resistance forces appearing during their motion. When $l/d \gg 1$ this preferred orientation is along the chord of the disc. The larger the ratio, the more parts are oriented in that way. Naturally, in this case the pockets should be made as shown in Figure 7.20a). For $l/d \cong 2$ the pockets are formed as in Figure 7.20b). To increase the number of pockets on the disc, they may be oriented radially (Figure 7.20c)), which increases the productivity of the device. However, to compel the parts to fall into radial pockets, the surface of the disc must be appropriately shaped with special radial bulges. The maximum rotational speed of the disc is determined by the falling speed of the parts into outlet tray 3. For this purpose the length of the pocket in case a) and its width in cases b) and c) must be great enough to provide clearance Δ. Thus, for the three types a), b), and c), respectively,

$$L = l + \Delta; \quad L = a + \Delta; \quad B = b + \Delta. \quad [7.23]$$

The peripheral speed V of the disc can be estimated from the formula

$$V \cong \Delta \sqrt{\frac{g \sin \alpha}{h}}. \quad [7.24]$$

Here, g is the acceleration due to gravity, and h is the height the part must fall to get free of the disc (obviously, h equals the thickness of the part or d, its diameter).

The next kind of feeder we consider is the so-called *sector hopper*. This device is shown in Figure 7.21 and consists of an oscillating sector 1 provided with slot 2, housing 3, outlet tray 4, and usually shut-off element 5. The parts 6 are thrown in bulk into the bowl of the housing. When the sector turns so that the slot is in its lower position, the slot is immersed in the parts and catches a certain number of them by chance as it is lifted by the sector. These then slide out along the slot and into tray 4. The shape of the slot must be suitable for the shape of the parts handled by the device (see Figure 7.22). To permit free movement of blanks in the slot and optimum feeding and orien-

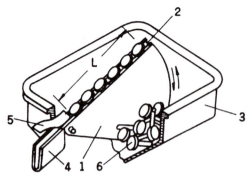

FIGURE 7.21 Sector-type hopper.

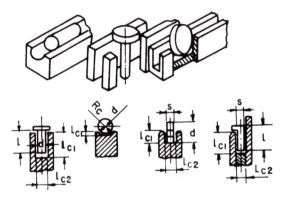

FIGURE 7.22 Shapes of slots for differently shaped details.

tation, the following empirical relationships between the dimensions of the parts and the slot parameters are recommended:

$$l_{c1} = l + (1-2)\text{mm}\,;\ l_{c1} = R_c\,;\ l_{c1} = 0.5d\,;\ l_{c1} = l + (1-2)\text{mm}\,;$$
$$l_{c2} = d + 1\text{mm}\,;\ R_c = (0.6-0.7)d\,;\ l_{c2} = s + 1\text{mm}\,;\ l_{c2} = s + 1\text{mm}\,.$$

[7.25]

A very similar feeding device is the *knife hopper*, a representative of which is shown in Figure 7.23. It consists of reciprocating knife 1 which slides vertically beside inclined plate 2, which has a slot on its upper edge. Bowl 3 also serves as a housing, and shut-

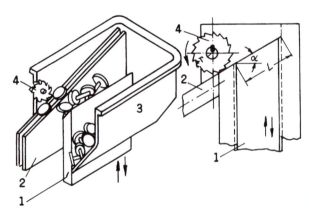

FIGURE 7.23 Knife-type hopper.

off wheel 4 rotates in the direction opposite to that of the parts movement. When the knife moves down it is immersed in the supply of blanks. In moving upward it catches some of them and, at the upper position of the knife, these blanks fall into the slot. Those that are successful in becoming oriented correctly will proceed in the slot under the shut-off wheel. The others will be resumed by this wheel back into the bulk for a new attempt.

The sliding time of an item along the slot in both the latter feeders can be estimated as shown in Chapter 3, Section 3.1. To provide the required productivity, the length L of the sector or the knife usually has the following relation to the blank's length l:

$$L \cong 10l. \tag{7.26}$$

Here l is the length of the blank in the direction of sliding when it is properly oriented.

The feeding rate of these devices is limited by the acceleration of the knife or sector as it reaches its upper position. Obviously, this acceleration a_0 must be smaller than g; otherwise the blanks will jump out of the slot or lose their orientation. It is easy to estimate the value of the acceleration of the knife or sector. Let us describe the displacements of the knife by the following expression:

$$s = s_0 \sin \omega t. \tag{7.27}$$

Thus, the acceleration a here has the form

$$a = -s_0 \omega^2 \sin \omega t, \tag{7.28}$$

and the maximum value of the acceleration a_0 has the value

$$a_0 = s_0 \omega^2. \tag{7.29}$$

We must ensure that

$$s_0 \omega^2 < g. \tag{7.30}$$

Here, s_0 is the amplitude of the knife or sector (at the point farthest from the axis of rotation), and ω is the frequency of oscillation in rad/sec—or in rpm we have

$$n = \frac{\omega}{2\pi} 60 \text{ rpm}.$$

These two feeders are examples of devices that issue parts in portions. The number of blanks fed per unit time is a statistical average and can be estimated experimentally to determine the productivity of the machine that the feeder serves. To avoid interruption of processing due to lack of blanks, the outlet tray should be long enough to hold about 25–30 blanks, to compensate for statistical deviation in the number of parts fed.

The third kind of feeding that provides a continuous flow of parts is *vibrofeeding*. We have already described the phenomenon of vibrotransportation in qualitative terms in Chapter 6, Section 6.4. A typical medium-sized vibrofeeder is illustrated in Figure 7.24. The device consists of bowl 1, whose internal surface is spirally grooved. The bowl is fastened to platform 2, which is supported by three slanted elastic rods 3. The rods are fastened to the platform and to base 4 by shoes 5 and 6, so that the projection of the rods on the horizontal plane is perpendicular to the bowl's radius. The platform is

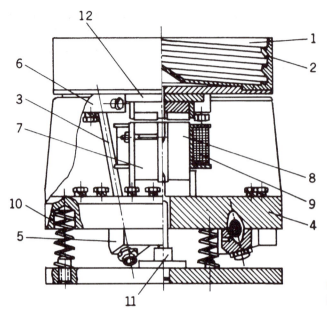

FIGURE 7.24 Vibrofeeder. General view.

FIGURE 7.24a) General view of a vibrofeeder with its controller. This device is driven by an electromagnet, like that shown schematically in Figure 7.24. This is an industrial device and can be used for feeding parts in concert with an automatic manufacturing machine. (Aylesbury Automation Ltd., Aylesbury, England)

vibrated by electromagnet 7 fastened in the middle of base 4. The electromagnet is made of core 8 and coil 9. To prevent transfer of vibrations to the system or machine on which the feeder is mounted, the latter can be installed on three springs 10, of relatively low stiffness. Pin 11 restrains the feeder from moving too much. When coil 9 of magnet 7 is energized by alternating current (usually the standard frequency of 50 Hz is used), an alternating force pulls armature 12. This force causes spiral oscillation of the bowl (because of inclined springs 3). Under certain conditions the alternating acceleration of this movement causes the parts in the groove to proceed, as we showed earlier for a vibrating tray (Figure 6.22).

Figure 7.25 shows a diagram of forces acting on an item located in the groove of a spirally vibrating bowl. The slope of springs 3 is indicated by angle γ, and that of the groove by α. Then we denote $\gamma - \alpha = \beta$. This diagram describes both straight and spiral vibrofeeding and differs from that shown in Figure 6.22 by the angle β between the groove and the direction of oscillation. Corresponding to the labels in Figure 7.25, the balance equations for the item in the groove have the following form:

$$-P\sin\alpha \pm m\ddot{x} = \pm F,$$
$$-P\cos\alpha \pm m\ddot{y} = N. \qquad\qquad [7.31]$$

where,

$P = mg =$ weight of the item,
$m =$ mass of the item,
$F =$ frictional force between the groove and the item,
$N =$ net force normal to the groove,
$x, y =$ displacement of the item along the x- and y-axes, respectively.

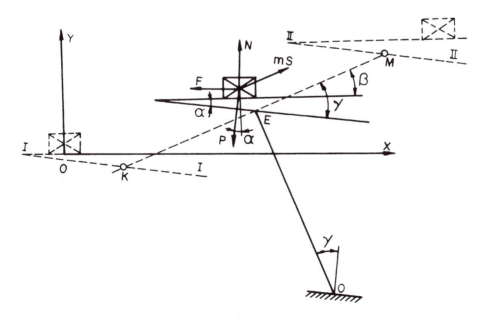

FIGURE 7.25   Forces acting on an item placed on the tray of a vibrofeeder.

If S is the actual displacement of the item, then $x = S \cos \beta$ and $y = S \sin \beta$. Obviously, $F = \mu N$ if μ is the frictional coefficient.

We now show the development of an expression for estimating the productivity of a vibrofeeder. We begin with considering the first half-period of the oscillation (section EM in Figure 7.25a), where $S \geq 0$ and $\ddot{S} \leq 0$. From (7.31) follows:

$$-mg \sin \alpha - \ddot{S} m \cos \beta = F,$$
$$-mg \cos \alpha - \ddot{S} m \sin \beta = N. \tag{7.32}$$

Substituting $F = \mu N$ into (7.32) and excluding N we obtain:

$$\ddot{S}_{cr} \geq g \frac{\sin \alpha - \mu \cos \alpha}{\mu \sin \beta - \cos \beta}. \tag{7.33}$$

For the second half-period (section EK in the same figure), where $S \leq 0$ and $\ddot{S} \geq 0$ we derive from (7.31) the following equations:

$$-mg \sin \alpha + \ddot{S} m \cos \beta = -F,$$
$$-mg \cos \alpha + \ddot{S} m \sin \beta = N, \tag{7.32a}$$

and correspondingly,

$$\ddot{S}_{cr}' \geq g \frac{\mu \cos \alpha + \sin \alpha}{\mu \sin \beta + \cos \beta}. \tag{7.33a}$$

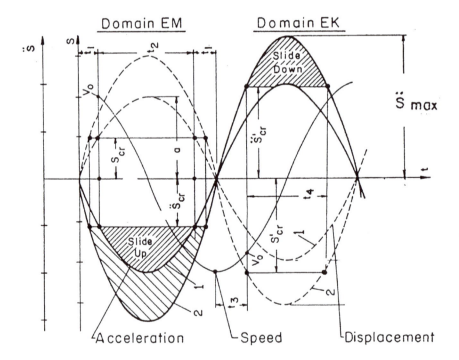

FIGURE 7.25a) Displacement, speed and acceleration of the vibrofeeder's bowl for two different oscillation amplitudes: 1) There is practially no backslide of the item on the groove; 2) There is backslide of the item.

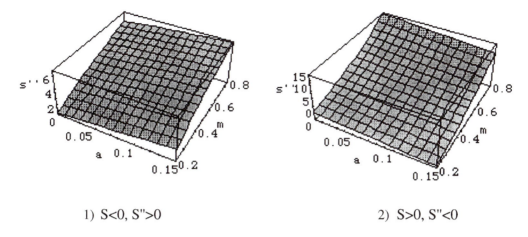

1) S<0, S">0   2) S>0, S"<0

FIGURE 7.25b)   Critical acceleration for 1) positive displacement and 2) Negative displacement of the bowl for different vibration amplitudes a and friction coefficients μ.

The relation between the concepts introduced here—displacement of the bowl S, its acceleration $\ddot{S}$, and critical values of the acceleration causing the body's slide relative to the bowl $\ddot{S}_{cr}$ and $\ddot{S}_{cr}'$—are shown in Figure 7.25a). Attention must be paid to the fact that these critical values depend only upon the geometry of the feeder and friction properties of the contacting materials.

Finally, we give visual representations of the dependences (7.33) and (7.33a). These representations are made for the case when the angle α changes from 0° to 5° and angle β changes from 30° to 35°.

The commands for the illustrations are given in MATHEMATICA language.

g1=Plot3D[9.8* (Sin[a]+m* Cos[a])/(m* .5+.866),
{a,0,.15},{m,.2,.8},AxesLabel->{"a","m","s'''"}]
g2=Plot3D[9.8* (Sin[a]-m* Cos[a])/(m* .5-.866),
{a,0,.15},{m,.2,.8},AxesLabel->{"a","m","s'''"}]

We can now proceed to calculations of the items displacement. From the curves in Figure 7.25a) it follows that the time t_1, when the slide begins (section EM), and the groove lags behind the item, is defined as

$$t_1 = \frac{1}{\omega}\arcsin\frac{\ddot{S}_{cr}}{\ddot{S}_{max}}. \qquad [a]$$

At this time the speed V_0 of the item (and the bowl) is defined correspondingly:

$$V_0 = \dot{S}(t_1) = a\omega\cos\left[\arcsin\frac{\ddot{S}_{cr}}{\ddot{S}_{max}}\right]. \qquad [b]$$

Thus, the slide begins with this speed and is under the influence of friction force $F \cong -\mu m(g-\ddot{y})$ acting backwards. We simplify this definition for our engineering purposes to a form $F \cong -\mu m(g-\ddot{S}_{cr}\sin\beta)$. This force causes deceleration $W \cong -\mu(g-\ddot{S}_{cr}\sin\beta)$. (This assumption gives a lower estimation of the displacement, while the higher estimation of it can be received when $W = 0$, neglecting friction.) This condition exists during time t_2, which is defined as

$$t_2 = \frac{1}{\omega}\left[\pi - 2\arcsin\frac{\ddot{S}_{cr}}{\ddot{S}_{max}}\right]. \qquad [c]$$

The displacement δ_1 then is

$$\delta_1 \cong V_0 t_2 - \frac{W t_2^2}{2}. \qquad [d]$$

Obviously, for the second half-period (section KM) we receive correspondingly (to expressions a, b, c, and d) a set of following formulas:

$$t_1' = \frac{1}{\omega}\arcsin\frac{\ddot{S}_{cr}'}{\ddot{S}_{max}} \qquad [a']$$

$$V_0' = \dot{S}'(t) = a\omega\cos\left[\arcsin\frac{\ddot{S}_{cr}'}{\ddot{S}_{max}}\right]. \qquad [b']$$

This speed value is negative while friction force acts in the positive direction:

$$t_2' = \frac{1}{\omega}\left[\pi - 2\arcsin\frac{\ddot{S}_{cr}'}{\ddot{S}_{max}}\right] \qquad [c']$$

$$\delta_2 \cong -V_0' t_2' + \frac{W t_2'^2}{2}, \qquad [d']$$

where $W \cong -\mu(g + \ddot{S}_{cr}'\sin\beta)$.

Summarizing these two slides we obtain the resulting displacement of the body relatively to the tray in the bowl. A more precise description of the relative movement of the body on the tray follows. We express the absolute coordinates of the body X and Y in the following form:

$$X = x + x_0(t), \quad Y = y + y_0(t),$$

where $-x_0(t)$ and $y_0(t)$ are coordinates of the tray; x and y are relative coordinates of the body. Then, rewriting, (7.31) we obtain

$$m\ddot{x} = -m\ddot{x}_0(t) - mg\sin\alpha + F,$$
$$m\ddot{y} = -m\ddot{y}_0(t) - mg\cos\alpha + N. \qquad [e]$$

For the case of dry friction we have $F = -\mu N \operatorname{sign}\dot{x}$, and for the normal reaction on the body correspondingly $N = m\ddot{y}_0(t) + mg\cos\alpha$.

Then the most important case of transportation without rebounds, the relative motion of the body along the tray, becomes as follows:

$$\ddot{x} = -\ddot{x}_0(t) - g\sin\alpha - \mu\left[\ddot{y}_0(t) + g\cos\alpha\right]\operatorname{sign}\dot{x}.$$

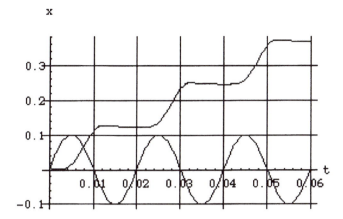

FIGURE 7.25c)   Relative body-tray displacement in comparison with the bowl oscillations.

A solution for this equation corresponding to the problem E7-1 made in MATHE-MATICA language is shown below. (To make the solution easier for the computer we propose to make a following substitution:

$$F = -\mu N \operatorname{sign} \dot{x} \cong \operatorname{th}(5\dot{x}).)$$

f2=x"[t]-8705.5* Sin[314* t]+342+
(-2777.3* Sin[314* t]+5876.4)* Tanh[5* x'[t]]
y2=NDSolve[{f2==0,x[0]==0,x'[0]==0},x[t],{t,0,.06},
MaxSteps->1500]
j2=Plot[Evaluate[x[t]/.y2],{t,0,.06},
AxesLabel->{"t","x"},GridLines->Automatic]

Expressions (7.33) and (7.34) permit calculating the accelerations that cause movement of the item on the tray. When the acceleration $\ddot{S}$ of the tray exceeds $\ddot{S}'_{cr}$ the item will move up the groove. When $\ddot{S}$ is greater than $\ddot{S}_{cr}$ the movement will be downward in the tray. By substituting $N = 0$ into the equations, we calculate the acceleration $\ddot{S}$ where rebound of the item occurs:

$$\ddot{S}_r > g \frac{\cos\alpha}{\sin\beta}. \qquad [7.35]$$

It follows from these expressions that, to obtain movement up the groove without bouncing (this is important for maintaining orientation of the parts), the acceleration $\ddot{S}$ must meet the requirements

$$\ddot{S}_{cr} < \ddot{S} < \ddot{S}'_{cr} \quad \text{and} \quad \ddot{S} < \ddot{S}_r.$$

Vibrofeeders possess certain advantages which explain their widespread use for automatic feeding. The advantages are:

- Motion of the parts along the tray does not depend on the masses. This means that, when the device is tuned appropriately, small and large items move at the same speed.

- Motion is due to inertia; therefore, there is less risk of damage to the parts.
- Constant and uniform speed of the parts is convenient for orientation (see next section).
- These devices are relatively simple, having no rotating links, and seizure of parts is less possible.
- The feeding speed can be easily tuned and controlled.

The vibrofeeder can be oscillated by an electromagnet (as mentioned above), pneumodrive, or mechanical means. Usually driving is done by a force field (electromagnetic or inertial); only the mechanical drive can function kinematically, thus ensuring a constant amplitude of vibration. Figure 7.26 shows some nonmagnetic vibrators for vibrofeeders. Case a) represents two masses 1 rotating in opposite directions. The resultant inertial force is a harmonically changing force F in the vertical direction (the horizontal components H cancel one another). The masses are driven by an electric motor, and a gear mechanism ensures rotation phase coincidence. Case b) illustrates a pneumatic vibrator consisting of housing I with a toroidal channel in which a massive ball 2 is driven by air flowing through nozzle 3. The mass develops a harmonic inertial force which is applied to the feeder. Case c) is a kinematic vibrator where eccentric cam 1 drives connecting rod 2 and slider 3. The slider's vibration is almost harmonic, with a constant amplitude that equals the eccentricity of the driving link.

This brief description of vibrofeeders permits us to reach some conclusions that can have important implications for designing vibrofeeders:

1. The peripheral acceleration of the tray of the spiral feeder is described in the form

$$\ddot{S} = -a\omega^2 \sin\omega t,$$

where a is the vibration amplitude, and ω is the frequency of the excitation force or movement.

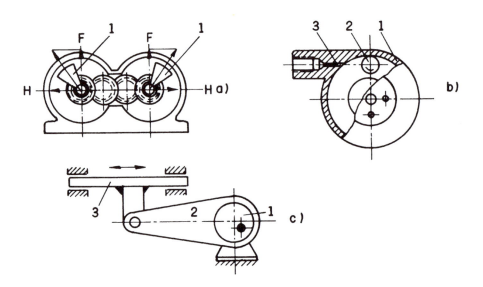

FIGURE 7.26 Nonmagnetic vibrators as drives for vibrofeeders (and vibroconveyors).

2. For kinematic excitation, the value of the amplitude stays constant and there-fore, to change the value of $\ddot{S}$, one must change the frequency ω (the rotation speed of eccentric cam 1 in Figure 7.26).
3. For the electromagnetic vibrator the frequency is usually determined by the stan-dard frequency of the alternating current in the network. Thus, changes in $\ddot{S}$ can be achieved only by voltage changes, which change the value of the amplitude a.
4. In the pneumatic vibrator, changes in the rate of air consumption cause both amplitude and frequency changes.

Let us now concentrate our attention on the electromagnetically driven vibrofeeder, which can be discussed in terms of a mechanical oscillator. Thus, this system has its resonance frequency which, when equal to the excitation frequency (remember, the magnet is powered by the network and therefore provides alternating force of 50 or 60 Hz), will cause relatively high vibrational amplitudes with minimal energy consump-tion. The resonance or natural frequency of the feeder depends on its oscillating mass and the stiffness of the springs. The device is usually designed so as to be close to res-onance conditions. However, this makes the feeder very sensitive to minor differences between the excitation and natural frequencies. These differences are caused by diffi-culties in making the actual parameters of the device exactly equal to their calculated values, by changes in the load or mass in the bowl of the feeder during its work, and by slight changes in frequency in the supply network. Therefore, the oscillation ampli-tude of the bowl varies over a wide range, resulting in variations in the feeding speed. To stabilize the work of the feeder, the voltage that energizes the magnet is tuned by a controller, such as a variable-ratio autotransformer. The sensitivity of the vibrofeeder can also be reduced by increasing frictional losses in the system. A better method, in my opinion, is to make an adaptive vibrofeeder that is automatically controlled so as to be always energized by alternating voltage, with the optimum frequency and value to maintain the desired bowl oscillation amplitude. Such a feeder is stable, has a con-stant part-feeding speed, and keeps the parts oriented. Figure 7.27 illustrates the depen-dence of the bowl's amplitude of vibration on voltage changes in the network and clearly indicates that the adaptive vibrofeeder is considerably more stable than the

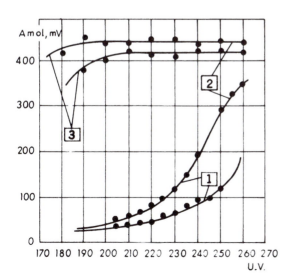

FIGURE 7.27   Influence of changes in network voltage on the bowl vibration amplitude: 1) Conventional controller; 2) Unloaded feeder; 3) Adaptive controller with amplitude stabilization.

conventional kind. Curves 1 belong to conventional design while curves 3 are for the adaptive feeder. It is almost independent of voltage changes and of the load in the bowl (curves 2). The energy consumed by a vibrofeeder working permanently at its mechanical resonance frequency is considerably less than that required by a conventional feeder, the energy savings being 60–80%. Thus, we see that both the stability and efficiency of vibrofeeding can be appreciably improved by using an adaptive self-tuning device. An automatic self-tuning system has been developed, built, and tested in the Engineering Institute of The Institutes for Applied Research of Ben-Gurion University of the Negev (Beersheva, Israel) by Dr. R Mozniker and myself.

7.6   General Discussion of Orientation of Parts

Solving orientation problems is one of the most important tasks in automatic production. Every item possesses six degrees of freedom in space. After orientation it is allowed only one degree of freedom, that for proceeding along the tray to the working position. For symmetrical items the number of restricted degrees of freedom can be less than five. For instance, orientation of a cylindrical item is unaffected by its rotation around the axis of symmetry. In contrast, a prism can move along a tray only on its face.

Items can be classified according to the number of planes, axes, and centers of symmetry that they possess. Figure 7.28 shows in a generalized form the principal shapes of items that automatic manufacturing deals with. The first group comprises nonsymmetrical hexahedral items which have 24 possible dynamically stable states on a

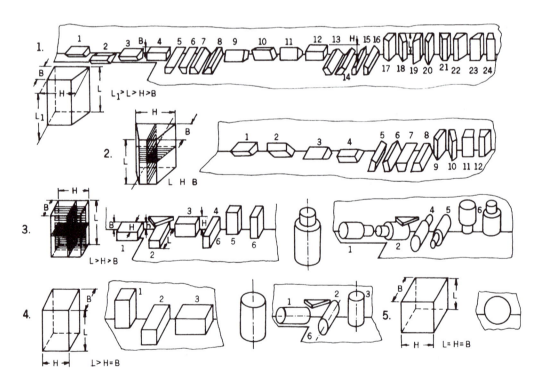

FIGURE 7.28 Generalized classification of item shapes (see text for explanation).

tray. The second group consists of hexahedral items possessing 2 planes of symmetry. The number of stable states that can be distinguished is reduced to 12. When a third plane of symmetry is introduced, there are only 6 distinguishable stable states. Items with one axis of symmetry also belong to this group. When an additional condition, $L > H = B$, is introduced, the number of distinguishable stable states equals 3. Cylindrical items having one axis of symmetry and one plane of symmetry also belong to this fourth group. The fifth group includes cubes and spheres which have only one distinguishable stable state on a tray. In other words, all 24 stable positions of a cube are perceived as identical by an observer, as are the infinite number of stable positions of a sphere.

What does "stable position" mean for an item located on a tray? Automatic orientation of parts in industrial applications is essentially a process of bringing parts from one stable state to another. This can be done in various ways, and the process can be described in qualitative and quantitative terms. Depending on its shape and other physical properties, an item on a tray may be found in unstable, stable, or indifferent states. Transition from some position into the desired position may require several intermediate changes in positions, which can be effected by applying forces to the item. The values of these forces depend on the specific shape and state of the item. Mathematical criteria are used to describe the degree of stability of an item (see Figure 7.29). The relative potential energy W of the item when moved from state 1 to state 2 relative to plane 1–1 is described by the expression

$$W = Rh = T\gamma .$$ [7.36]

Here,

> R = net external force acting on the item (in Figure 7.29, gravitation),
> T = net torque of external forces acting on the item relative to point A,
> h = vertical change in the center of gravity,
> γ = the angle of inclination of the item in state 2.

Taking the derivative of Equation (7.36) we obtain

$$dW = R\,dh = T\,d\gamma$$ [7.37]

or

$$T = R\frac{dh}{d\gamma}.$$ [7.38]

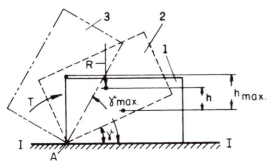

FIGURE 7.29   Criteria describing the item stability on a feeder tray.

Now we can introduce a parameter called the *specific stability torque m* that describes the stability of an item on a plane and is defined as the ratio $m = T/R$. This parameter, as follows from Expression (7.38), can be written in the form

$$m = \frac{dh}{d\gamma}. \qquad [7.39]$$

By rotating the part relative to point A, we obtain its limit state 3 where force R creates zero restitution torque (i.e., no torque is needed to return the item to its stable state 1). The angle of inclination γ_{max} also serves as a stability parameter, describing the so-called stability field (or domain). Obviously, the larger the value of γ_{max} the more stable is the state of the item on a flat surface. To reach this limit state, a certain amount of work W_{max} must be done. The amount of this turnover work obviously equals

$$W_{max} = Rh_{max}. \qquad [7.40]$$

As an exercise, the reader can calculate the values of the specific stability torque, the stability field, and the turnover work for a cylindrical body with height H and radius r, standing on its base.

To help solve orientation problems correctly and in the optimum way, we present here a brief classification of shapes that are often dealt with in manufacturing processes. First, we will classify conical and cylindrical items (see Figure 7.30). Four classes of shapes can be distinguished.

I. Symmetrical parts having one axis of symmetry and one plane of symmetry perpendicular to the axis, i.e., cylindrical items such as shafts or bushings without collars (examples 1, 4, 5, 6) or with collars or otherwise shaped, identical ends (examples 2, 3, 7). These parts have only one degree of freedom orientation. Indeed, one must bring the symmetry axis of the part in line with one of the coordinate axes.

II. Parts possessing only one axis of symmetry. To this class belong cylindrical parts with different ends (examples 1–4), parts with asymmetric necks (example 5), discs with ring-like grooves (example 7) or bevels (example 8) on one face, details with ring-like protuberances or heads (screws, rivets, bolts, example 6), conically shaped parts (examples 9–13), caps (examples 14–17), turned details with undercuts or recesses on one end (examples 18–22), and details with knurling or threading on one end (example 23). Orientation requires aligning the symmetry axis of the part with the coordinate axis (as in 1), as well as an additional rotation of the detail by 180° around the Z-coordinate axis.

III. Details whose cylindrical shape is disturbed by slots, grooves, flats, drilled holes, etc. These parts possess two planes of symmetry. One passes through the rotation axis; the other is perpendicular to the first. To this class belong shafts with a flat (example 1) or a slot (example 2), bushings with a slit (example 3), shafts with a perpendicular hole (example 4), rings, bushings, or discs with holes parallel to the rotation axis (example 5), or lengthwise slots (example 6). These details, after orientation as in case 1, require rotation around the X-coordinate axis by some angle.

IV. By subtracting one plane of symmetry from the class 3 items, we obtain class 4 details. Orientation of these parts generally requires two actions: rotation around the Z-coordinate axis by 180°, and rotation around the X-coordinate axis by some angle.

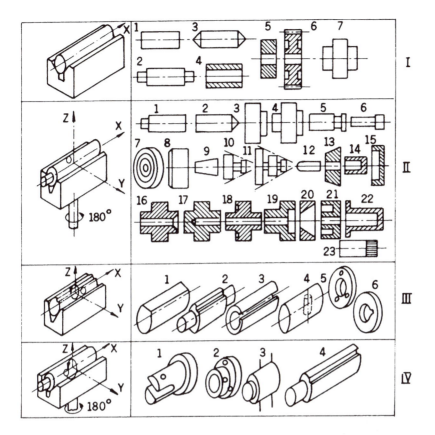

FIGURE 7.30 Classification of conically or cylindrically shaped items to be oriented (see text for further explanation).

The first stage in all these cases (in the first case it is the only action) is easily achieved in feeders, especially in vibrofeeders. All other orientation stages need additional approaches, which will be described and discussed in the following sections.

Flat parts allow more possibilities, and we consider six classes, which are represented in Figure 7.31. The simplest class, I, relates to flat items possessing three planes of symmetry and three different dimensions, say $L > H > B$. These sorts of details are led out of the tray by simple means (see next section) into position with the longest dimension L along the route on the tray. However, when dimensions L and B are close but still different (class II), the part may require additional rotation around the Z-axis by 90° to bring the dimension L into the right direction. Analogously, details with only two planes of symmetry (class III) require an additional rotation around the Z-axis by 180°. To class III belong items shaped like examples 3, 4, and 5.

To class IV belong parts with two planes of symmetry and $L = B$, as shown in examples 6 and 7. This time, rotation around the Z-axis is needed, but the angle must be chosen automatically (or manually) as 90°, 180°, or 270°.

To class V belong details with one plane of symmetry and $L > B > H$. Thus, after the previous orientation, four different positions are possible. Therefore the final orientation may require two additional rotations around the Z- and X-axes, both by 180°.

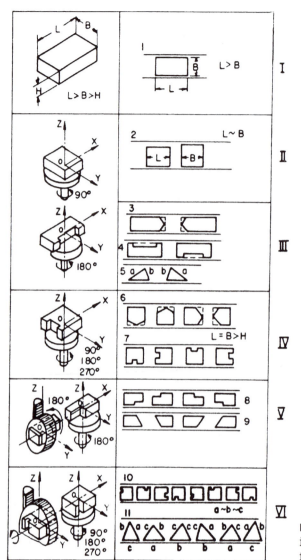

FIGURE 7.31 Classification of flat items to be oriented (see text for further explanation).

Class VI includes items possessing one plane of symmetry and in addition $L = B$. Eight different positions can be obtained after the first orientation. To bring the part into the desired orientation, two rotations around the X- and Z-axes are also required. However, this time the rotation angles around both axes maybe 90°, 180°, or 270°. Triangular parts having approximately equal sides also belong to this class. Here, the rotation angles needed for orientation depend on the angles of the triangle. These latter two classes are relatively difficult to orient automatically and require sophisticated methods. The design of an orientation device is a kind of art. Usually, the designer gives an idea of how to solve the specific problem, while the craftsman finds the correct proportions, dimensions, and sometimes principles of the devices. There is no substitute for the craftsman's skills, experience, and intuition in the field of automatic orientation.

7.7   Passive Orientation

Passive orientation is based on the idea of keeping on the tray only those parts or details that are oriented as desired. The wrongly oriented parts are simply thrown off the tray for additional trials. The working principles of this orientation are explained on the basis of examples. Figure 7.32 shows the behavior of a cylindrical part of the first class (Figure 7.30), running along a chute (Figure 7.32a). If the length L of the cylinder is longer than its diameter d, a shutoff element 1 is put in the way of the part in the feeder, thus throwing every part not in position 2 off the tray. In Figure 7.32b) the device ensures the output only of parts in a standing position (especially when L is only slightly longer than d). The tray has a cutout 3, so that the remaining width b of the tray is $b < d/2$. Above the tray a catch 1 is located. Thus, parts proceeding in position 2 fall when they reach the cutout; otherwise they safely continue the journey along the tray in orientation 4.

Figure 7.33 shows the handling of cylindrical parts belonging to class 11. The parts have diameters much greater than their heights. Case a) uses hook 1 shaped so as to catch the part by its head and take it over the cutout to continue on tray 3. Parts in position 2 fall back into the supply bin for the next trial. Case b) uses cutout 1 with an inclined surface. Parts in position 2 fall back into the bin. The dimensions of the part dictate the dimensions of the cutout.

Cup-shaped items of the same class can be handled by the passive means illustrated in Figure 7.34. Parts shaped as in case a) have two possible stable positions (the opening facing upward or downward) on tray 1. Shutoff 2 prevents stacking up of parts. The curved opening 3 causes parts with the opening facing downward to turn over and fall onto the lower level of the tray (we are dealing with a spiral vibrofeeder, whose tray is shaped like a rising spiral) and the part has a chance to land with the opening facing upward. In this position the parts cross cutout 3 successfully. The modification shown in Figure 7.34b) throws the "wrongly" oriented parts into the bin of parts at the bottom of the feeder. For more massive parts of the same class (thicker walls), a cutout (shown in Figure 7.34c)), with the addition of guide 1, ensures that parts 2 oriented with the

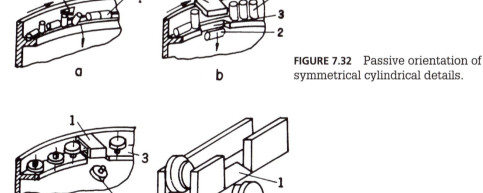

FIGURE 7.32   Passive orientation of symmetrical cylindrical details.

FIGURE 7.33   Passive orientation of symmetrical cylindrical details with one axis of symmetry.

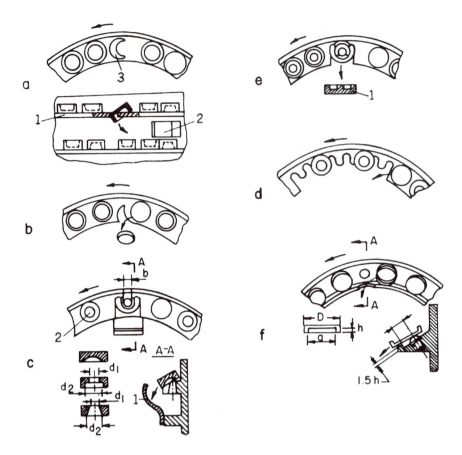

FIGURE 7.34   Passive orientation of cup-shaped details.

opening upward continue on their way. From the figure it follows that the dimensions b, d_1, and d_2 must be related so that $d_2 > b > d_1$. The guide puts the wrongly oriented parts on the next lower level of the tray, in the right orientation. Case d) shows the same separating idea as in e), except that the wrongly oriented parts fall into the bin and begin their way again. The structure shown in Figure 7.34e) works analogously. In this case part 1 is more complicated—it has a protuberance in the middle of the indentation. The shape of the cutoff in the tray permits the parts oriented with the open side upward to proceed. The other parts are removed from the tray. The last case (Figure 7.34f)) is for parts with small h values. Here, the part succeeds when the opening is downward. These parts remain on the tray while those oriented differently fall back.

Next, Figure 7.35 illustrates passive orientation for some representative class III parts. Cylindrical parts moving along a vibrating tray rotate. We use this phenomenon. In case a) the rotation of part 1 brings it to the position where slot A is caught by tooth 3. From this position the oriented detail can be taken by a manipulator for further handling. To ensure that rotation to the proper orientation is complete, electric contact 2 (insulated from the device) closes a circuit through the part. Case b) is for a part having a flat. Shutoff 2 lets only details in position I pass. The rare position II, which can also go through the shutoff, can be checked by another shutoff. A tray with the profile shown in Figure 7.35c) orients cylindrical parts having a flat. A tray with a rail orients parts

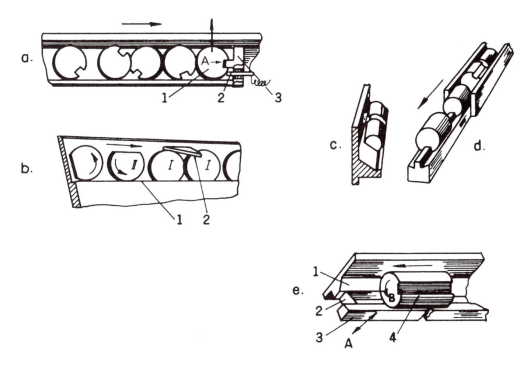

FIGURE 7.35 Passive orientation of almost-cylindrical details with one plane of symmetry:
a), d), and e) Details with slot; b) and c) Details with flat.

having a slot (Figure 7.35d)). Details which are not oriented properly fall from the tray
at the end of the side supports. The design shown in case e) is useful for details having
a diameter greater than 5 mm. A section of the tray is composed of an immobile element
1 and vibrating element 3 fastened by springs 2. The direction A of vibration causes
rotation of detail 4 in direction B until it is stopped by its slot.

It can be difficult to distinguish positions of cylindrical parts having slightly different ends, as shown in Figure 7.36a). For this purpose special devices are sometimes
designed, as in Figure 7.36b). Here, a mechanism moving with two degrees of freedom
consists of lug 5 rotating around horizontal axle 4. The latter is fixed in shackle 3, which
rotates around vertical axle 2. Spring 1 keeps shackle 3 in position. Tail 6 on lug 5 keeps
the latter in its normal position. In the scheme in Figure 7.36c), the response of lug 5, as
it depends on the orientation of the part on the tray, is shown. When the part moves to
the right with the bevelled face forward, lug 5 twists upwards around axle 4; when the
part moves with the straight edge forward, the system rotates around vertical axle 2. As
a result of this latter rotation, bulge 7 of shackle 3 removes the part from the tray. To facilitate this action, the tray is made as shown in Figure 7.36d). This idea is very effective
and can be adapted for flat details with insignificant differences, as shown in Figure 7.37.
Here, the device must sense the small chamfer at one of the corners. When the part
moves with the chamfer ahead, lever 1 together with strip 4 twists around horizontal
axes 3 and the part passes the checkpoint. When the chamfer is in another place, the
detail turns lever 1 around vertical axle 2, and bulge A removes the part from the tray.

Let us now consider more examples of passive orientation of rectangular parts. In
Figure 7.38a) a part with four possible positions on the tray is shown. The shape and

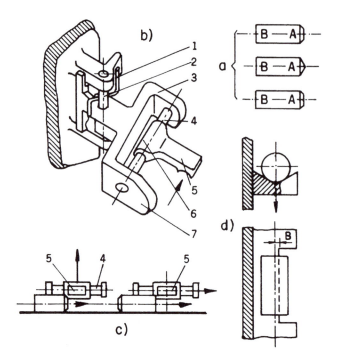

FIGURE 7.36 Device for passive orientation of cylindrical details with slight differences between their ends. a) Examples of parts having slightly different ends; b) Layout of a device able to distinguish slightly different ends of the parts; c) Front view of the device at work; d) Shape of the tray providing removing of the part when needed.

dimensions of the tray allow only one stable oriented position of the part, namely, that marked I. The other three possibilities will be extracted from the tray. Positions II and IV are unstable because of the location of the mass center relative to the edge of the tray. The part oriented as shown in III falls from the tray when it reaches cutouts 1. Asymmetrical angle pieces are conveniently oriented by the method presented in Figure

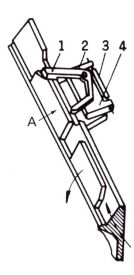

FIGURE 7.37 Device for passive orientation of flat details with insignificant asymmetry.

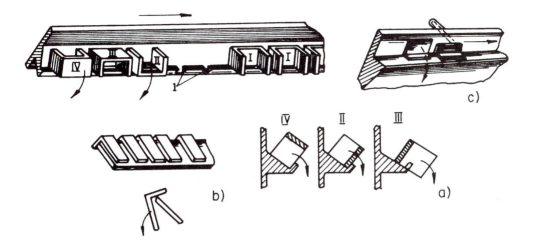

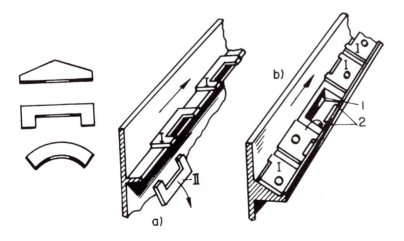

FIGURE 7.38 Passive orientation of rectangular details: a) and b) Due to force of gravity; c) Due to air flow.

7.38b). These parts are brought onto the tray in two possible positions. Obviously, when suspended by its narrow side on the vibrating tray's edge, the part falls back into the bin. Another position selection method for asymmetrical angle pieces is based on the use of blowing air, as shown in Figure 7.38c). The part placed with the wide side vertically is blown away from the tray when it reaches the nozzle.

Oblong asymmetrical flat details shaped like the examples in Figure 7.39 are easily oriented as shown in case a) when the asymmetry is strong enough to cause loss of balance on the tray. When the asymmetry is not strong enough, the idea shown in case b) can be used. The parts positioned as I pass cutout 1 successfully since they are supported by bulge 2, which is a bit smaller than the cutout in the detail. Details positioned with their cutout downward (II) fall from the tray when they reach cutout 1 in the tray.

Slotted details can be oriented as illustrated in Figure 7.40. Details shown in section a) of the figure are oriented by a rail, when the slot should be underneath, or by the device shown in section b), when the slot must stay on top. Details moving from left

FIGURE 7.39 Passive orientation of asymmetrical flat details.

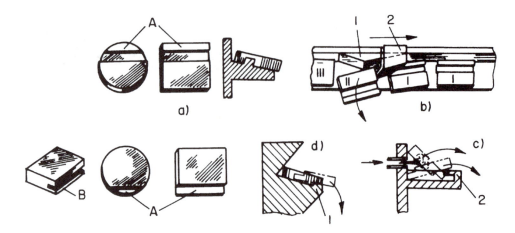

FIGURE 7.40 Passive orientation of flat slotted details: a) The slot must remain underneath; b) The slot must remain on top; c) and d) The slot is on the edge of the detail.

to right are caught by knife 2 when oriented like I (the knife fits the slot). When oriented otherwise, for example, as in II, they are pushed away from the tray by protuberance 1 and the knife does not catch the slot. The same happens when details are oriented with the slot downwards (case III). When the details are shaped as in Figure 7.40c) (the slot is on the edge of the detail as in case a or case b), orientation is done as shown in section d) by the edge of tray 1 and the force of gravity or by the edge 2 of the tray and an air stream. This latter (pneumatic) case is useful for detail B.

Details with protuberances as shown in Figure 7.41a) can be oriented by the approach shown in this figure. Details with the protuberance facing upwards are caught by hook 3, so that they do not fall from the tray. Details oriented with the protuberance downwards are extracted from the tray by slot 1, which leads them out of tray 2. Details which have passed the orientation device continue their movement in position 5, held by edge 4 of the tray. We leave it to the reader to analyze the orientation devices and processes shown in Figures 7.42–7.44.

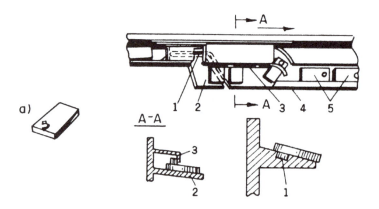

FIGURE 7.41 Passive orientation of a flat detail with a protuberance.

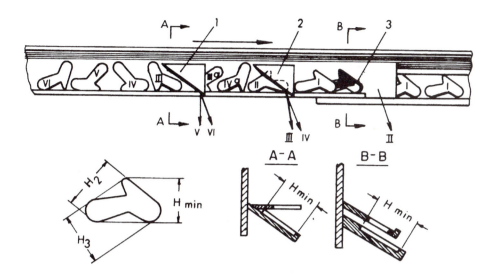

FIGURE 7.42   Exercise. Explain the process of passive orientation.

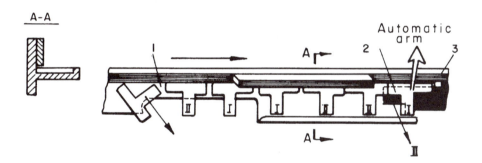

FIGURE 7.43   Exercise. Explain the process of passive orientation.

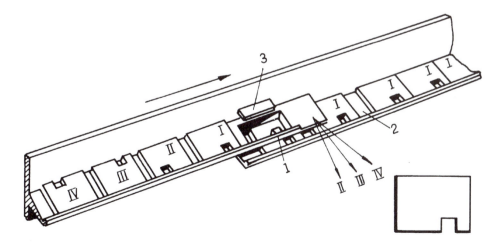

FIGURE 7.44   Exercise. Explain the process of passive orientation.

7.8 Active Orientation

Active parts orientation consists of actions which bring every part on the feeder's tray into position, oriented as required. No parts are thrown back into the hopper. Some general methods for this purpose are described briefly in this section.

To begin with, we consider a method for orientation of a square part with an asymmetric cutout A (see Figure 7.45a)). This part can have eight different positions on the tray. To bring it into the desired position IV, which is selected by openings 1 (appropriately shaped), the part is moved along the tray. When part 2 is not properly oriented and passes opening 1 it is (by the shape of the tray) turned by 90° and checked by the next opening 1. Obviously, the part will be selected after three or fewer turns if it is moving on its correct side. If not, it passes a turnover device as shown in Figure 7.45b). Here the part is forced to slide down from tray 1 via inclined guide 4. Screen 3 turns it by 180° to its other side. Thus, every part is handled and sooner or later achieves the desired orientation.

Often the difference between the geometrical center and the center of mass is used for active orientation (see Figure 7.46). Here a hollow cylindrical part closed on one end is moving along the tray of a vibrofeeder. It approaches opening 1 in one of two possible states: the closed end faces either the front or the back of the part. The length of the part is l, the center of mass is located near point e, and the width of opening 1 in the tray equals t. Because of the difference in locations of the geometrical and mass centers, the value of t can be chosen so as to satisfy the following inequality:

$$x_1 < t < x_2 ,$$

where

$$x_1 < l/2 \quad \text{and} \quad x_2 > l/2 . \qquad [7.41]$$

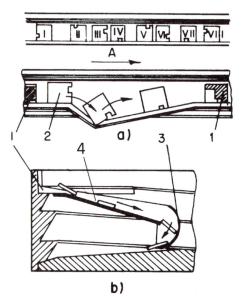

FIGURE 7.45 Active orientation of a flat, square part: a) Turning in the plane of the part; b) Turning over to the second side.

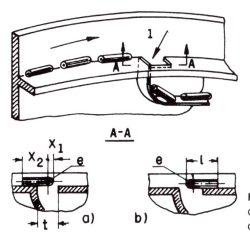

A-A

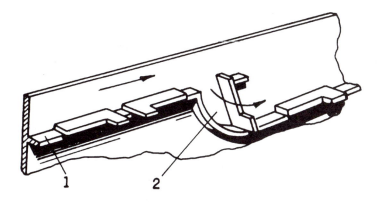

FIGURE 7.46   Active orientation of cylindrical details due to the difference between the center of mass and the geometric center.

Thus, if the part approaches opening 1 with the closed end first (Figure 7.46a)), it falls before the end of the part proceeds across the opening by the distance $l/2$ and continues with the closed end in front to the outlet of the feeder. If the part approaches the opening 1 with the open end in front, it passes it, as shown in Figure 7.46b), and flips over as it falls with the closed end first.

The same idea is used for orientation in the example in Figure 7.47. A modified form of this idea is illustrated by examples presented in Figures 7.48 and 7.49. Here we use both the differences between the mass and geometrical centers of the details and their specific shapes. These details possess one axis or plane of symmetry. A shaft with a neck is first oriented along its axis of symmetry (Figure 7.48) and then moved through cylindrical guide 2. If the neck is in front, the shaft moves up to support 4, passes gap 3, and flips over when freed from the guide, thus falling onto tray 6 with the neck toward the rear. If the neck already faces backward when the part moves though guide 2, the shaft does not reach support 4 because cutout 1 permits the shaft to fall before it passes gap 3. Again, the part falls onto tray 6 with the neck facing backward. Threshold 5 forces parts to fall from tray 6 when the latter is overfilled. The same explanation applies to

FIGURE 7.47   Active orientation of flat details due to the difference between the center of mass and the geometric center.

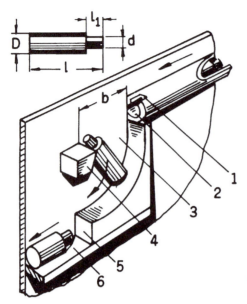

FIGURE 7.48 Active orientation of cylindrical details with an appropriately shaped guide.

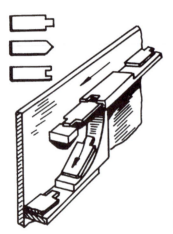

FIGURE 7.49 Active orientation of flat details with an appropriately shaped guide.

the case shown in Figure 7.49, where feeding of a flat detail is illustrated. Obviously, for differently shaped details the device must have the appropriate dimensions and proportions. The reader can try to design such devices for the details shown in Figure 7.50 (the dimensions can be chosen arbitrarily).

The location of the center of mass is widely used in automatic orientation. For instance, details having a large head such as screws, bolts, and rivets, can easily be brought into a position as shown in Figure 7.51a) by means of a through slot. Analogously, flat forked details, as in Figure 7.51b), are oriented.

If the slot is not deep, Figure 7.52 shows reorientation of parts with heads, so that they continue their movement along the tray with the heads forward. Figure 7.53 schematically illustrates a device for active orientation of needle-like details. Whichever the direction of the point, the cutout forces the needle to fall with the point forward.

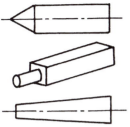

FIGURE 7.50 Exercise. Try to design guide shapes for these details. (Use the same idea as in Figures 7.48 and 7.49.)

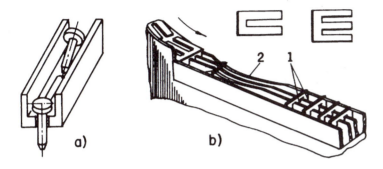

FIGURE 7.51 Active orientation of: a) Nail-like details; b) Flat, forked details.

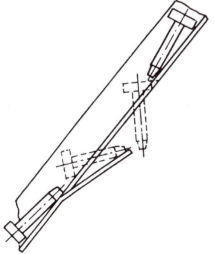

FIGURE 7.52 Turning over of nail-like details.

Figure 7.54 illustrates three methods for active orientation of caplike details. Case a) is based on the difference between the center of mass of the detail and its geometric center. Knife 1 supports the part under its geometric center while gravity turns the detail over so that it always falls with the heavier end forwards. Case b) uses hook 1. The parts move in the tubular guide 2 in two possible positions. When approaching hook 1 with the open end forward, the detail, under pressure of the line of details in the guide, comes into the position shown by dotted lines and falls with the closed end

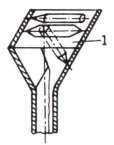

FIGURE 7.53   Active orientation of needle-like details.

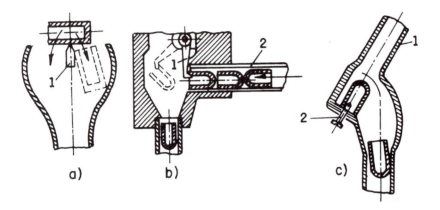

a)   b)   c)

FIGURE 7.54   Active orientation of cup-like cylindrical details by the use of:
a) Balancing support; b) Hook; c) Pin.

downward. When approaching the hook with the closed end forward, the part immediately falls down, again with the closed end forward. In case (c) the shape of guide 1
and auxiliary pin 2 fulfills the same function: whatever the direction of the detail in
the upper part of the guide, when it meets the pin it falls with the closed end forward.
The pin does not catch the detail when it approaches with the closed end forward.
Otherwise, the pin catches the detail, and it flips over.

We continue our analysis of active orientation by mechanical means with an
example dealing with conical details (see Figure 7.55). The details roll along inclined
plane 3 and turn because of the difference in radii, becoming sorted into two lines of
details, depending on the side to which radius r faces. The curvative of the trajectory
is L, which can be calculated from the following formula:

$$L = \frac{hR}{R-r}.$$ [7.42]

The two rows can be merged later, when the parts are oriented.

Sometimes the device for active orientation can require a certain degree of sophistication; for instance, the orientation of rings with internal bevels on one side, as in Figure
7.56. Rings 3 are placed automatically in the channel and feeler 4 is brought in contact
with each ring. Feeler 4 is driven by lever 10 and bushing 6, which slides in guide 7. If
the bevel faces the right side of the ring, feeler 4 penetrates deeper into it and screw
8, which is fastened onto feelers rod 12, presses microswitch 9, thus energizing elec-

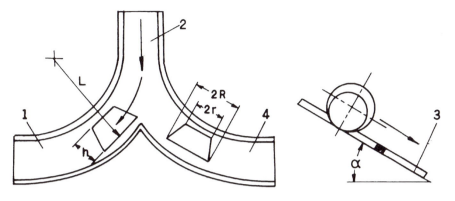

FIGURE 7.55 Active orientation of rolling conical details (orientation due to difference in radii).

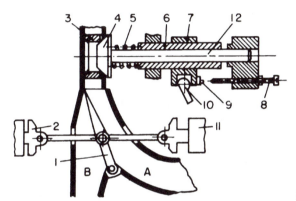

FIGURE 7.56 Active orientation of rings with one internal bevelled face.

tromagnet 2 and putting the directing blade into the appropriate position, say B. If the bevel faces left (as in Figure 7.56), feeler 4 penetrates less and compresses spring 5 while screw 8 does not reach microswitch 9. Magnet 11 is then activated, changing the position of blade 1 so as to direct the ring into channel A.

7.9 Logical Orientation

As was mentioned earlier, every detail we deal with has a certain number of stable positions on the tray. Usually, one position is desired and the others must be driven into the desired position by forcing the detail to turn around coordinate axes. The desired position of an item can be described by some events which must happen. For instance, an asymmetrical item must lie on a certain side (event a), with a certain cutout facing in a given direction (event b). The correctly oriented item is, thus, an event c which is a logical function of two (or more) logical variables a and b. This statement can be written in terms of Boolean algebra in the following forms:

$$c = f(a, b) = a \cdot b = a \wedge b.$$

[7.43]

This operation is called conjunction (logical multiplication) and means: statement c is true when and only when both statements a and b are true. When a and b take place we write

$$a = 1 \quad \text{and} \quad b = 1.$$

Thus:

$$c = a \cdot b = 1 \cdot 1 = 1. \tag{7.44}$$

When one or both of the variables are not true ($a = 0$ and/or $b = 0$), the result also is 0 and event c does not exist ($c = 0$). It is convenient to use inversion (another operation in Boolean algebra), i.e., the opposite of the variable's value. Thus, denial $\bar{a}$ means not a. When $a = 1$ the denial $\bar{a} = 0$. We can write

$$c = \bar{a} \cdot b = a \cdot \bar{b} = \bar{a} \cdot \bar{b}. \tag{7.45}$$

In performing active orientation, we often deal with only two possibilities, one of them $c = a$ and the other $c = \bar{a}$. For example, a device for active orientation of disclike details, which have one smooth side I and another side II with a certain degree of roughness, is shown in Figure 7.57. The sensor is pneumatic. Its nozzle 4 is placed at distance h from detail 3. The pressure to which sensor 5 responds depends on the smoothness of the detail's surface under the nozzle (the smoother the surface, the lower the pressure in the sensor). In effect, the control unit solves the logical task:

- The pressure in sensor 5 is low: then the detail continues from table 2 to tray 7.
- The pressure in sensor 5 is high: then the detail is lowered by means of an electromagnet (controlled by unit 6) and the detail continues to tray 8.

Thus, the details on trays 7 and 8 are oriented oppositely.

Another example is shown in Figure 7.58. Here a flat detail with an asymmetric cutout is actively oriented. The details move in positions I or II along tray 3. Light source 1 and lens 2 project an image of these parts onto screen 5, which is placed behind diaphragm 4, and actuate photocells 6 or 7. In accordance with these signals, the logical decision is made and rotary gripper 8 brings the detail into the desired position.

A more complicated example is illustrated in Figure 7.59. Again, a flat detail is considered; however, here four positions are possible. As it follows from Figure 7.59, three

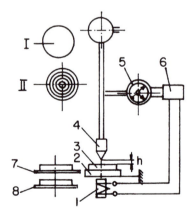

FIGURE 7.57 Pneumatic device for active orientation of flat detail with one rough side.

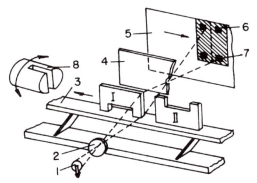

FIGURE 7.58   Photoelectric device for active orientation of asymmetrical, flat detail.

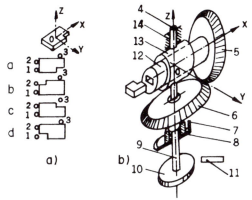

FIGURE 7.59   Device for logical active orientation of asymmetrical flat detail: a) Possible positions of the detail; b) The repositioning device.

contacts 1, 2, and 3 make the following four connections when the detail being oriented touches them in the various orientations:

$$1-2-3 \quad \text{events } a, \bar{b}, \bar{c}, \bar{d}$$
$$1-3 \quad \text{events } b, \bar{a}, \bar{c}, \bar{d}$$
$$1-2 \quad \text{events } c, \bar{d}, \bar{a}, \bar{b}$$
$$2-3 \quad \text{events } d, \bar{a}, \bar{b}, \bar{c}$$

The logical orientation system must be able to bring the detail into the desired position from any other. This means, for instance, that if the desired position is "a" and the detail is in state "b," the contact with point 2 is lacking. This can be corrected by rotating 180° around the Z-axis. For state "c" to be brought into the desired state, one must rotate the detail around the X-axis 180°. State "d" can be brought to situation "a" by rotating around the Y-axis 180°. Alternatively, the same effect can be achieved by two consecutive rotations around the Z- and X-axes (both for 180°). Figure 7.59b) shows a plan of a device for this kind of manipulation. The detail is inserted into the pocket in shaft 12. This shaft is installed in bushing 13 and rotates around the X-axis. The bushing, due to shaft 4 supported by bearing 14, rotates around the Z-axis. Shaft 12 is driven by bevel gears 5 and 6. Wheel 6 is connected to pinion 8. Stop 7 provides rotation for exactly 180°. The continuation 9 of shaft 4 has pinion 10, which is braked by means of stop 11. According to the logical function defined by contacts 1, 2, and 3, the control unit processes commands to motors (not shown in the figure), driving pinions 8 and 10 so as to bring the detail into the required position.

7.10 Orientation by Nonmechanical Means

We discuss here some ideas based on the use of electromagnetic fields for orientation. Figure 7.60 shows a classification of the different combinations of materials and electromagnetic fields that are used. We will present some examples relevant to electrostatic, magnetostatic, and alternating magnetic fields (some other special cases are omitted) in combination with parts made of ferromagnetic or nonmagnetic conductors or dielectric materials. The diagram in Figure 7.60 indicates that:

- An electrostatic field is useful for orientation of oblong items made of any material;
- A magnetostatic field is useful mainly for orientation of items made of ferromagnetic materials;
- An alternating magnetic field can be used for orienting items made of nonmagnetic electricity conductors.

Where do we use these orientation approaches? What are their main properties? These are noncontact methods of orientation. Theoretically, orientation could be carried out in a vacuum, manipulating the detail while it is suspended by the forces set up by the field.

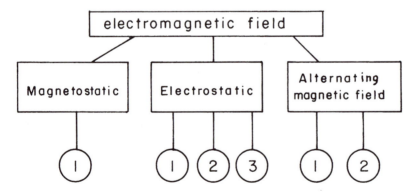

FIGURE 7.60 Classification of electromagnetic fields in combination with materials of different natures.

- These are active methods of orientation. If properly designed, the devices creating the field can bring the details or elements into the desired position regardless of their previous position.
- This orientation method is convenient for details with slight or negligible asymmetries in external shape or with specific internal features.

Let us begin our consideration with an illustration describing the use of a magnetostatic field (permanent magnetic field). As was mentioned earlier, this kind of field is effective for ferromagnetic items. Figure 7.61 shows examples of details with internal and external asymmetry, and Figure 7.62 shows a coil 2 creating a permanent mag-

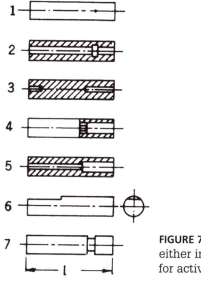

FIGURE 7.61 Examples of ferromagnetic details with either internal or slight external asymmetry suitable for active orientation by a magnetostatic field.

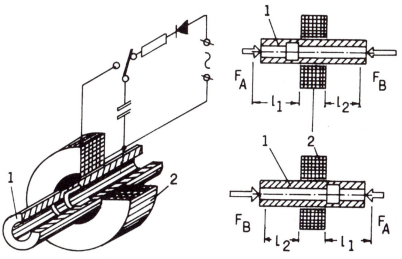

FIGURE 7.62 Diagram of an active orientation device for the details shown in Figure 7.61.

netic field suitable for such details. The diagram also shows the balance of forces appearing when the detail is put into the coil. There is a difference between the forces F_A and F_B when the detail is placed symmetrically in the coil. This results in displacement of the detail so that the distances l_1 and l_2 are not equal:

$$l_1 > l_2 . \qquad\qquad [7.46]$$

This difference is enough to be used for active orientation.

Figure 7.63 illustrates a dielectric part (clamp-shaped, with a slight difference between the ends, as shown in part c) of the figure) being oriented by means of an electrostatic field. The field is created between the pair of electrodes 1. The parts 2 move from the left to the right and, whatever their orientation I before they enter the field, the interaction between the field and the parts brings them into position II, where the thick end of the part faces forward. Section b) of Figure 7.63 shows the creation of a torque T due to the difference between forces F_1 and F_2 appearing because of the nonuniformity of the electrostatic field at the entrance into the space between the electrodes. By appropriately designing the shapes and relative locations of the electrodes, active orientation of dielectric parts can be achieved.

Another example of this sort is presented in Figure 7.64. Here electrodes 1 create a nonuniform field because of the wedge-like space between them. Such a field causes rotation of parts 2 from position I into position II. In addition, the parts will stop in position III in the narrowest section 3 of the field, which situation provides coincidence of the maximum value of dielectric permittivity with the direction of maximum field intensity.

The forces developed by the field are usually small; therefore, resistance caused by friction or other factors must be minimal when using this kind of orientation. The smaller the part, the higher are the effects. One specific example can emphasize this point. This is a process for producing imitation velvet, which is widely used for decoration. The process consists of glueing short-cut silk fibers to the surface of paper, fabric, board, etc., and is carried out by an automatic machine, the design of which is presented in Figure 7.65. The fabric (or paper or whatever) 2 is transported through

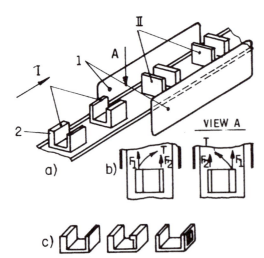

FIGURE 7.63 Active orientation in an electrostatic field: a) Behavior of details on the tray; b) Forces and torques acting on the detail; c) Representative details.

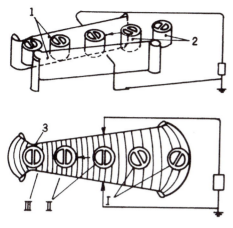

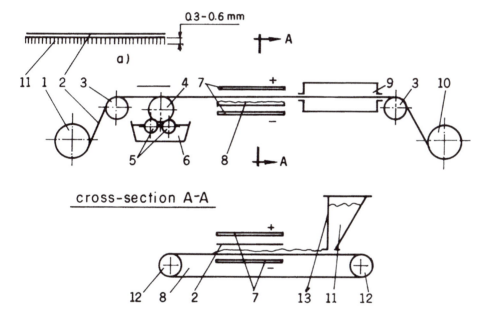

FIGURE 7.64   Active orientation in an electrostatic field of a cylindrical detail having internal features.

FIGURE 7.65   Design of an automatic machine for producing imitation velvet. (Electrostatic field is used for orientation of short silk threads.)

the machine from roll 1 to roll 10, supported by rollers 3. Roller 4 deposits glue on the lower surface of the fabric. The glue is applied to roller 4 by rollers 5, which take it from glue pan 6. The cut silk fiber (previously dried) 11 is supplied from hopper 13 by conveyor belt 8, which is driven by rollers 12. The main purpose of this machine is to stick the cut fibers upright to the surface of the fabric (paper, board, etc.), as shown in Figure 7.65a). It seems that the use of an electrostatic field is the only industrially relevant solution. The field orients every single piece of fiber along the field lines and moves them from one electrode 7 toward the other. After the fiber is stuck, the product is dried in dryer 9. The voltage used between the electrodes 7 is about 10,000–15,000 V.

An alternating magnetic field is a means for orienting nonferromagnetic parts made of electrically conducting materials. The physical phenomenon exploited here is the interaction between the magnetic field and eddy currents induced in the parts by the alternating field. Figure 7.66 shows the orientation of metal details having slight differences in shape at their ends, such as threading, small holes, or cutouts, etc. The electrodynamic forces F_1 and F_2 appearing here create a torque T which turns the part on the tray. The inequality $F_1 > F_2$ is caused by the difference in the currents $i_1 > i_2$, which is due to the slight differences in shape between the two ends of the part. As a result, the part is brought from position I into position II, with the rough (thread or knurling) end forwards.

It is worthwhile to mention here that orienting of these kinds of parts by mechanical means requires a lot of effort (if it is even possible). To illustrate this, we show here the means to orient a part (a stud) of the type shown in Figure 7.66. A mechanical solution is presented in Figure 7.67. The detail is caught by two clamps 2. Knife 3 strikes the middle of the stud. Because the friction in the clamp where the threatened end of the stud is located is higher, this end will be freed later. Thus, the stud falls with the smooth end downward into guide 4.

An alternating magnetic field acting on angle pieces develops forces as shown in Figure 7.68. The resulting force $\Sigma\Delta F$ rotates the part from every position I into the desired position II, with the vertical side facing backward. The values of the forces in such cases depend upon:

- The intensity of the field,
- The frequency of the alternating field,
- The shape of the detail being manipulated,
- The material of which the detail is made, and
- The dimensions and shape of the detail.

This chapter is largely based on the valuable material presented in the excellent book by Prof. A. Rabinovich, *Automatic Orientation and Feeding of Piecelike Details* (Technika, Kiev, 1968) (in Russian), and the book by J. Grinshtein and E. Vaisman, *Auto-*

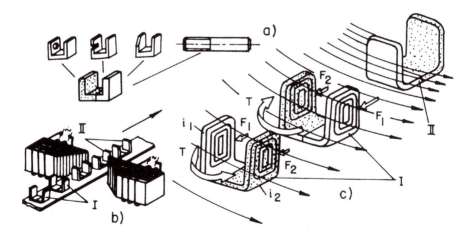

FIGURE 7.66 Active orientation of asymmetrical details in an alternating magnetic field: a) Representative details; b) Behavior of the details on the tray; c) Forces and torques acting on the details.

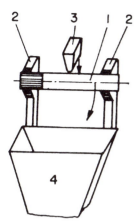

FIGURE 7.67 Mechanical orientation of a stud.

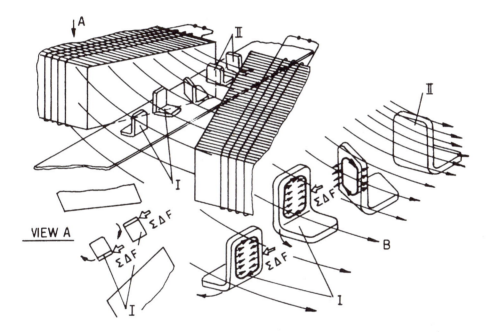

FIGURE 7.68 Active orientation of an angle piece by an alternating magnetic field.

matic Feeding Systems in the Instrument Construction Industry (Mashinostrojenie, Moscow, 1966) (in Russian). A book relevant to the subject of orientation by means of electromagnetic fields is that by Dr. B. Yoffe and R. K. Kalnin: *Orientation of Parts by Electromagnetic Fields* (Zinatne, Riga, 1972) (in Russian).

Exercise 7E-1

A strip-feeding device is shown in Figure 7E-1. The thickness of the strip $h = 0.004$ m, and the force needed to move it $F = 100$ N. Other dimensions indicated in the figure have the following values: $L = 0.1$ m, $l = 0.05$ m, and $H = 0.06$ m. What is the force Q

that the spring must develop to provide reliable functioning of the device? What are the reactions R_x and R_y at point O? The friction coefficient $\mu = 0.15$.

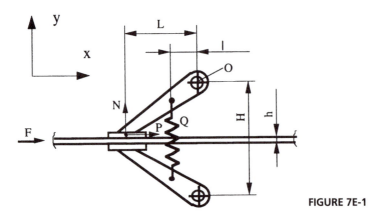

FIGURE 7E-1

Exercise 7E-1a)

One of the two elements of a ribbon feeder is shown in Figure 7E-1a). The spring in it develops a force $F = 20$ N. The spring acts on two rollers which, due to the shape of the device, create a friction force between the ribbon (point B) and the rollers and the inner inclined surface of the housing (point A). The inclination angle $= 15°$, and the coefficient of friction $\mu = 0.3$. What is the pulling force Q that this device is able to develop?

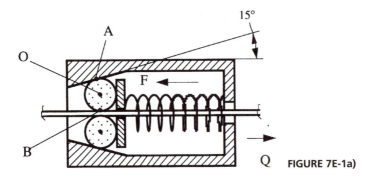

FIGURE 7E-1a)

Exercise 7E-1b)

A vertical rod-feeding mechanism is shown in Figure 7E-1b). The mechanism acts as a result of the friction forces developing between the fed rod and the gripping jaws. The weight of the levers holding the jaws $P = 0.8$ N, the weight of the feed rod $Q = 40$ N, and the friction coefficient $\mu = 0.4$. Find the value A that provides the normal feeding process of the mechanism if $H = 80$ mm and $h = 20$ mm.

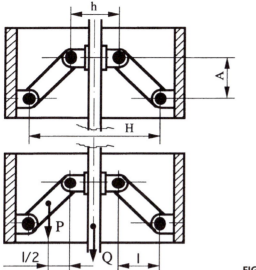

FIGURE 7E-1b)

Exercise 7E-2

Calculate the displacement H per second of a part placed on the groove of a spirally vibrating bowl, such as in Figure 7E-2, of a vibrofeeder. Pertinent data for the feeder are clear from Figure 7E-2:

Inclination angle of the groove $\alpha = 2°$,
Slope angle of the springs $\gamma = 30°$,
Coefficient of friction between the groove and the feed part $\mu = 0.6$,
Frequency of vibration $f = 50$ Hz, and
Amplitude of the harmonic vibration a = 0.1 mm.

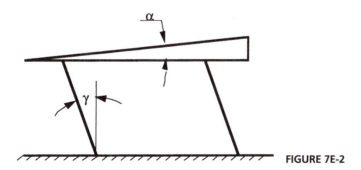

FIGURE 7E-2

Exercise 7E-3

This is the same exercise as that in the previous case (Exercise 7E-2) except that the amplitude of vibration is increased to a value of $\alpha = 0.15$ mm.

Exercise 7E-4

How many stable modes on the tray of a feeder can the part shown in Figure 7E-4 have when:

$H = B$;
$H \neq B$;
$h = H/2$; and
$h \neq H/2$?

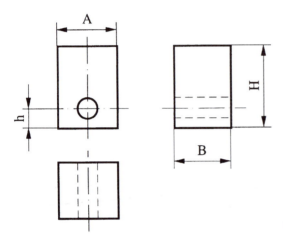

FIGURE 7E-4

Exercise 7E-5

How many stable modes on the tray of a feeder can the part shown in Figure 7E-5 have when:

$H = B = L$;
$H \neq B = L$; and
$H \neq B \neq L$?

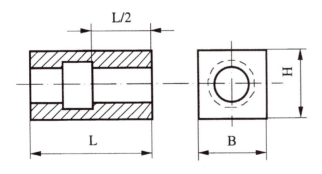

$L \neq B \neq H$

FIGURE 7E-5

8

Functional Systems and Mechanisms

8.1 General Concepts

In the previous six chapters we have discussed the problem of automatic manufacturing of products, describing common features, approaches, systems, devices, and mechanisms typical for the kind of equipment used. Now we must turn to the specific elements that make every machine useful for a specific task (or a group of specific tasks). The means that carry out these tasks or processes will be called functional systems and mechanisms. The list of such processes is endless, as is the list of systems and mechanisms to carry them out. Some processes used in automatic manufacturing are:

- Metal cutting,
- Molding of metals and plastics,
- Stamping,
- Assembling,
- Coloring,
- Galvanic coating and plating,
- Measuring, controlling, and sorting, and
- Chip technologies.

The processes listed here can be subdivided further. For instance, metal cutting includes the following operations:

- Turning,
- Milling,

- Drilling,
- Threading,
- Counterboring, countersinking, and
- Reaming.

Further classification is possible even at this level. For example, there are several methods for carrying out the process of threading. Indeed, we can distinguish between:

- Chase-threading or thread-chasing,
- Single-point-tool threading, and
- Threading by screw tap.

The analysis can often continue to even lower sublevels. Our purpose in this listing, however, is to show that an attempt to cover the ocean of automatic means of accomplishing all the possible manufacturing tasks within the limits of a chapter (or even a book) is not realistic. Therefore, we discuss here only some selected examples, with the emphasis on assembling because this is an important stumbling-block in automatic manufacturing today.

8.2   Automatic Assembling

Assembly accounts for about 50 to 60% of the workload in machine building and apparatus building. In some fields this percentage is even higher. For instance, in the electronics industry assembly includes chip production, circuit production, and manufacture of the final product. The high relative importance of assembly in manufacturing makes the need for automation of these processes crucial. Every success in automating the assembling process results in considerable profit.

There are many kinds of assembly techniques used in industry, and, obviously, the methods for automation for each of them must also differ. A brief list of these techniques follows:

1. Mechanical assembly with fastening by: screws, rivets, stamping, binding, expanding, and forge-rolling.
2. Welding: arc welding of various kinds, gas welding, seam resistance welding, butt resistance welding, and resistance spot welding.
3. Soldering and brazing: ultrasonic soldering, flow soldering or brazing, salt-bath drip brazing, and metal dip brazing.
4. Bonding with glues, resins, and adhesives.
5. Sewing or stapling with: threads, wires, clips, clamps, etc.
6. Magnetic mounting, twisting, curling, coiling, interference fit, slide fit, wedge-insertion, spring catch (latch, pawl, trip).

The assembly process, whatever its nature, consists of a number of operations. Two operations that are almost always needed in assembly are alignment and control—for instance, control for the presence of needed parts, quality control, etc. The next operation is completion of the assembly, which requires appropriate tools and actions. The operations needed for direct preparation for assembly (orientation of details, coating with glue, tin-plating, etc.) also depend on the nature of the whole process. A simplified example of assembly, illustrating this general description, is shown in Figure 8.1.

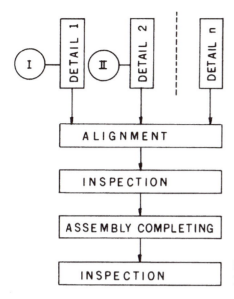

FIGURE 8.1 Flow diagram of assembly process.
I and II: preparation operations.

The question of whether one, two, or no inspection operations must be included, as well as that of the number and location of preparation operations, is answered according to each specific case. One of the criteria taken into account is the time needed for every operation. For instance, inspection for the presence of a part in the correct position takes about 0.5 second; completing an assembly by driving a screw can take about 1 second; assembling that is done simultaneously with feeding of the parts takes about 0.3 second, etc. We illustrate this in Figure 8.2. Here screws 1 slide down along tray 2 while washers 3 slide down along tray 4. Guide 5 controls the behavior of the screw heads. The slopes of trays 2 and 4 are shaped so as to promote insertion of the

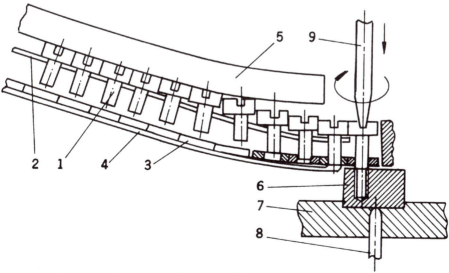

FIGURE 8.2 Process of screw-washer assembly.

screws into the washers during their movement. When the screw and washer come to the end of their travel they rest against the threaded opening of detail 6, which is held at this moment in a pocket of holder 7. The presence of part 6 is checked by feeler 8. Now screwdriver 9 can complete the assembly.

This example helps in further explanation of typical assembly problems. To achieve high reliability during screw-washer assembly, special measures can be taken, as illustrated in Figure 8.3. Case a) is adequate for manual assembly but is less suitable for automatic assembly because of the highly precise alignment required. Human hands, fingers, and eyes can provide this level of precision, but automatic devices find it difficult. Humans unconsciously correct for deviations of dimensions, locations, forces, etc., but an automatic tool is unable to. So one must look for a compromise, such as in case b). Tail 1 of the screw helps the latter to find the washer's opening. Additional help is provided by facet 2, resulting in a considerable improvement in the reliability of the assembly process.

We formulate here some principles one must pay attention to when automatic assembly is under consideration.

Principle I

Avoid assembling as much as possible. In other words, the design of the product must minimize the number of assembly steps. Of course, this relates chiefly to automated mass production. Let us consider the lever shown in Figure 8.4a) which consists of two parts: lever 1 itself and bushing 2. These two parts are connected by expanding the bushing into the opening made in part 1. To save effort in automatic assembly of this product, one can consider another design, for instance, that shown in Figure 8.4b). This lever is made by stamping and consists of one single piece of metal. (The openings are processed separately in both cases.) In another example in Figure 8.5a, shaft 1 and pinion 2 are made separately and require assembly. It is worthwhile to weigh the alternative shown in Figure 8.5b), where the detail is made as one piece. No assembling is needed; however, either a larger-diameter blank material or forging is used in the manufacturing process. An additional example appears in Figure 8.6. Here a riveted bracket consisting of two simple parts is shown in a). The alternative presented in b) is made by cutting slices from a rolled or extruded strip.

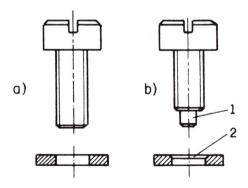

FIGURE 8.3 Examples of a) conventional screw and washer and b) those suited for automatic assembly.

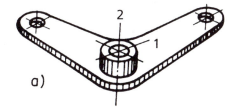

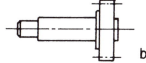

FIGURE 8.4   Examples of a) an assembled and b) a one-piece lever.

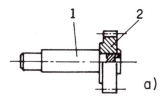

FIGURE 8.5   Examples of a) an assembled and b) a one-piece shaft-pinion unit.

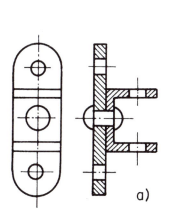
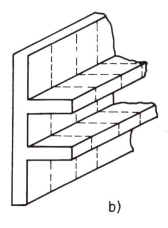

FIGURE 8.6   a) Conventional bracket; b) Design suited for mass production.

Principle II

Try to combine the assembly process with the production of one or more parts. This principle is usually applicable when the parts are produced by stamping or molding. Figure 8.7 shows the process of assembling relay contact 1 onto flat contact spring 2 (the figure shows a cross section of the spring at the point where the silver contact is fastened). The process of assembly is divided into four stages. In the first stage I, the section of silver wire 3 is inserted into the stamp. In the second stage II, the wire is plastically deformed so as to form the fastening to contact spring 2. In the next stage III, contact 4 is formed. And in the last stage IV, both the contact and its assembly on the spring are completed. (This process can accompany the process for flat spring production shown in Figure 7.11.) Another example is presented in Figure 8.8, which shows a plastic handle. This handle consists of metallic nut 1 and plastic body 2 made by molding. The parts are assembled during molding of the plastic body by inserting the nut into the mold.

This principle is implemented to some extent when self-threading screws are used, in that these screws create the thread in the fastened parts as they are screwed in. In Figure 8.9 detail 1 has holes permitting free passage of screw 3, and detail 2 has corresponding holes of smaller diameter. When the threads of the screws are forced to pass through these smaller holes, the threads cut their way into the material of detail 2, thus providing both hole-formation and assembly.

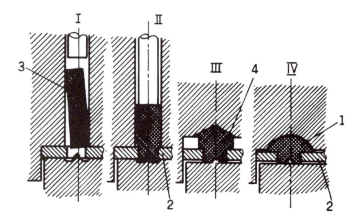

FIGURE 8.7 Relay contact assembly during manufacture of the contact.

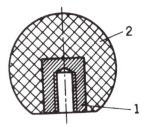

FIGURE 8.8 Handle produced simultaneously with its assembly on a metal nut for fastening.

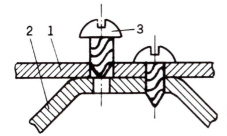

FIGURE 8.9 Screw that creates a thread during assembly.

Principle III

Try to avoid assembly of separated parts. Design the assembly process so as to complete it before the assembled components are separated. This principle serves to simplify the orientation problem and to reduce the accuracy needed for component alignment before assembly. The example shown in Figure 8.10 illustrates this. Here we consider automatic assembly of spring 1 with detail 2. Spring 3 is produced and fed into the device continuously through guide 4. Parts 2 are moved along tray 5 so as to stop with opening 6 opposite the guide, thus permitting spring 3 to easily enter the opening. Afterwards, knife 7 cuts the spring at a certain level, finishing both assembly and production.

Assembly of the relay contacts discussed in Chapter 7 and illustrated in Figures 7.11 and 8.7 also serves as an example of implementation of this principle. Indeed, the relay springs are formed so as to remain in a continuous band, and only after the contact is fastened onto the spring is the spring separated finally and the part completed (see Figure 7. 11, section c). We can also imagine a chain of rivets (screws, nails, etc.) as shown in Figure 8.11, which can be convenient for assembly with other components before these components are fastened by the rivets. Feeding this chain is obviously

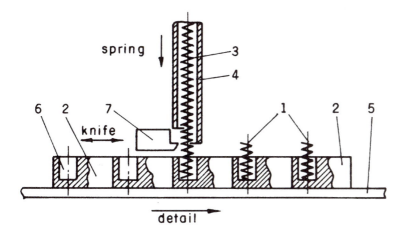

FIGURE 8.10 Spring assembly by cutting a section from a continuously fed spring. Orientation and separate handling of the spring are avoided with this approach.

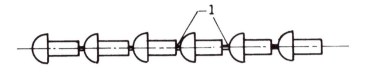

FIGURE 8.11   Example of continuous chain of rivets for more effective assembly.

simpler than feeding separate rivets; therefore, assembly will be more reliable. After the rivet is put into the appropriate opening, it is cut from the chain at neck 1.

In Chapter 7 (see Figure 7.20) we also considered the idea of transforming essentially separate units into continuous form, for example, details used in electronic circuit assembly. Sometimes it is worthwhile to expend some effort in making this transformation (e.g., gathering resistors into a paper or plastic bond) to increase the effectiveness of automatic assembly.

Principle IV

Design the component for convenient assembly. This principle is actually a particular case of a more general principle which reads: Design the product so it is convenient for automatic production. We have already met one relevant example in Figure 8.3. One of the most important features required in components one intends to assemble is convenience for automatic feeding and orientation. And here two recommendations must be made:

- Design parts so as to avoid unnecessary hindrances;
- Design parts so as to simplify orientation problems: with fewer possible distinct positions or emphasized features such as asymmetry in form or mass distribution.

Some examples follow. Figure 8.12a) shows a spring that is not convenient for automatic handling. Its open ends cause tangling when the springs are placed in bulk in a feeder. The design shown in Figure 8.12b) is much better (even better is the solution discussed earlier and shown in Figure 8.10). Tangling also occurs with details such as those shown in Figure 8.13. Rings made of thin material and afterwards handled automatically must be designed with a crooked slit to prevent tangling. Analogously, thin flat details with a narrow slot, as illustrated in Figure 8.14, should be designed so that $\Delta < \delta$. This condition obviously protects these details from tangling when in bulk. An additional example appears in Figure 8.15, where a bayonet joint is used for a gasket-

a)   b)

FIGURE 8.12   a) Spring design not recommended for automatic handling; b) Design of a spring more suitable for automatic handling.

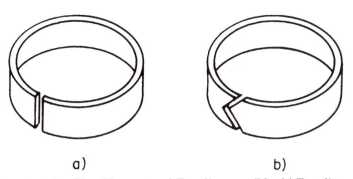

FIGURE 8.13   Ring-like parts: a) Tangling possible; b) Tangling almost impossible during automatic handling.

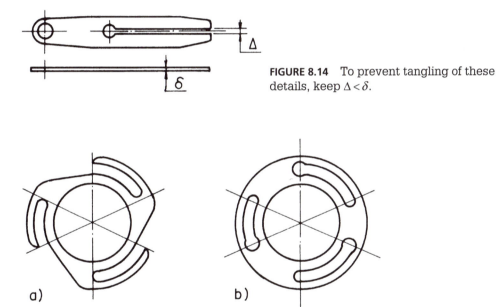

FIGURE 8.14   To prevent tangling of these details, keep $\Delta < \delta$.

FIGURE 8.15   To avoid tangling, design b) is better than design a).

like detail. Case a), with open horns, is dangerous from the point of view of automatic handling. Obviously, these horns cause tangling, they may be bent, and so on. The alternative shown in case b) is much more reliable. The behavior of details shaped as in Figure 8.16a) is clearly much worse than those in case b). The screws with cylindrically shaped heads behave more consistently on the tray than those with conical heads. The latter override one another, where the cylindrical screws stay in order.

Reducing the number of stable positions on the orientation tray will simplify the orientation process and increase its reliability. For example, the part presented in case a) of Figure 8.17 is preferred over that in case b) because of the symmetry around the y-axis. This is true even if the design of the product requires only two openings (as in case b)). (Of course, the cost of making two additional openings must be taken into

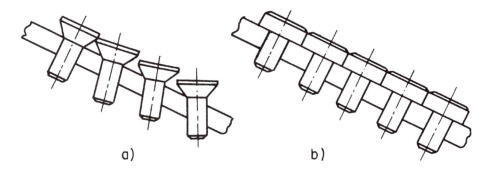

FIGURE 8.16 Details with the shape shown in case a) are less reliable on the feeding tray than those in case b).

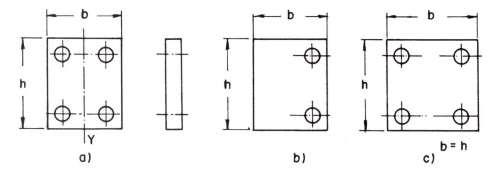

FIGURE 8.17 Orientation conditions of the part in case b) are worse than for those in case a), and those in case c) are best of all.

consideration, in addition to the concurrent simplification of orientation and assembly.) We should also consider the dimensions b and h. As one can see, in cases a) and b), the difference between b and h is rather small. It is worthwhile to redesign the part so that $b = h$ (see case c)) or, on the contrary, to increase their difference. In the first case ($b = h$) we obtain four indistinguishable positions of the detail on the tray, thus considerably simplifying the requirements for orientation. In the second case, making the dimensions b and h very different, for instance $b \ll h$, also facilitates orientation.

The same idea, of exaggerating the difference in some feature of the part is useful in cases where a shift in the center of mass is used in orientation. Figure 8.18 illustrates this for a stepwise-shaped roller. In cases a) and b) the difference Δ between the center of mass (m.c.) and the geometrical center (g.c.) of the detail is insignificant and difficult to detect and exploit reliably. To make this detail more suitable for automatic handling and assembling, use either cases c) or d), where the design is symmetrical, or case e), where the asymmetry is emphasized to make the difference Δ large enough for convenient and reliable orientation.

For convenient assembly the details must be designed so as to decrease the requirements for accuracy. For instance, as shown in Figure 8. 19, it is much more difficult to assemble the design shown in case b) than that in case a), where the right-hand opening has an oblong shape. The latter design provides the same relative location between

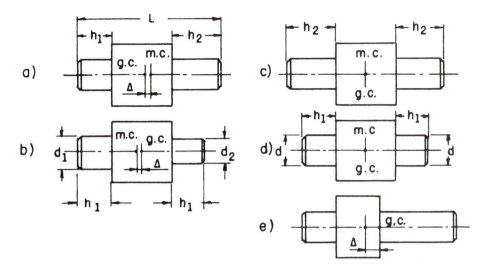

FIGURE 8.18   Effect of the relative location of the center of mass (m.c.) of a part with respect to its geometric center (g.c.). (See text for explanation.)

parts 1 and 2 after assembly as case b) does; however, the effort in carrying out this assembly step is less for case a) because one can pay less attention to the accuracy of dimension *l*. The relations between the dimensions of components of an assembly are important in various ways. In addition to the previous example, Figure 8.20 illustrates the general subprinciple: do not try to fit two mounting surfaces simultaneously; do it in series: first one, then the other. The mounting surfaces in Figure 8.20 are denoted A and B. In case a) the pin (dimension L_1) is designed so that it must be fitted simultaneously to openings A and B during assembly, while in case b) the proper choice of value L_2 makes the assembly process sequential: first the pin is fitted to opening B and then guided by this opening toward completion of assembly, i.e., penetration of the thicker part of the pin into opening A.

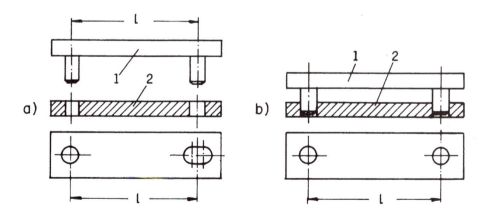

FIGURE 8.19   Use design a) for automatic (and even for manual) assembly; avoid the situation shown in b).

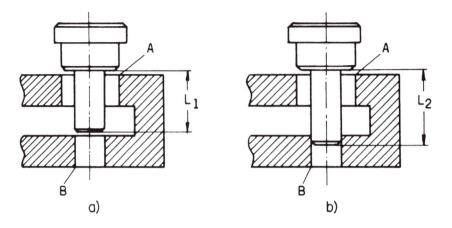

FIGURE 8.20 Do not try simultaneous fitting of a pin into two openings. This kind of assembly must be done in series.

Another subprinciple says: for automatic assembly the components must possess a certain degree of accuracy (which is correlated with their cost). A simple example based on automatic screwing of an accurate screw (Figure 8.21) is obvious. Case a) is normal, while in cases b) and c) the slot or the head is not concentric on the body of the screw. Cases d) and e) show defective screws: the first not slotted, the second not threaded. All the abnormal screw types should of course be prevented from arriving at the assembly position, or never be supplied in the first place.

Even when all conditions are met, automatic assembly remains a serious problem, and its reliability influences the effectiveness of the whole manufacturing process.

Reliability of Assembly Process

Let us now suppose that some product consists of n components which are brought in sequence to the assembly positions, with the end result that a certain product is obtained (see Figure 8.22). Each position is characterized by reliability $R_1, R_2, R_3, ..., R_n$

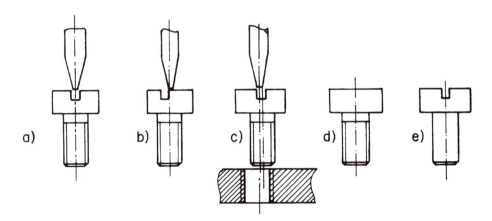

FIGURE 8.21 a) Normal screwdriver and screw in position; b) and c) Eccentricity of the slot or screw head. Defective screws: d) Without slot; e) Without thread.

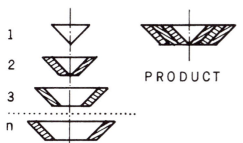

FIGURE 8.22   Simple model of an assembly process.

of assembly. We define the values R_i (where $i = 1, ..., n$) as the ratio between the number of successful assemblies N_{vi} and the total number of attempts N_i; that is:

$$R_i = N_{vi} / N_i .$$ [8.1]

The reliability of an automatic system R can be calculated as follows:

$$R = R_1 \cdot R_2 \cdot R_3 \cdot ... \cdot R_n .$$ [8.2]

For instance, if $n = 4$ and

$$R_1 = R_2 = R_3 = R_4 = 0.99 ,$$

we have for R,

$$R = 0.99^4 = 0.96 .$$

The reasons for the appearance of defective assemblies have different sources:

- Defective components, as shown in Figure 8.21, for example,
- Defective operation of the assembly mechanism.

Both types of reasons occur randomly.

To increase the reliability, special approaches can be taken, some of which will be considered in the following section.

8.3   Special Means of Assembly

In this section we consider some possibilities for increasing the efficiency of automatic assembly. As a criterion for estimating the efficiency, we use the reliability R, which we defined above as

$$R = N_v / N .$$

Here N_v = the number of successful assemblies, and N = the number of assembly attempts.

We also stated that, when an assembly or some other process requires a series of operations, the overall reliability is defined by Expression (8.2). The more components the whole assembly includes, the higher will be the number of failed assemblies and the smaller will be the estimated reliability. To improve this value we can propose duplicating some of the mechanisms comprising the assembly machine. A diagram of such an assembly machine of improved reliability is shown in Figure 8.23. This machine

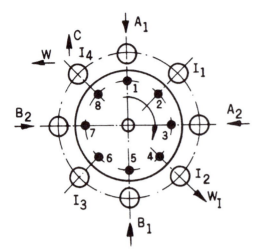

FIGURE 8.23 Diagram of high-reliability assembly machine.

must put together two components, A and B. However, each of these components is fed twice: A at both positions A_1 and A_2, and B at positions B_1 and B_2. Thus, if feeding fails at positions A_1 or B_1 the inspection devices placed at positions I_1 and I_3 give a command to operate the feeding devices at positions A_2 and B_2, respectively. The concept of failure includes:

- Lack of a part in pocket 1 or 5,
- A defective part, or
- Defective orientation of a part.

Let us compare the final reliability of this machine with one lacking duplicate feeding. Assuming that the reliability of each position in this machine equals R_i, we obtain the estimation of the probability P that the feeding of component A (or B) fails, from the following expression:

$$P = \prod_{i=1}^{2}(1 - R_i), \quad i = 1, 2; \tag{8.3}$$

here i = the number of the feeding positions A or B.

And thus the reliability of the whole machine equals

$$R = (1 - P)^2 = \left[1 - \prod_{i=1}^{2}(1 - R_i)\right]^2. \tag{8.4}$$

For example, for $R_i = 0.90$ (for both A and B) we obtain

$$R = \left[1 - (1 - 0.90)^2\right]^2 = 0.98.$$

For the same R_i value, a machine without duplication has the following reliability:

$$R = R_i^2 = 0.81.$$

Inspection position I_2 serves to stop feeding B_1 if, despite the duplication, something is wrong with part A, and to remove defective part A from pocket 4. Position I_4 directs correctly assembled products into collector C and wrongly made products into collector W.

We mentioned above that reliable assembly requires high accuracy in handling components. There is a method based on vibration that can increase the reliability of assembly. To explain the principal idea of this method, let us consider the following model for assembling two components, as shown in Figure 8.24. Here, bushing 1 represents one component and pin 2 the other component of the assembly being put together. Bushing 1 is kept in pocket 3 while the pin is guided by part 4. The mating diameters of the bushing and pin are D_1 and D_2, respectively. Because of various kinds of deviations in these dimensions and in assembly-tool displacements, an error δ_0 in alignment occurs. The assembling force P can complete the process as long as the value δ_0 is within certain limits $[\delta_0]$. To increase the chance of achieving satisfactory alignment, relative vibration between the components in the plane perpendicular to the force P may be helpful. The real situation existing during the alignment process is, of course, more complicated than that shown in Figure 8.24. Bevels on both details create an inclination angle α at the contact point A between the two details (this is helpful), as shown in Figure 8.25. The skew between the axes, designated γ in the figure (this is harmful), is an obstacle in assembling. When vibrating, say, guide 4 (Figure 8.24) relative to part 1, the chances of creating better conditions for the penetration of pin 2 into the hole of part 1 are improved. Of course, the amplitude of vibration, the speed of relative displacement between the two parts (in the horizontal plane), the force P, the deviation δ, and the dimensions of the bevels are mutually dependent. The value of the vibration amplitude A should be estimated from the following formula:

$$A = \frac{\sqrt{P\delta}}{100\pi\sqrt{rm\cos\alpha}}. \tag{8.5}$$

This dependence is derived for the frequency 50 Hz (electromagnetic vibrators fed by the industrial AC supply). Here,

δ = manufacturing tolerance of the conjugate parts,
m = mass of the parts including pin 1 moved by force P,
r = radius of the bevels, both inner and outer.

The rest of the symbols are clear from Figure 8.25.

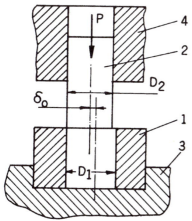

FIGURE 8.24 Model of assembling two components.

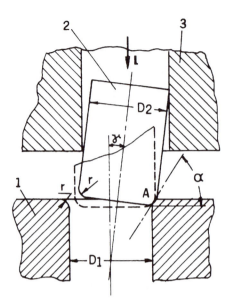

FIGURE 8.25   Skew phenomenon appearing during assembly of a pin into a hole.

Figure 8.26 shows a plan for a specific device for vibration-assisted assembly. Bushing 1 and pin 2 are in the assembly device. The bushings are fed into pocket 3 and the pins are placed in guide 4. Pusher 5 presses the pin against the bushing with force P. Guide 4 is vibrated by magnets 6 and springs 7. As is clear from the cross section A-A, the magnets are energized from the main supply by coils 8 and, due to rectifiers 9, they produce a 50-Hz force. This force actuates armature 10 of guide 4. Tray 11 serves to lead parts 2 from the feeder into the assembly device.

Another idea for increasing the effectiveness of assembly is based on rotation of the pin relative to the bushing, as presented in Figure 8.27. Pin 1 is placed in rotating cylindrical guide 3 and pressed towards the hole in part 2 by pusher 4 with force P. The angle γ between the device's axis of rotation and the pin's axis of symmetry must be less than 2°. (The use of vibration and rotation for improving assembly has been investigated and recommended by K. J. Muceniek, B. A. Lobzov, and A. A. Stalidzan, Riga Politechnic, USSR.)

It is interesting to mention here that an electromagnetic field is a powerful means for assembly. A diagram of its effects is presented in Figure 8.28. The components we want to put together are placed in an alternating magnetic field so that the vector of induction is directed along the assembly axis. Here, the components are three rings 1, 2, and 3 of different sizes. The rings can be scattered, in which case no other method can gather them together (part a) of the figure). This scattering may reach about 80–90% of the ring diameters. It is interesting to note that the gathering of the rings is done in the shortest way by this electrodynamic method. At the end of the process the three rings are assembled, as shown in line e) of Figure 8.28. This phenomenon has the following explanation: the alternating magnetic field results in the appearance of alternating currents i_1, i_2, and i_3 in the rings (part b) of the figure). The latter induce circular magnetic fields B_1, B_2, and B_3 (part c) of the figure). The interactions between these fields move the rings together in the manner shown in part d) of Figure 8.28 until they come into the assembled state, as in part e). The proper choice of frequency of the magnetic field can even heat one of the rings and thus help to carry out assembling

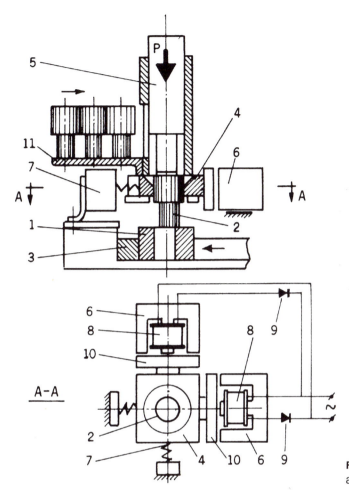

FIGURE 8.26   Vibrating assembly device.

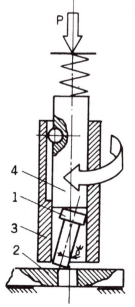

FIGURE 8.27   Rotating assembly device.

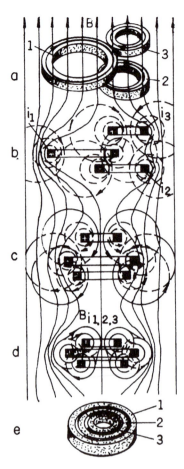

FIGURE 8.28 Assembly of three ringlike metal parts in an alternating magnetic field.

with tension. These kinds of methods are described in USSR patents 38008 (1972), R. K. Kalnin and others; 434699 (1972), B. Joffe and others; and 413724 (1972), B. Joffe and R. K. Kalnin.

8.4   Inspection Systems

Three kinds of inspection devices are widely used in automatic production. The purpose of one kind is to check the manufacturing process at various stages as the product is made. This kind of inspection must prevent idle strokes or operations of tools when a component, blank, or material is missing for some reason; it must save materials and components (by not completing assembly if something is missing); and it must warn the operator that the process is out of order.

The second kind of inspection relates to the tools and the system. Its purpose is to be aware of the wear of tools (for example, cutting tools), thus being able to change, tune, or sharpen the tools (automatically or manually). This sort of inspection ensures the quality of the product and saves time losses due to unexpected damages and their reappearances.

The third type of inspection is connected with sorting and checking the finished product. Its purpose is to separate damaged products from the bulk and (no less important) to sort the products into several groups according to parameters measured during inspection. This is a very powerful approach when selective assembly is to be carried out later, using the sorted products. This is the way ball-bearings are assembled. The example of sorting bushings of roller chain (Chapter 6, Figure 6.21) also belongs to this kind of inspection.

To carry out these inspection operations, appropriate sensors are used (see Chapter 5). The types of inspections can be arranged in several groups according to their level of complexity. For instance, very often during assembly, the presence of the proper component at the correct place at the right moment is important (discussed in Chapter 5). For this purpose, sensors of the "on-off" type can be used. These are not too sophisticated, and prevent the production of incomplete products, which is especially dangerous when the outside of the product does not indicate this defect. This solution is almost the only possibility when nonmetallic details and products are being handled. It seems worthwhile to make a brief digression here to mention an electromagnetic device that can reveal defective metallic assemblies among finished products. Such a device is diagrammed in Figure 8.29. Here, assemblies 1 are falling through an alternating magnetic field created by coil 2, which is fed by an alternating voltage of a certain frequency. The eddy currents induced in assemblies 1 are a function of the mass and shape of their components. Thus, the energy absorbed by bodies 1 depends on their perfection and so does the current in the coil. This results in a voltage drop U_{output} across the resistance R. This voltage is used for sorting out defective assemblies.

Another level of inspection takes place when, for example, the dimensions of cut, ground, etc., parts are checked. In this cases the sensors must provide continuous measurement within a certain range of values. This inspection level is useful in two cases:

- When the product (either some detail, part, or piece of material) must be sorted and, say, collected into separate groups according to its dimensions or other parameter;
- When the dimension or other parameter measured during production reflects the state of the instrument, tool, or process, and serves as a feedback for correcting, retuning, or replacing the tool or process.

Figure 8.30 shows a system belonging to the second case. Here, grindstone 1 processes rotating cylindrical part 2. These parts are automatically fed and turned by

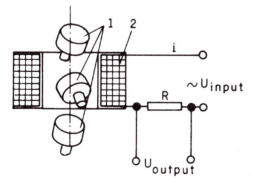

FIGURE 8.29 Scheme of a device for checking assembly completeness.

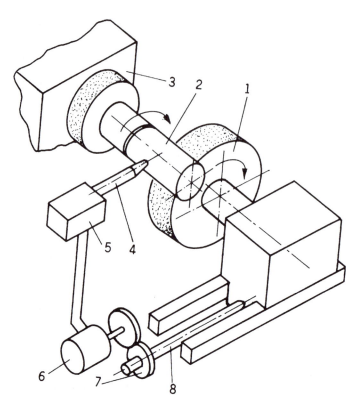

FIGURE 8.30 Device for examining and correcting grindstone wear.

lathe 3. Pick-up 4 (which can be pneumatic) measures the gap between its tip and the cylindrical surface of the detail. Its signal is processed in unit 5 and transmitted to motor 6, which moves grindstone 1 appropriately, by means of gears 7 and lead screw 8. Other, partial examples of this inspection level are presented in Figure 6.13 and 6.20 (see Chapter 6), where inspection is combined with transportation and is done on a discontinuous basis.

The next inspection level is concerned with more complicated handling of the measured results. In general, the results must be remembered, compared, processed, etc. Some algorithm governing the sequence and logic of handling the data obtained by the system must also reach a certain conclusion and control the action of the machine. We discuss here one example of this kind of equipment: an automatic machine for sorting aneroids (Chapter 2, Section 2.1) according to their sensitivity, and for checking their linearity. When the aneroids have been sorted in different groups, it helps to assemble them into blocks of four or five aneroids so as to obtain approximately uniform sensitivity of these pressure sensors even when the sensitivity of every single aneroid may differ significantly. However, the characteristics of each aneroid must be sufficiently linear. This means that the maximum deviation of the measured deformations of the aneroid resulting from pressure changes must not fall outside a certain range of allowed values (see Figure 8.31). When the aneroid is subjected to changing pressure (in our example, the pressure changes from the atmosphere value P_0 to zero), its thickness S in the center also changes. By changing the pressure from P_0 through

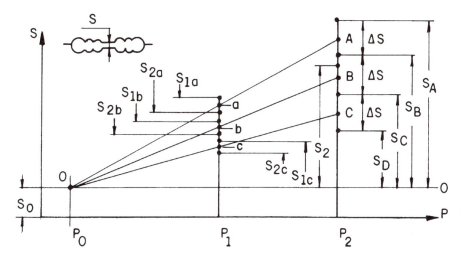

FIGURE 8.31 Deformation versus pressure for aneroids.

P_1 to P_2, we obtain increases in S from S_0 to values corresponding to points a, b, c and A, B, C, respectively. Point O is a floating one; thus, the value S_0 for each aneroid differs and we are forced to create a system of coordinates for measuring S from the floating axis 0–0. By applying pressure to the aneroid, we obtain its characteristics in the form of curves passing through the points mentioned above. The S values corresponding to points A, B, C, ... define the sensitivity of the aneroid and the group it belongs to. The width of the ranges ΔS may be made equal for each group, so that we have:

$$S_A - S_B = S_B - S_C = S_C - S_D = \ldots = \Delta S.$$ [8.6]

When, as a result of measuring, the value of the deformation S_2 corresponding to the pressure P_2 falls within a certain range, the aneroid possesses a certain sensitivity and belongs to a certain group. All aneroids that have S_2 in this range must be collected into one box or compartment. The first task of the machine is thus completed.

The second task—checking the linearity of response—is based on measuring the deformation S for some intermediate pressure P_1. For ideally linear characteristics, these deformations are described by points on a straight line, as shown in Figure 8.31. In reality, there is usually some deviation of points a, b, c from the ideal locations. Some range of deviations is allowed, and the linearity check consists of examining whether points a, b, c, ... lie in the allowed range of deviations, between S_{1a}, and S_{2a}, S_{1b}, and S_{2b}, S_{1c}, and S_{2c}, ..., respectively.

First we describe the mechanical layout of this automatic machine (see Figure 8.32). The aneroids are loaded into a magazine-type hopper 11, from which rotating index-ing table 12 brings them into test position 10. The table is driven by electric motor 16 and wormgear speed-reducer 17. In the test position, the aneroid is closed in hermet-ically sealed chamber 20 (sealing is provided by the super-finished surfaces of housing 7, cover 8, and rings 21). Then electromagnet 5 raises pick-up 6 until the aneroid is caught between the pick-up and upper support 9. Now the coils of inductance-type displacement sensor 22 come into action, and the distance between pick-up 6 and upper support 9 creates the measured value S. Three different pressures P_0, P_1, and P_2

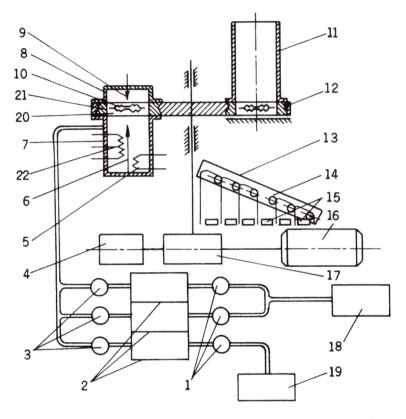

FIGURE 8.32 Layout of an automatic machine for sorting and checking the linearity of aneroids.

are applied in turn in the chamber, by compressor 19 and vacuum pump 18. Control valves 1 provide certain constant pressures in receivers 2 so that $P_0 = 800$ mm of mercury column, $P_1 = 350$ mm, and $P_2 = 5$ mm. Valves 3 connect the receivers to the test chamber in sequence. Thus, when pressure P_0 is applied to the chamber the value S_0 is measured. The values S_1 and S_2 are similarly obtained for pressures P_1 and P_2. These S-values are stored in the memory of the machine and processed, so that all aneroids that do not meet the linearity requirements are removed via tray 13. When the linearity test is passed satisfactorily, the aneroids are sorted according to sensitivity. The bottom of tray 13 is made of gates 14 actuated by magnets 15. The first gate at the upper end of the tray is used to remove the defective aneroids. Only aneroids that pass this gate reach the selection process. Here, according to remembered value S_2, the appropriate gate opens and the aneroid falls into a box meant for this specific S_2. The memory of the machine must hold the data until table 12 makes its next 90° rotation, bringing the measured aneroid to the top of the tray (not shown in Figure 8.32), where it falls onto the tray. The machine is operated by master cam system 4 driven by main motor 16.

Now we pass over to the "brains" of this machine, the layout of which is shown in Figure 8.33. We should note that the number of aneroids per year undergoing this inspection and sorting is close to 1.5 million. Thus, no special flexibility is needed for the machine. In addition, the response of an aneroid to pressure changes is relatively

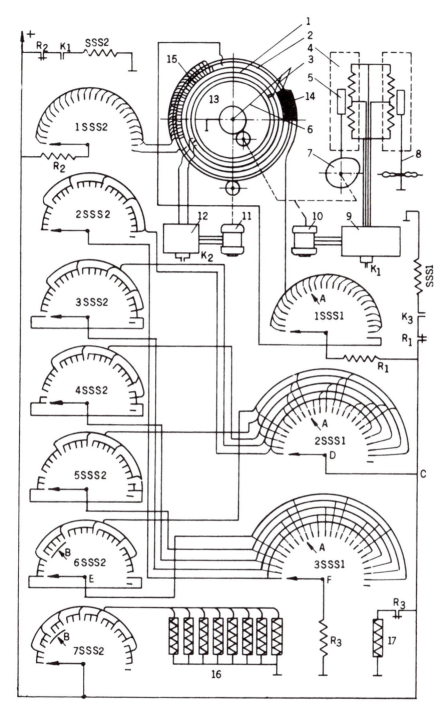

FIGURE 8.33 Memory and computation circuit for the machine shown in Figure 8.32 (see text for explanation).

slow. Thus, no rapid computational means are justified. For these reasons, the control and computation layout uses step-by-step selectors (SSS). In addition, this machine was designed when electronics had not yet reached its present level. In my opinion, this scheme is more appropriate for explanation and understanding; and, of course, a purely electronic analog of this layout also can be designed and used.

For those who do not know what SSS are, a brief description follows. Figure 8.34 shows a diagram of this device, which consists of several contact rows 1, each of which includes, say, 25 contacts located around an arc, slide contact 2 driven by ratchet wheel 3, and pawl 4 actuated by electromagnet 5 and spring 6. The system is designed so that the slide contact passes over to the next contact for each current pulse in the magnet coil. As was stated earlier, the device consists of several rows of contacts 1, while slide contacts 2 are also fitted to each row. The slide contacts move and stop synchronously as a solid body. One of the rows serves for control of the SSS device. A circuit can be included that will stop the slide contact at each desired contact in the contact row.

Figure 8.33 shows two SSS devices: SSS1 and SSS2. The rows ISSS1 and ISSS2 are used for self-control, while the other two rows in SSS1 and six rows in SSS2 are used to remember the measured results and for control. The circuit shown in Figure 8.33 works as follows. The aneroid is held between the support and pick-up 8. The latter actuates inductive bridge 4 and amplifier 9, which cause the motor to drive cam 7 until armature 5 balances bridge 4. In addition, motor 10 drives slide contact 3. Thus, when the aneroid is under pressure P_0 and dimension S_0 is measured which, as mentioned earlier, serves as an initial coordinate for further measurements, slide contact 3 moves to a certain place. At this moment, amplifier 12 is brought into operation (due to contacts K_2) and actuates motor 11 driving disc 13. This disc carries two contact rings. Ring 1 is fed from amplifier 12. Thus, a feedback is created between the location of slide contact 3 and motor 11. The latter rotates disc 13 until the slot in ring 1 finds contact 3. Value S_0 is thus set and remembered. When the aneroid is brought under pressure P_1 and its deformation becomes S_1, inductive bridge 4 (through amplifier 9 and motor 10) brings contact 3 into the position corresponding to value S_1. If the aneroid under inspection is normal, slide contact 3 stops on ring 2 somewhere in the first domain of contacts 14. Each of these contacts is connected with the first row of ISSS1. When the slide contact of SSS1 reaches this specific contact, relay R_1 disconnects the normally closed contact R_1 and stops SSS1 at this position, thus remembering value S_1. Analo-

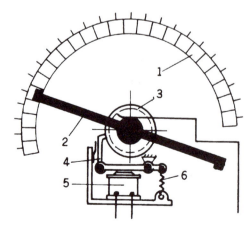

FIGURE 8.34   Step-by-step selector (one-motion rotary switch) or miniselector.

gously, pressure P_2 applied to the aneroid brings motor 10 into action, resulting in rotation of contact 3, which stops somewhere in the second domain of contacts 15 (ring 2). Thus, the second SSS2, the first row (1SSS2) of which is connected to each contact 15, is stopped at the corresponding position, due to relay R_2 and normally closed contact R_2. As a result, SSS2 remembers value S_2. The seventh row (ISSS2) is for sorting. The contacts of this row constitute eight groups, each of which (when energized) actuates corresponding electromagnet 16. As was explained earlier (and illustrated in Figure 8.32), the sorting tray consists of eight gates actuated by magnets 16 according to value S_2. The reader can see in Figure 8.33 that contacts in ISSS2 are united in eight groups, each of which actuates a corresponding magnet. (The wiring is done in bundles.)

Now let us consider the decision-making process in this circuit when the linearity of an aneroid is tested. For this purpose we must explain the wiring between the SSS1 and SSS2 contacts. The contacts of 3SSS1 (third row of SSS1) are connected in groups so that the 1st, 6th, 11th, 16th, and 21st contacts are connected to the slide contact of 2SSS2 (second row of SSS2); contacts number 2, 7, 12, 17, and 22 are connected to slide contact 3SSS2; contacts number 3, 8, 13, 18, and 23 are connected with slide contact 4SSS2; contacts number 4, 9, 14, 19, and 24 with slide contact 5SSS2; and contacts 5, 10, 15, 20, and 25 with slide contact 6SSS2. In addition, the contacts in 2SSS1 are wired to contact rows of SSS2 so that numbers 1, 6, 11, 16, and 21 of 2SSS1 are connected to five contact groups in 2SSS2; contacts 2, 7, 12, 17, and 22 to five groups in 3SSS2; contacts 3, 8, 13, 18, and 23 to five groups in 4SSS2; contacts 4, 9, 14, 19, and 24 to five groups in 5SSS2; and contacts 5, 10, 15, 20, and 25 to five groups in 6SSS2. Now, say, value S_1 has brought the slide contact of SSS1 to the tenth position (arrow A in Figure 8.33); then if the aneroid is linear, value S_2 must bring the slide contact of SSS2 into position between the 5th and 9th contacts (arrow B). Only then will current flow from C through D, A, B, E, A, and F, to energize relay R_3 and open normally closed contact R_3; thus, coil 14, which controls the first gate on the sorting tray, stays closed. (This gate leads aneroids into the box for defective parts, as was stated earlier.) Thus, the aneroid continues its movement along the tray until it reaches open gate 2 or 3 (because contacts 5 to 9 belong to these two groups), depending on the specific one that the value S_2 will indicate. (See USSR Patent by B. Sandler & A. Strazdin, Automatic machine for pressure sensing elements linearity measurement and sorting, 1964, No. 168507.)

Of course, modern digital circuits can be used for the same purpose, especially when flexibility is desired in the decision-making algorithm.

8.5   Miscellaneous Mechanisms

In this section we consider several mechanisms used for some common manufacturing operations. For bending or cutting external and internal shapes in sheet material, flattening material, caulking, making riveted joints, etc.—stamping, forging, or forming mechanisms are used. For relatively small forces (about 1,000–2,000 N), electromagnetic heads are suitable. Figure 8.35 shows a design for an electromagnetic stamping press. Here a linkage is used for amplifying the stamping force. The design consists of an immobile core provided with coil 1, and armature 2 which, through connecting rod 3, actuates links 4 and 5. The latter drives punch 6 sliding in guides 7 and clip 8 and, due to die matrix 9, the sheet of material 10 is punched. The punch is

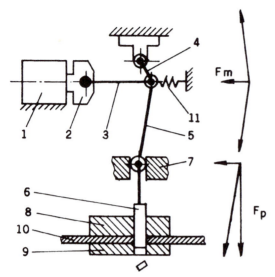

FIGURE 8.35 Electromagnetic punching head.

returned upwards after completing the task by spring 11. The resolved force vectors are shown in the diagram attached to the design. One can easily see how the relatively small force F_m of the magnet is transformed into the large punching force F_p.

When higher forces must be developed, a pneumatic stamping head is convenient. Such a press is shown in Figure 8.36 and consists of a pneumatic drive composed of two parts (1 and 2), between which flexible diaphragm 3 is clamped around the edges, creating a hermetically sealed chamber above this diaphragm. Punch rod 5 is fastened to the diaphragm by means of two flat disks 4. Due to spring 6, the diaphragm is kept in the upper position when resting. To energize the press, pressure is introduced through inlet 7 into the upper chamber. This force presses against the diaphragm, thus moving rod 5 and punch 8 (the example presented here shows a bending stamp) bending strip 9 against matrix 10.

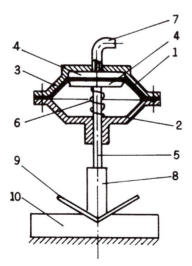

FIGURE 8.36 Pneumatic diaphragm punching press.

Both mechanisms illustrated here can carry out other operations, besides those shown in Figures 8.35 and 8.36; for instance, riveting with hollow or solid rivets (Figure 8.37). Here, the parts 1 to be joined are shown before riveting (position 2) and after the riveting is completed in position 3. Also, punch 4 and matrix 5 are shown. These kinds of stamping heads allow flexibility in design and construction of the machine, as well as in the method for control and timing.

Of course, pneumatic and hydraulic drives can be based on conventional cylinders and pistons. When high forces are required, hydraulic means are more suitable (see Chapter 3). Forces of tens of tons can be developed in this way (much higher forces are available, too, but are seldom used in automatic machinery), which are needed for heavy bending or other kinds of plastic deformation of materials, for example, to cut or bend thick rods or strips. Figure 8.38 gives an idea of how such a hydraulically driven pipe-bending head acts. Cylinder 1 and its piston 2 are engaged with pinion 4, by means

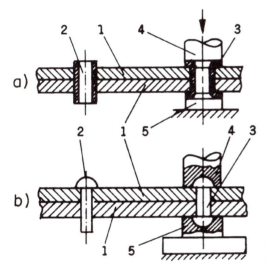

FIGURE 8.37 Examples of riveting: a) Hollow rivet; b) Ordinary rivet.

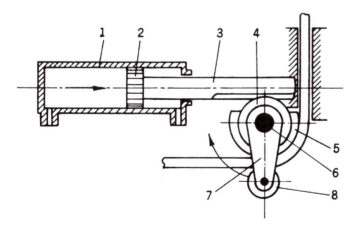

FIGURE 8.38 Hydraulic bending head based on a reciprocating plunger mechanism.

of piston rod 3, which is made in the form of a rack. Cylindrical-former accessory 5 is mounted concentric with the pinion (in another plane). On pinion 4, shaft 6 and lever 7 are fastened. An adjustable roller 8 is mounted on the latter. Thus, when rotating, pinion 4 drives lever 7, and roller 8 carries out the bending of pipe (or rod) 9 around former 5. The same rotational movement of the bending tool (roller 8) can be obtained with the rotary hydraulic motor shown in Figure 8.39. Here, lug 1 is placed in a sector-like chamber created by two checkpieces 3 and 4 and curved part 2. Due to inlets-outlets 5 and 6, the pressurized working liquid is introduced into the motor and rotates the lug (according to the inlet-outlet connection mode). The torque developed in the motor is transferred by splined shaft 7, which drives the roller around the cylindrical former (see Figure 8.38), with the action of the bending head being as in the previous case.

Another operating head is a hydraulically driven drilling head (Figure 8.40). This device must provide a certain rotating speed for the drill and also the torque required for cutting the material. An axial force must also be developed for feeding the drill. Thus, the device consists of electric motor 1 which, by means of pinion 2 and speed-reducer 3, transfers rotation to gear wheel 4, which is built as one body with bushing 5. The latter is engaged by key 6 with shaft 7, which can slide along the bushing while transferring torque. Shaft 7 is mounted by bearings 8 in sleeve 9, which has a rack that is engaged with pinion 10. This pinion is also engaged with plunger 11 (also provided with a rack). The working stroke of the plunger is powered by liquid pressure and the return stroke by spring 12. The flow rate of the liquid into chamber 13 of the cylinder controls the feed of the drill 14; analogously, the outlet flow from the chamber controls the rate of the return stroke. All these elements are installed in housing 15. This design is not appropriate for cases where the advance of the tool must be strictly matched with its rotation, as needed for threading, boring, and some milling operations.

A possible design for such a boring head is presented in Figure 8.41. It includes two electromotors 1 and 2. Motor 1 is provided with pinion 3 and, through a gear trans-

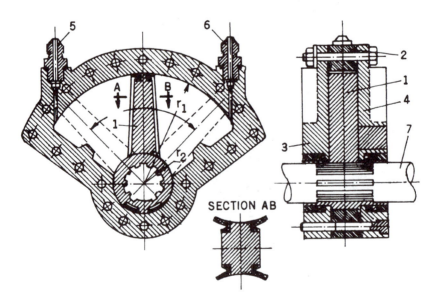

FIGURE 8.39 Rotary hydraulic motor.

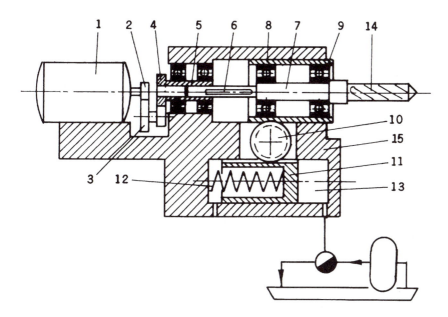

FIGURE 8.40 Drilling head.

mission, drives hollow bushing 4, which is engaged via key 5 with boring shaft 6, in which cutter 7 (adjusted according to the bore's diameter d) is fastened. This shaft is mounted by bearings in sleeve 8. Thus, the drive, described above, provides the boring shaft with the torque needed for cutting the material while keeping the tool free for axial displacement. To provide the desired axial displacement, which must be in a definite ratio with the rotation of the tool, the second motor 2 is used. The latter transmits its rotation via coupling 9 to threaded shaft 10, which is engaged with nut 11. Nut 11 and sleeve 8 are connected through gear wheel 12 by means of racks on the surfaces

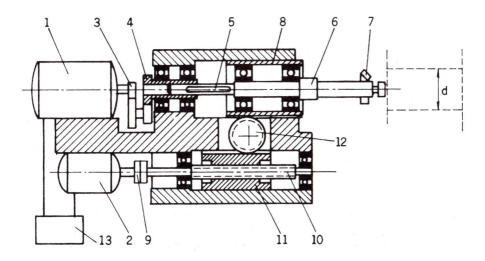

FIGURE 8.41 Boring (or threading) head.

of the sleeve and nut. Rotation of shaft 10 drives the nut, resulting in movement of the sleeve. The speeds of motors 1 and 2 are regulated by controller 13, which governs the advance of the cutter in each revolution.

A question arises here: could we use here a rigid-ratio mechanism providing a strictly constant feed of the cutter per revolution? Of course, then one motor 1 with a specially designed transmission, linking bushing 4 with shaft 10, could serve the same purpose. The point, obviously, is that in this case the lower cost of this simplified design leads to almost no flexibility for carrying out different cutting regimes. Which approach to choose depends on constraints at the designer's disposal.

When looking at automated assembly, we must include a consideration of a powerful method of assembly that relies on programmable manipulators (robots). We will illustrate this method by using a puzzle assembly process. This puzzle is shown in Figure 8.42a) in a nonassembled condition, while in Figure 8.42b) it appears completely assembled. It is clear from these figures that the puzzle is composed of three kinds of parts: three units 1, 2, and 3 with two cylindrical slots, two units 4 and 5 with two slots on one side and one slot in a perpendicular plane, and finally, one smooth, locking part 6. To carry out the assembly process a device called a "cradle" was designed and built. In Figure 8.43. this "cradle" is shown and the sequence of the parts involved in assembly is denoted by numbers (mentioned also above) indicating the order of the process.

The previously programmed manipulator (or, as some people say, robot) takes the separate parts from the pallets where they are placed in an oriented position in the order mentioned above, and brings them into the corresponding place of the "cradle," thus carrying out the assembly of the puzzle. (This work was supervised by Dr. V. Lifshits in the CIM laboratory of the engineering faculty of the Ben-Gurion University of the Negev, Beersheva, Israel).

In our case, at this stage of development, the puzzle parts orientation procedure was a manual one. In Figure 8.44 some consequent situations demonstrating the process are shown.

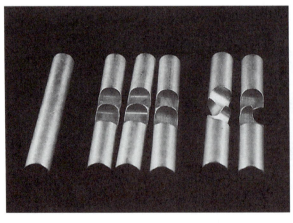 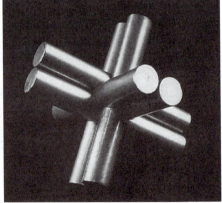

FIGURE 8.42   a) Parts constituting the puzzle; b) Assembled puzzle.

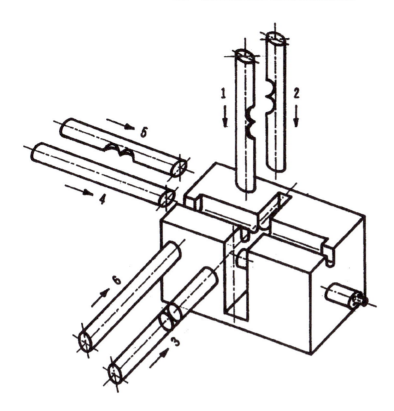

FIGURE 8.43   Assembly "cradle" and assembly sequence.

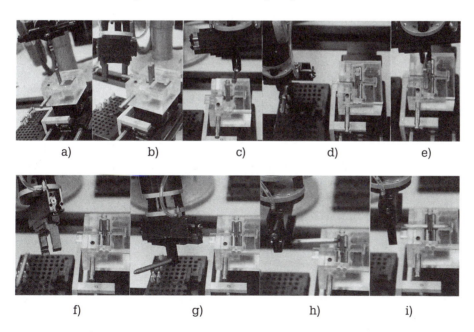

FIGURE 8.44   a) Assembly process of the puzzle; b) First three components placed at the corresponding places; c), d), and e) Components 4 and 5 next to the three previous parts; from f) to i) The locking action.

9

Manipulators

9.1 Introduction

This chapter is devoted to a discussion, in greater depth and breadth, of the aspects of manipulators mentioned in Section 1.3. The following problems that arise when a manipulator is being designed are considered here:

- Dynamics of manipulators;
- Location of drives and kinematics for motion transfer to the links of the manipulator;
- Grippers used in manipulators, their properties, kinds, design;
- Methods for achieving the required level of accuracy in displacement.

Industrial manipulators are basically arm-like devices for handling parts, tools, blanks, etc. This handling consists of a defined sequence of motions or operations. These devices are good at tasks that require a considerable amount of repetition within close tolerances. To become a part of advanced robotics, the device must be reprogrammable, or even able to make simple decisions. After the program is introduced into the device, it must operate automatically, without human guidance.

Some accuracy problems that appear in mechanical systems of automatically acting machines have been discussed in Chapter 4. Here we will consider some design approaches for achieving the required accuracy level in manipulators. A well-designed, manufactured, and assembled mechanical system of a manipulator must be provided with a well-developed control system to utilize the accuracy level attainable with the mechanics. After all, badly designed mechanisms will not perform accurately even if provided with sophisticated electronics or other controls.

In industry, manipulators are used for a wide range of tasks, most commonly:

- Taking parts or blanks from a feeder tray or conveyor and transferring them to a machine, e.g., an indexing table, another conveyor, chuck, vise, etc.;
- Taking parts from chucks, vises, etc., and transferring them for further handling;
- Assembly operations, which include taking parts from trays, hoppers, etc., and putting them into some other assembly positions (e.g., in electronic circuit assembly);
- Taking tools from magazines or hoppers and putting them into spindles, or vice versa; bringing the proper tool into the correct position at the right moment;
- Moving welding electrodes or burners in accordance with the welding trace; moving and activating resistance-point welding heads;
- Moving dye or paint sprayers;
- Functioning as X–Y coordinate tables for different purposes, for instance, in all manufacturing stages of integrated-circuit wafer production;
- Operating in hostile environments such as radioactive, hot or extremely cold, or poisonous atmospheres (devices acting in outer space and under water may also be included in this category).

9.2 Dynamics of Manipulators

We have already mentioned that manipulators usually work in concert, creating automatic production machines of several levels of flexibility. We have also emphasized the importance of time, when the productivity of a process is under consideration. Thus, we begin the discussion of manipulator design problems with a consideration of their dynamics.

We consider this problem on the basis of some examples, and try to show the approach for answering a nontrivial question: What are the unconstrained trajectories of the links, for a manipulator with multiple degrees of freedom, to bring a gripper from point A to point B (arbitrarily chosen) in minimal time? The trajectory providing this condition is the so-called optimal-time trajectory. The example under discussion presents a cylindrical manipulator with three degrees of freedom, as in Figure 9.1. Here

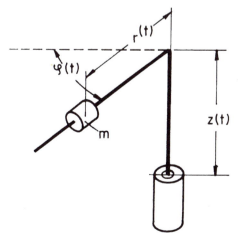

FIGURE 9.1 Layout of cylindrical-type manipulator.

the gripper with the grasped part is represented by sliding mass m. The distance of this mass from the axis of rotation is $r(t)$, which is a function of time. So also is the angle of rotation $\phi(t)$, the azimuth. The coordinate $z(t)$ describes the height of the manipulator's arm. The angular motion is caused by torque T, and the radial motion by force F.

The variables are

$$
\begin{aligned}
x_1(t) &= r(t), & u_1(t) &= F(t), \\
x_2(t) &= \dot{r}(t), & u_2(t) &= T(t). \\
x_3(t) &= \phi(t), \\
x_4(t) &= \dot{\phi}(t),
\end{aligned}
$$

[9.1]

Neglecting frictional forces, the motion equations are

$$
\dot{x}_1(t) = x_2(t); \qquad \dot{x}_3(t) = x_4(t);
$$

$$
\dot{x}_2(t) = \left[u_1(t) + mx_1(t)x_4^2(t)\right]/m;
$$

[9.2]

$$
\dot{x}_4(t) = \left[u_2(t) - 2mx_1(t)x_2(t)x_4(t)\right]/\left[J_0 + mx_1^2(t)\right].
$$

Here, J_0 = moment of inertia of the unloaded arm (without the mass m), and the control constraints are

$$
\begin{aligned}
\left|u_1(t)\right| &\leq F_{\max}, \\
\left|u_2(t)\right| &\leq T_{\max}.
\end{aligned}
$$

[9.3]

Here u_1 and u_2 are control functions describing the behavior of the force and torque. In this case the $z(t)$ degree of freedom is not considered, because it does not influence the azimuth and radial displacements of the mass m.

The motion of the manipulator is considered from an arbitrary initial position A (at standstill):

$$
\begin{aligned}
x_1(0) &= r_A, \\
x_3(0) &= \phi_A, \\
x_2(0) &= x_4(0) = 0,
\end{aligned}
$$

[9.4]

to an arbitrary final position B (at standstill) at time τ:

$$
\begin{aligned}
x_1(\tau) &= r_B, \\
x_3(\tau) &= \phi_B, \\
x_2(\tau) &= x_4(\tau) = 0,
\end{aligned}
$$

[9.5]

in such a way as to complete this motion in the shortest time. Let us suppose that the system shown in Figure 9.1 is described by the following data

$$
\begin{array}{lll}
m = 8.5 \text{ kg}, & T_{\max} = 5 \text{ Nm}, & \phi_A = 0, \\
J_0 = 0.4 \text{ kg m}^2, & r_A = 0.75 \text{ m}, & \phi_B = 90°. \\
F_{\max} = 15 \text{ N}, & r_B = 0.75 \text{ m},
\end{array}
$$

[9.6]

Computation of Equations (9.1) with control constraints (9.2) and boundary conditions (9.3) and (9.4) brings us to the results presented in Figure 9.2. Here, movement of the arm and mass from point A to point B took about 1.2 seconds, and their position was calculated every 0.1 second. To achieve this minimal time, both the arm and the mass m along the arm were moved, so that the force changes direction twice during the operation while the torque T changes once (line c). The control functions obtained for this case have the following form:

$$u_1 = \left\{ -F_{max}, \; +F_{max}, \; -F_{max} \right\},$$
$$u_2 = \left\{ +T_{max}, \; -T_{max} \right\}.$$

[9.7]

The changes in the force occur at about 0.4 second and 0.8 second, while the torque changes once in the middle of the operating time at about 0.6 second. It is interesting to note that when the mass is not moved during this operation and stays at the end of the arm, the operating time (when the same T_{max} is applied) equals 1.5 seconds (line d). Thus, 20% of the time can be saved by introducing optimal dynamic control of the manipulator.

The calculations for this example (and later ones in this section) are taken from the paper "Time-Optimal Motions of Robots in Assembly Tasks" by H. P. Geering, L. Guzzella, S. A. R. Hepner, and C. H. Onder, presented at the 24th IEEE Conference on Decision and Control, Fort Lauderdale, Florida, December 11–13, 1985.

Obviously, the solution presented in Figure 9.2 for the manipulator's optimal behavior is not the only possibility. For instance, when the value F_{max} is considerably greater than in the above case, the mass m will rapidly reach the axis 0 and its further movement (if the design of the manipulator permits it) will not be desired (to minimize the moment of inertia of the system). This leads to the necessity of keeping the control function $u_1 = 0$ for a certain time.

The next example we consider is presented in Figure 9.3. This kind of device is often called a serpent-like manipulator. It consists of arm 1 and lever 2. The location of

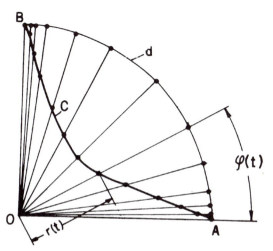

FIGURE 9.2 Optimal-time trajectory of the gripper providing fastest transfer from point A to point B.

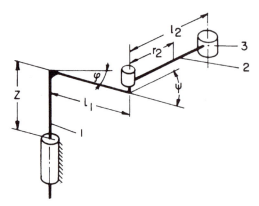

FIGURE 9.3 Serpent-like manipulator.

gripper 3 can be spatially defined by coordinate z and two angles ϕ and ψ. In this case also, vertical displacement $z(t)$ does not influence the dynamics of the mechanism in the horizontal plane. (We do not consider here the remaining degrees of freedom of the gripper.) Such manipulators are convenient for automated assembly. The variables describing the motion of this system are

$$
\begin{aligned}
x_1(t) &= \phi(t), \\
x_2(t) &= \dot{\phi}(t), \\
x_3(t) &= \psi(t), \\
x_4(t) &= \dot{\psi}(t).
\end{aligned}
\qquad [9.8]
$$

The movement of the manipulator is provided by applying torques T_ϕ and T_ψ; thus, the control variables are

$$
\begin{aligned}
u_1(t) &= T_\phi(t), \\
u_2(t) &= T_\psi(t).
\end{aligned}
\qquad [9.9]
$$

We denote the parameters as follows:

l_1, l_2 = lengths of links 1 and 2, respectively;
r_2 = distance from the joint to the center of mass of the second link;
m_2, m_3 = masses of link 2 and gripper 3;
J_1 = moment of inertia of link 1 with respect to the first axis;
J_2 = moment of inertia of link 2 with respect to the second axis.

Then,

$$
\begin{aligned}
J_3 &= J_2 + l_2^2 m_3, \\
J_4 &= J_1 + l_1^2 (m_2 + m_3), \\
J_5 &= l_1 (r_2 m_2 + l_2 m_3).
\end{aligned}
\qquad [9.10]
$$

FIGURE 9.3a) General view of a cylinder-type (serpent-like) robot with hydraulic drives. This device is used in teaching and is produced by Cybernetic Application Company (Partway Trading Estate Andover Hants SP103LF, England).

We can now rewrite the motion equations in the following form:

$$\dot{x}_1 = x_2; \qquad \dot{x}_3 = x_4;$$

$$\dot{x}_2 = \frac{\left[J_3 \left\{ u_1 - u_2 + J_5 (x_2 + x_4)^2 \sin x_3 \right\} - J_5 \left\{ u_2 - J_5 x_2^2 \sin x_3 \right\} \cos x_3 \right]}{\left[J_3 J_4 - J_5^2 \cos^2 x_3 \right]}; \qquad [9.11]$$

$$\dot{x}_4 = \frac{\left[(J_4 + J_5 \cos x_3)(u_2 - J_5 x_2^2 \sin x_3) - (J_3 - J_5 \cos x_3) \left\{ u_1 - u_2 + J_5 (x_2 + x_4)^2 \sin^2 x_3 \right\} \right]}{\left[J_3 J_4 - J_5^2 \cos^2 x_3 \right]}.$$

The control constraints are

$$|u_1(t)| \le T_{\phi \max},$$

$$|u_2(t)| \le T_{\psi \max}. \qquad [9.12]$$

Note that, because of changing moments of inertia, even constant forces or torques applied to the links will cause variable accelerations.

The computation was carried out for the following set of data:

$$J_1 = 1.6 \text{ kgm}^2, \qquad l_1 = 0.4 \text{ m}, \qquad m_2 = 15 \text{ kg}, \qquad T_{\phi max} = 25 \text{ Nm},$$

$$J_2 = 0.44 \text{ kgm}^2, \qquad l_2 = 0.25 \text{ m}, \qquad m_3 = 6 \text{ kg}, \qquad T_{\psi max} = 9 \text{ Nm}, \qquad [9.13]$$

$$r_2 = 0.125 \text{ m}.$$

Figure 9.4 shows the optimal trajectory traced by the gripper of the manipulator for equidistant time moments when the boundary conditions are

$$\phi_A = 0; \qquad \phi_B = \pi; \qquad \dot{\phi}_A = \dot{\phi}_B = 0. \qquad [9.14]$$

The time needed to complete this transfer is 1.25 seconds and the torques to carry out this minimal (for the given circumstances) time have to behave in the following manner:

$$u_1 = \begin{cases} T_{\phi max} > 0 & \text{for } 0 < t \leq \dfrac{\tau}{2}, \\[2mm] T_{\phi max} < 0 & \text{for } \dfrac{\tau}{2} < t \leq \tau, \end{cases} \qquad [9.15]$$

$$u_2 \begin{cases} T_{\psi max} > 0 & \text{for } 0 < t \leq 0.278 \text{ seconds}, \\[2mm] T_{\psi max} < 0 & \text{for } 0.278 \text{ seconds} \leq t < \dfrac{\tau}{2} = 0.625 \text{ seconds}, \\[2mm] T_{\psi max} > 0 & \text{for } \dfrac{\tau}{2} < t \leq 0.974 \text{ seconds}, \\[2mm] T_{\psi max} < 0 & \text{for } 0.974 \text{ seconds} < t \leq 1.25 \text{ seconds} = \tau. \end{cases}$$

The meaning of these functions is simple. Arm 1 is accelerated by the maximum value of torque $T_{\phi max}$ half its way (here, until angle ϕ reaches $\pi/2$); afterwards arm 1 is decelerated by the negative torque $-T_{\phi max}$ until it stops. Obviously, this is true when friction can be ignored. Link 2 begins its movement, being accelerated due to torque $T_{\psi max}$ for 0.278 second. Then it is decelerated by torque $-T_{\psi max}$ until 0.625 second has elapsed, and then again accelerated by torque $T_{\psi max}$. After 0.974 second the link is decelerated by negative torque $-T_{\psi max}$ until it comes to a complete stop after a total of 1.25 seconds.

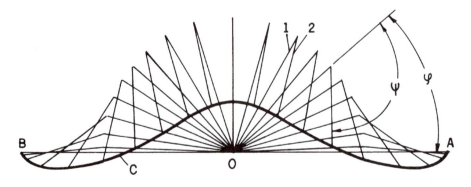

FIGURE 9.4 Optimal-time trajectory C of the gripper shown in Figure 9.3, providing fastest travel from point A to point B for link 1; rotational angle of $\phi = \pi$.

It is interesting (and important for better understanding of the subject) to compare these results with those for a simple arc-like trajectory connecting points A and B, made by straightened links 1 and 2 so that the length of the manipulator is constant and equals $l_1 + l_2$. To calculate the time needed for carrying out the transfer of mass m_3 from point A to point B under these conditions, we have to estimate the moment of inertia J of the moving masses. This value, obviously, is described in the following form:

$$J = J_1 + J_2 + m_2(l_1 + r_2)^2 + m_3(l_1 + l_2)^2 \cong 8.52 \text{ kgm}^2. \qquad [9.16]$$

Applying to this mass a torque $T_{\phi\max}$ we obtain an angular acceleration α

$$\alpha = \frac{T_{\phi\max}}{J} \cong 2.93^1 / \text{seconds}^2. \qquad [9.17]$$

Considering the system as frictionless, we can assume that for half the way, $\pi/2$, it is accelerated and for the other half, decelerated. Thus, the acceleration time t_1 equals

$$t_1 = \sqrt{\pi/2\alpha} \cong 0.73 \text{ seconds}, \qquad [9.18]$$

which gives, for the whole motion time T,

$$T = 2t_1 \cong 1.46 \text{ seconds}.$$

The previous mechanism gives a 17% time saving (although the more complex manipulator is also more expensive).

The mode of solution (the shape of the optimal trajectory) depends to a certain extent on the boundary conditions. The examples presented in Figures 9.5, 9.6, and 9.7 illustrate this statement.

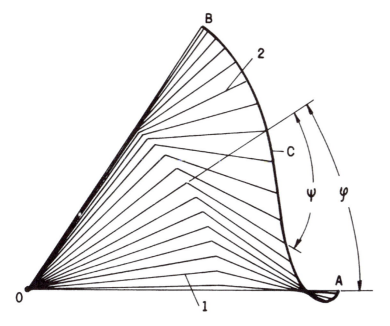

FIGURE 9.5 Optimal-time trajectory C of the gripper providing fastest travel from point A to point B for link 1; rotational angle of $\phi = 1$.

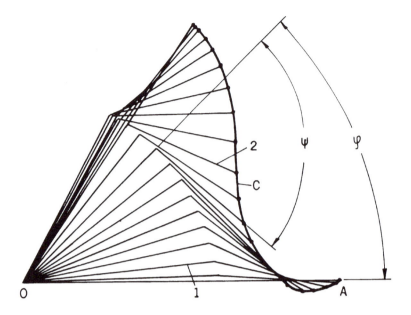

FIGURE 9.6 Optimal-time trajectory C of the gripper providing fastest travel from point A to point B for link 1; rotational angle of $\phi = 1$.

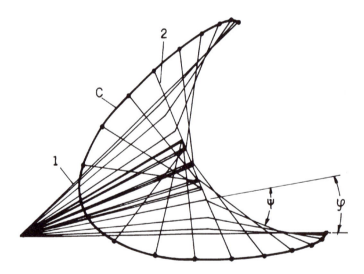

FIGURE 9.7 Optimal-time trajectory C of the gripper providing fastest travel from point A to point B for link 1; rotational angle of $\phi = 0.76$.

For the conditions:

$$\phi_A = 0, \qquad \phi_B = 1, \qquad \dot{\phi}_A = \dot{\phi}_B = 0, \qquad [9.19]$$

we have, for example, a motion mode shown in Figure 9.5.

The transfer time $T = 1.085$ seconds and the control functions have the following forms:

$$u_1 = \begin{cases} +T_{\phi\max} & \text{for } 0 < t \leqslant 0.5425 \text{ second,} \\ -T_{\phi\max} & \text{for } 0.5425 \text{ second} < t \leqslant 1.085 \text{ seconds,} \end{cases}$$

$$u_2 = \begin{cases} +T_{\psi\max} & \text{for } 0 < t \leqslant 0.088 \text{ second,} \\ -T_{\psi\max} & \text{for } 0.088 \text{ second} < t \leqslant 0.588 \text{ second,} \\ +T_{\psi\max} & \text{for } 0.588 \text{ second} < t \leqslant 1.085 \text{ seconds.} \end{cases} \qquad [9.20]$$

In Figure 9.6 we see another path of motion of the links for the same conditions (9.19) and here the control functions are

$$u_1 = \begin{cases} +T_{\phi\max} & \text{for } 0 < t \leqslant 0.513 \text{ second,} \\ -T_{\phi\max} & \text{for } 0.513 \text{ second} < t \leqslant 1.056 \text{ seconds,} \\ +T_{\phi\max} & \text{for } 1.056 \text{ seconds} < t \leqslant 1.085 \text{ seconds,} \end{cases}$$

$$u_2 = \begin{cases} +T_{\psi\max} & \text{for } 0 < t \leqslant 0.113 \text{ second,} \\ -T_{\psi\max} & \text{for } 0.113 \text{ second} < t \leqslant 0.634 \text{ second,} \\ +T_{\psi\max} & \text{for } 0.634 \text{ second} < t \leqslant 1.085 \text{ seconds.} \end{cases} \qquad [9.21]$$

Note that in Figure 9.5, link 1 does not pass the maximum angle, while in the trajectory shown in Figure 9.6, link 1 passes this angle a little and then returns.

By decreasing the maximum angle ϕ_B, we obtain another very interesting mode ensuring optimal motion time for this manipulator. Indeed, for

$$\phi_A = 0, \qquad \phi_B = 0.76, \qquad \dot{\phi}_A = \dot{\phi}_B = 0, \qquad [9.22]$$

the result is as shown in Figure 9.7. Link 2 here moves in only one direction, creating a loop-like trajectory of the gripper when it is transferred from point A to point B. The control functions in this case are

$$u_1 = \begin{cases} +T_{\phi\max} & \text{for } 0 < t \leqslant 0.19 \text{ second,} \\ -T_{\phi\max} & \text{for } 0.19 \text{ second} < t \leqslant 0.488 \text{ second,} \\ +T_{\phi\max} & \text{for } 0.488 \text{ second} < t \leqslant 0.785 \text{ second,} \\ -T_{\phi\max} & \text{for } 0.785 \text{ second} < t \leqslant 0.9755 \text{ second,} \end{cases}$$

$$u_2 = \begin{cases} +T_{\psi\max} & \text{for } 0 < t \leqslant 0.488 \text{ second,} \\ -T_{\psi\max} & \text{for } 0.488 \text{ second} < t \leqslant 0.9755 \text{ second.} \end{cases} \qquad [9.23]$$

Comparing these results (the time needed to travel from point A to point B for the examples shown in Figures 9.5, 9.6, and 9.7) with the time T' calculated for the conditions (9.16), (9.17), and (9.18) (i.e., links 1 and 2 move as a solid body and $\psi = 0$), we obtain the following numbers:

Figure	$T_{optimal}$	T' for $\psi = 0$	Time saving
9.5	1.085 sec	1.17 sec	~13%
9.6	1.085 sec	1.17 sec	~13%
9.7	0.9755 sec	1.02 sec	~10%

The ideal motion described by the Equation Sets (9.1) and (9.2) does not take into account the facts that: the links are elastic, the joints between the links have backlashes, no kinds of drives can develop maximum torque values instantly, the drives (gears, belts, chains, etc.) are elastic, there is friction and other kinds of resistance to the motion, or there may be mechanical obstacles in the way of the gripper or the links, all of which do not permit achieving the optimal motion modes. Thus, real conditions may be "hostile" and the minimum time values obtained by using the approach considered here may differ when all the above factors affect the motion. However, an optimum in the choice of the manipulator's links-motion modes does exist, and it is worthwhile to have analyzed it.

Note: The mathematical description here is given only to show the reader what kind of analytical tools are necessary even for a relatively simple—two-degrees-of-freedom system—dynamic analysis of a manipulator. We do not show here the solution procedure but send those who are interested to corresponding references given in the text and Recommended Readings.

Another point relevant to the above discussion is that, in Cartesian manipulators (see Chapter 1), such an optimum does not exist. In Cartesian devices the minimum time simply corresponds with the shortest distance. Therefore, if the coordinates of points A and B are X_A, Y_A, Z_A and X_B, Y_B, Z_B, respectively, as shown in Figure 9.8, the distance $\overline{AB}$ equals, obviously,

$$\overline{AB} = \sqrt{\left(X_B - X_A\right)^2 + \left(Y_B - Y_A\right)^2 + \left(Z_B - Z_A\right)^2} \, . \qquad [9.24]$$

Physically, the shortest trajectory between the two points is the diagonal of the parallelepiped having sides $(X_B - X_A)$, $(Y_B - Y_A)$, and $(Z_A - Z_H)$. Thus, the resulting force F acting along the diagonal must accelerate the mass half of the way and decelerate it during the other half. Thus, the forces along each coordinate cause the corresponding accelerations

$$a_X = \frac{F_X}{m_X}; \qquad a_Y = \frac{F_Y}{m_Y}; \qquad a_Z = \frac{F_Z}{m_Z} \, . \qquad [9.25]$$

Here,

a_X, a_Y, a_Z = accelerations along the corresponding coordinates,
F_X, F_Y, F_Z = force components along the corresponding coordinates,
m_X, m_Y, m_Z = the accelerated masses corresponding to the force component.

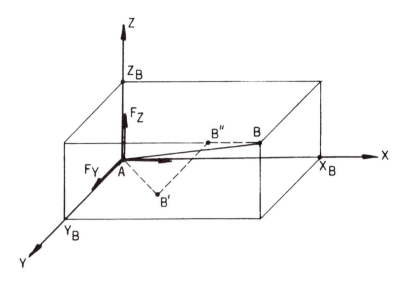

FIGURE 9.8 Fastest (solid line) and real (dotted line) trajectory for a Cartesian manipulator.

Thus, the time intervals needed to carry out the motion along each coordinate component are

$$T_X = 2\sqrt{\frac{X_B - X_A}{a_X}} \ ; \quad T_Y = 2\sqrt{\frac{Y_B - Y_A}{a_Y}} \ ; \quad T_Z = 2\sqrt{\frac{Z_B - Z_A}{a_Z}} \ . \tag{9.26}$$

To provide a straight-line trajectory between points A and B, the condition

$$T_X = T_Y = T_Z \tag{9.27}$$

must be met. Obviously, this condition requires a certain relation between forces F_X, F_Y, and F_Z. For arbitrarily chosen values of the forces (i.e., arbitrarily chosen power of the drive), the trajectory follows the dotted line shown in Figure 9.8. In this case, for instance, the mass first finishes the distance $(Y_B - Y_A)$ bringing the system to point B' in the plane Y_B = constant; then the distance $(Z_B - Z_A)$ is completed and the system reaches point B''; and last, the section of the trajectory lies along a straight line parallel to the X-axis, until the gripper reaches final point B. Sections AB' and B'B'' are not straight lines. The duration of the operation, obviously, is determined by the largest value among durations T_X, T_Y, and T_Z. (In the example in Figure 9.8, T_Z is the time the gripper requires to travel from A to B.) This time can be calculated from the obvious expression (for the case of constant acceleration)

$$T_Z = 2\sqrt{\frac{Z_B - Z_A}{a_Z}} \ ; \tag{9.28}$$

substituting expression (9.25) into (9.28) we obtain:

$$T_Z = 2\sqrt{\frac{(Z_B - Z_A)m_Z}{F_{Z\,max}}} \ . \tag{9.29}$$

Here, $F_{Z\,max}$ = const.

Because of the lack of rotation, neither Coriolis nor centrifugal acceleration appears in the dynamics of Cartesian manipulators. The idealizing assumptions (as in the previous example) make the calculations for this type of manipulator much simpler.

9.3 Kinematics of Manipulators

This section is based largely on the impressive paper "Principles of Designing Actuating Systems for Industrial Robots" (Proceedings of the Fifth World Congress on Theory of Machines and Mechanisms, 1979, ASME), by A. E. Kobrinkskii, A. L. Korendyasev, B. L. Salamandra, and L. I. Tyves, Institute for the Study of Machines, Moscow, former USSR.

This section deals with motion transfer in manipulators. We consider here mostly Cartesian and spherical types of devices and discuss the pros and cons mainly of two accepted conceptions in manipulator design. The conceptions are:

- The drives are located directly on the links so that each one moves the corresponding link (with respect to its degree of freedom) relative to the link on which the drive is mounted;
- The drives are located on the base of the device and motion is transmitted to the corresponding link (with respect to its degree of freedom) by a transmission.

Obviously, in both cases the nature of the drives may vary. However, to some extent the choice of drive influences the design and the preference for one of these conceptions. For instance, hydraulic or pneumatic drives are convenient for the first approach. A layout of this sort for a Cartesian manipulator is shown in Figure 9.9. Here 1 is the cylinder for producing motion along vertical guides 2 (Z-axis). Frame 3 is driven by cylinder 1 and consists of guides 4 along which (X-axis) cylinder 5 drives frame 6. The latter supports cylinder 7, which is responsible for the third degree of freedom (movement along the Y-axis). By analyzing this design we can reach some important conclusions:

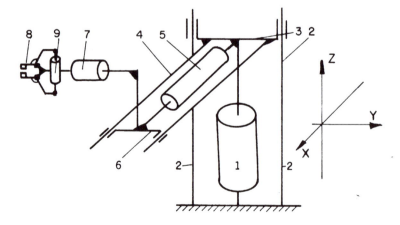

FIGURE 9.9 Cartesian manipulator with drives located directly on the moving links.

- More degrees of freedom can easily be achieved by simply adding cylinders, frames, and guides. In Figure 9.9, for example, gripper 8 driven by cylinder 9 constitutes an additional degree of freedom;
- The resultant displacement of the gripper does not depend on the sequence in which the drives are actuated;
- The power or force that every drive develops depends on the place it occupies in the kinematic chain of the device. The closer the drive is to the base, the more powerful it must be to carry all the links and drives mounted on it; every added drive increases the accelerated masses of the device;
- The drives do not affect each other kinematically. In the above example (Figure 9.9), this means that when a displacement along, say, the X-coordinate is made, it does not change the positions already achieved along the other coordinate axes.

These conclusions are, of course, correct regardless of whether the drives are electrically or pneumohydraulically actuated.

Let us consider the second conception. Figure 9.10 shows a design of a Cartesian manipulator based on the use of centralized drives mounted on base 1 of the device. Motors 2, 3, and 4 are responsible for the X, Y, and Z displacements, respectively. These displacements are carried out as follows: motor 2 drives lead screw 5, which engages with nut 6. This nut is fastened to carriage 7 and provides displacement along the X-axis. Slider 8, which runs along guides 9, is also mounted on carriage 7. Another slider 10 can move in the vertical direction (no guides are shown in Figure 9.10). The position of slider 10 is the sum of three movements along the X-, Y-, and Z-axes. Movement along the Y-axis is due to motor 3, which drives shaft 11. Sprocket 13 is mounted on this shaft via key 12 and engaged with chain 14. The chain is tightened by another sprocket 15, which freely rotates on guideshaft 16. The chain is connected to slider 8,

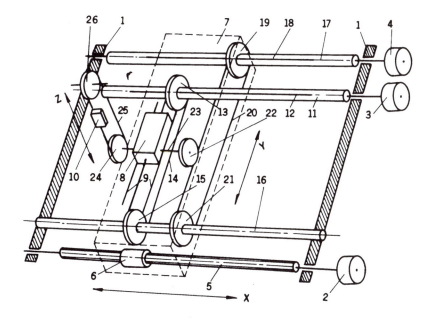

FIGURE 9.10 Cartesian manipulator with drives located on the base of the device and transmissions for motion transfer.

so that the latter is driven by motor 3. Motor 4 drives shaft 17 which also has key 18 and sprocket 19. The latter is engaged with chain 20, which is tightened by auxiliary sprocket 21 that freely rotates on guideshaft 16. Chain 20 is also engaged with sprocket 22 which, due to shaft 23, drives another sprocket 24. Shaft 23 is mounted on bearings on slider 8. Sprocket 24 drives (due to chain 25) slider 10, while another sprocket 26 serves to tighten chain 25. Sprockets 13 and 19 can slide along shafts 11 and 17, respectively, and keys 12 and 18 provide transmission of torques. Sprockets 15 and 21 do not transmit any torques since they slide and rotate freely on guideshaft 16. Their only task is to support chains 14 and 20, respectively. The locations of sprockets 13, 14, 19, and 21 are set by the design of carriage 7.

The following properties make this drive different from that considered previously (Figure 9.9), regardless of the fact that here electromotors are used for the drives. Here,

- The masses of the motors do not take part in causing inertial forces because they stay immobile on the base;
- One drive can influence another. Indeed, when chain 14 is moved while chain 20 is at rest, sprocket 22 is driven, which was not the intention. To correct this effect, a special command must be given to motor 4 to carry out corrective motion of chain 20, so as to keep slide 10 in the required position;
- The transmissions are relatively more complicated than in the previous example; however, the control communications are simpler. The immobility of the motors (especially if they are hydraulic or pneumatic) makes their connections to the energy source easy;
- Longer transmissions entail more backlashes, and are more flexible; this decreases the accuracy and worsens the dynamics of the whole mechanism.

The two conceptions mentioned in the beginning of this section are applicable also to non-Cartesian manipulators. Figure 9.11 shows a layout of a spherical manipulator, where the drives are mounted on the links so that every drive is responsible for the angle between two adjacent links. Figure 9.12 shows a diagram of the second approach; here all the drives are mounted on the base and motion is transmitted to the corresponding links by a rod system. Here, for both cases, each cylinder C_1, C_2, C_m, and C_{n-1} is responsible for driving its corresponding link; however, the relative positions of the links depend on the position of all the drives.

Let us consider the action of these two devices. First, we consider the design in Figure 9.11. The cylinders C_1, C_2, C_3, and C_4 actuate links 1, 2, 3, and 4, respectively. The cylinders develop torques T_1, T_2, T_3, and T_4 rotating the links around the joints between

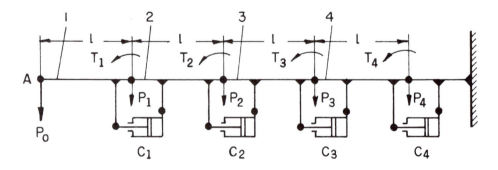

FIGURE 9.11 Spherical manipulator with drives located on the moving links.

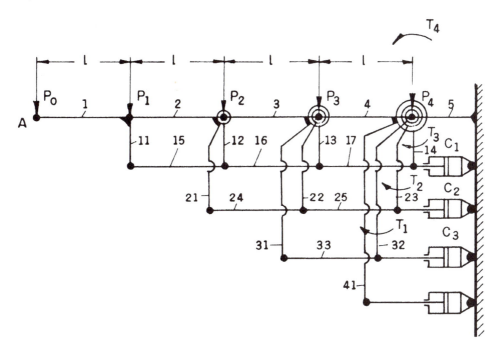

FIGURE 9.12 Spherical manipulator with drives located on the base and transmissions transferring the motion to the corresponding links.

them. To calculate the coordinates of point A (the gripper or the part the manipulator deals with), one has to know the angles ϕ_1, ϕ_2, etc., between the links caused by the cylinders (or any other drive). In Figure 9.13 we show the calculation scheme. Thus, we obtain for the coordinates of point A the following expressions:

$$x_A = l\left[\cos\phi_3 + \cos(\phi_3 - \phi_2) + \cos(\phi_3 - \phi_2 - \phi_1)\right],$$
$$Y_A = l\left[\sin\phi_3 + \sin(\phi_3 - \phi_2) - \sin(\phi_3 - \phi_2 - \phi_1)\right].$$

[9.30]

(These expressions are written for the assumption that the lengths of all links equal l.)

The point is that, to obtain the desired position of point A, we have to find a suitable set of angles $\phi_1, \phi_2, \dots \phi_n$, and control the corresponding drives so as to form these angles.

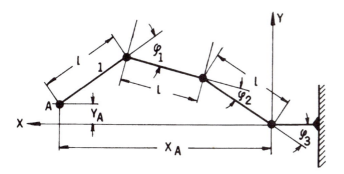

FIGURE 9.13 Kinematics calculation scheme for the design shown in Figure 9.11.

The design considered in Figure 9.12 acts in a different manner. Here cylinders C_1, C_2, C_3, and C_4 move the links relative to the bases via a system of rods and levers which create four-bar parallelograms. Thus, cylinder C_1 pushes rods 17, 16, and 15. The latter, through lever 11, moves link 1, while rods 12, 13, and 14 serve kinematic purposes, as a transmission. The latter are suspended freely on joints between links 2-3, 3-4, and 4-5. In the same manner, cylinder C_2 pushes rods 25 and 24. The latter moves link 2, through lever 21, while suspensions 22 and 23 form the kinematic chain. Cylinder C_3 pushes rod 33, actuating link 3 via lever 31, while rod 32 is a suspension. Link 4 is driven directly by cylinder C_4.

In Figure 9.14 we show the computation model describing the position of point A through the input angles ψ_1, ψ_2, and ψ_3, and intermediate angles ϕ_1, ϕ_2 and ϕ_3. Obviously, the intermediate angles describe the position of point A in the same manner as in the previous case because these angles have the same meaning. Therefore, Equations (9.30) also describe the position of point A in this case. However, these intermediate angles must be expressed through the input angles ψ_1, ψ_2, and ψ_3 which requires introducing an additional set of equations. In our example this set looks as follows:

$$\psi_3 = \phi_3 ,$$
$$\psi_2 = \phi_3 - \phi_2 , \qquad [9.31]$$
$$\psi_1 = \phi_3 - \phi_2 - \phi_1.$$

The position of point A is then described as

$$X_A = l\left[\cos\psi_3 + \cos\psi_2 + \cos\psi_1\right],$$
$$Y_A = l\left[\sin\psi_3 + \sin\psi_2 - \sin\psi_1\right]. \qquad [9.32]$$

Of course, this form of equations is true for links of equal lengths and transmissions equivalent to a parallelogram mechanism. Figure 9.15 shows a device with equivalent kinematics. Here, motors 1, 2, and 3 drive wheels 4, 5, and 6, respectively. The ratios of the transmissions are 1:1. Wheel 4 is rigidly connected to shaft 7 which drives link

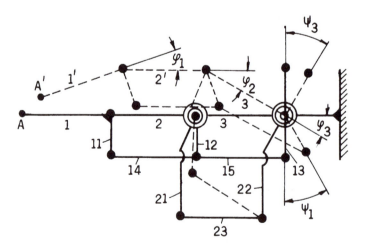

FIGURE 9.14 Kinematic calculation scheme for the design shown in Figure 9.12. In this figure, link 2 stays horizontal, which gives $\psi_2 = 0$.

I of the manipulator (i.e., frame 8). Wheels 5 and 6 rotate freely on shaft 7. Each of these wheels is rigidly connected with other wheels 9 and 10, respectively. Wheels 9 and 10 transmit motion to wheels 11 and 12, respectively; here also the ratios are 1:1. Wheel 11 is rigidly connected to shaft 13 and the latter drives frame 14 which constitutes link II. Wheel 12 rotates freely on shaft 13 and drives wheel 15, from which the motion goes to wheel 16, which drives shaft 17, i.e., link III.

Dependencies (9.31) can be rewritten in the following forms:

$$\psi_3 = \phi_3 \, ,$$

$$\psi_2 = \psi_3 - \phi_2 \, ,$$ [9.33]

$$\psi_1 = \psi_2 - \phi_1.$$

This means that changing either angles ψ_3 or ψ_2 changes the values of the other angles. This fact entails the necessity to correct these deviations and mutual influences by special control means. The latter connection is described in a general form by a matrix C' as follows:

$$
C' = \begin{vmatrix}
\dfrac{\delta\psi_1}{\delta\phi_1} & \dfrac{\delta\psi_1}{\delta\phi_2} & \cdots & \dfrac{\delta\psi_1}{\delta\phi_n} \\[2mm]
\dfrac{\delta\psi_2}{\delta\phi_1} & \dfrac{\delta\psi_2}{\delta\phi_2} & \cdots & \dfrac{\delta\psi_2}{\delta\phi_n} \\[2mm]
\vdots & \vdots & & \vdots \\[2mm]
\dfrac{\delta\psi_n}{\delta\phi_1} & \dfrac{\delta\psi_n}{\delta\phi_2} & \cdots & \dfrac{\delta\psi_n}{\delta\phi_n}
\end{vmatrix}
$$ [9.34]

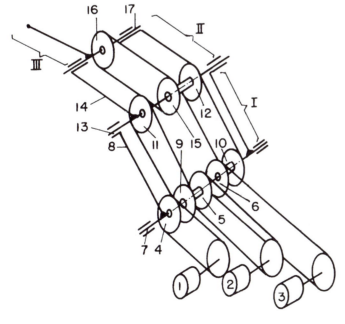

FIGURE 9.15 Design of kinematics for a manipulator according to the scheme shown in Figure 9.14.

Here, for instance, the elements of column number m ($m = 1, 2, \ldots, n$) represent kinematic ratios in the mechanism of a manipulator when all other angles ϕ (except ϕ_m) are held constant.

The dependence between increments of angles $\Delta\phi$ and drive angles $\Delta\psi$ for the approach in Figure 9.11 is

$$
\begin{vmatrix} \Delta\psi_1 \\ \Delta\psi_2 \\ \vdots \\ \Delta\psi_n \end{vmatrix} = \begin{vmatrix} 1 & 0 & \cdots & 0 \\ 0 & 1 & \cdots & 0 \\ \vdots & \vdots & & \vdots \\ 0 & 0 & \cdots & 1 \end{vmatrix} \begin{vmatrix} \Delta\phi_1 \\ \Delta\phi_2 \\ \vdots \\ \Delta\phi_n \end{vmatrix}. \tag{9.35}
$$

For the approach in Figure 9.12 this dependence has the form:

$$
\begin{vmatrix} \Delta\psi_1 \\ \Delta\psi_2 \\ \vdots \\ \Delta\psi_n \end{vmatrix} = \begin{vmatrix} 1 & 1 & \cdots & 1 \\ 0 & 1 & \cdots & 1 \\ \vdots & \vdots & & \vdots \\ 0 & 0 & \cdots & 1 \end{vmatrix} \begin{vmatrix} \Delta\phi_1 \\ \Delta\phi_2 \\ \vdots \\ \Delta\phi_n \end{vmatrix}. \tag{9.36}
$$

We will now illustrate an approach for evaluating the optimum choice of drive location: on the joints or on the bases. We make the comparison for the worst case when the links are stretched in a straight line (in this case, obviously, the torques the drives must develop are maximal) for the two models given in Figure 9.11 and Figure 9.12. We assume:

1. All links have equal weights P_0 and lengths l.
2. For the model in Figure 9.11 the weight of each driving motor P_d is directly proportional to the torque T developed by it in the form

$$
P_d = kT. \tag{9.37}
$$

3. The weight P_m of link number m together with the drive can be applied to the left joint.
4. The energy W or work consumed or expended by the whole system can be estimated as

$$
W = \sum_{m=1}^{n} T_m \Delta\phi_m. \tag{9.38}
$$

Here, $\Delta\phi_m$ = rotation of a link at joint m, for $m = 1, 2, \ldots, n$. We assume

$$
\Delta\phi_m = \Delta\phi.
$$

T_m = torque needed to drive link m.

Thus, we can write for (9.35),

$$
W = P_0 l \Delta\phi \sum_{j=0}^{n-1} C_{2n+1-j}^{j} (kl)^{n-1-j}. \tag{9.39}
$$

The weights P_m are described in terms of the above assumptions in the following form:

$$P_m = P_0 \left[1 + \sum_{j=0}^{m-1} C_{2m-j}^{j} (kl)^{m-j} \right].$$ [9.40]

Here, C = number of combinations.

For the second model (Figure 9.12) we assume, in addition:

5. That the weights P_i of each link together with the kinematic elements of transmission are proportional to the torque the link develops. Thus,

$$P_i = k|T_i| \quad i = 1, 2, \ldots, n-1.$$ [9.41]

Then the weight P_m of the link between joints m and $(m+1)$ is

$$P_m = P_0 + k \sum_{i=1}^{m} |T_i|.$$ [9.42]

6. The sum of the torques T for all n drives (n = the number of degrees of freedom) gives an indication of the power the system consumes for both approaches, and this sum can be expressed for the model in Figure 9.11 as:

$$T_\Sigma = \sum_{i=1}^{n} |T_i| = P_0 l \sum_{j=0}^{n-1} C_{2n+1-j}^{j} (kl)^{n-1-j},$$ [9.43]

and for the model in Figure 9.12 as

$$T_\Sigma = \sum_{i=1}^{n} |T_i| = P_0 l \sum_{j=0}^{n-1} C_{2n-j}^{j} (kl)^{n-1-j}.$$ [9.44]

We derive two conclusions from this last assumption:

- These sums of torques depend on the values (kl).
- The ratio $T_\Sigma / P_0 l$ describes the average specific power consumed by one link, and this ratio can serve as a criterion for comparing the two approaches.

Figure 9.16 shows a diagram of the relations between the ratio $T_\Sigma / P_0 l$ and the number of degrees of freedom n for different (kl) values. Curves 1 and 2 are for the layout in Figure 9.11, with stepping motors mounted on every joint (for this case $k \cong 0.2 - 0.35^1/\text{cm}$). It follows (high torque per link value) from these curves that it is not worthwhile to have more than three degrees of freedom in this type of device. Curves 3 and 4 belong to designs where hydro- or pneumocylinders are mounted at each joint. These solutions are suitable even for 6 to 8 degrees of freedom. Thus, for this number of degrees of freedom the designer has to use either the first approach with hydraulic or pneumatic drives or the second approach (Figure 9.12) with electric drives, which is more convenient for control reasons. (For pneumo- and hydraulic drives the value of $k \cong 0.004 - 0.13^1/\text{cm}$.) Curves 5 and 6 illustrate the limit situations for the first and second approaches, respectively, when $k = 0$.

The curves shown in Figure 9.16 also reflect the easily understandable fact that pneumo- or hydraulic cylinders develop high forces in relatively small volumes while

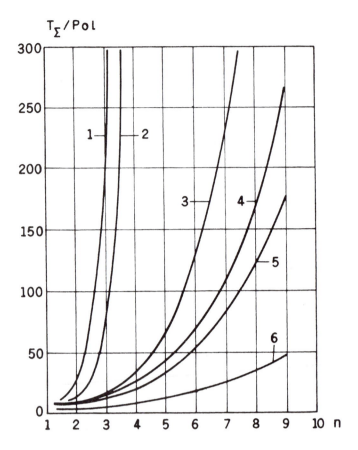

FIGURE 9.16   Specific driving torque versus the number of degrees of freedom of the manipulator being designed (see text for explanation).

electromotors usually develop high speeds and low torques. To increase the torque, speed reducers must be included, and this increases the masses and sizes of the devices. Special kinds of lightweight but expensive reducers are often used, such as harmonic, epicyclic or planetary, or wobbling reducers. Introduction of reducers into the kinematic chain of a manipulator entails the appearance of backlashes, which decrease the accuracy of the device. Recently, motors for direct drive have been developed. These synchronous-reluctance servomotors produce high torques at low speed. Thus, they can be installed directly in the manipulator's joints. This type of motor consists of a thin annular rotor mounted between two concentric stators and coupled directly with the load. Each stator has 18 poles and coils. The adjoining surfaces of the rotor and both stators are shaped as a row of teeth. When energized in sequence, these teeth react magnetically and produce torque over a short angle of about 2.4°. The high performance of these motors is a result of thin rotor construction, a high level of flux density, negligible iron losses in the rotor and stators, and heat dissipation through the mounting structure.

Thus, comparing the two approaches used for drive locations in manipulator systems, we can state that the first one has a simple 1:1 ratio between the angles ϕ and ψ, while the second approach suffers from the mutual influences of the angles ψ on the angles ϕ (see Equations (9.33–36)). On the other hand, the second approach is preferable from the point of view of the inertial forces, torques, and powers that the whole system consumes. How can we combine these two advantages in one design? Such a solution is presented schematically in Figure 9.17. The layout of the manipulator here copies that in Figure 9.12; however, special transmissions are inserted between cylinders C_1, C_2, C_3, and C_4. These transmissions consist of connecting rods 18, 26, 34, and 42, which transfer motion from cranks driven by wheels I, II, III, and IV. The latter are in turn driven by racks 19, 27, 35, and 43. These racks are actuated by cylinders C_1, C_2, C_3, and C_4, respectively, via differential lever linkages. This mechanism operates as follows: for instance, when cylinder C_1 is energized, its piston rod 120 pushes lever 121, through which connecting rod 122 is actuated. Obviously, the position of rack 19 depends not only on the displacement of the piston in cylinder C_1 but also on the position of rack 27, which dictates the position of lever 123. (The linkages we deal with here are spatial, as are the joints in them.) Thus, the motion of every piston rod affects the racks relative to other rods' positions.

As mentioned above, the solution given in Figure 9.17 is essentially only schematic. To make it more realistic, the design shown in Figure 9.18 is considered. Here, a seven-degrees-of-freedom spherical manipulator is presented. Link 1 rotates around axis X–X, and links 2, 3, and 5 rotate around axes 8, 9, and 10, respectively. In turn, links 4, 6, and 7 also rotate around axis X–X. Links 1 to 7 are driven by sprockets 11 to 17, due to chains C_1 to C_7 and sprockets 21 to 27, respectively. The latter are driven by motors (DC or stepping) M11 to M17, respectively, through differentials and gear transmissions shown in more detail in Figure 9.18b. Motor number M_n, drives bevel differential D_n. As this figure shows, the position of sprocket $2n$ in this case is determined by the positions of motor $M(n-1)$, via gears $3(n-1)$ and $4(n+1)$, which ensures that the positions of motors $M(n-1)$ and $M(n+1)$ also affect link n. Obviously, this is true for each value of n (from 1 to 7). Thus, sprocket 11 is responsible for link 1's rotation; sprocket 12, by means of a pair of bevel gears, drives link 2 (due to bevel gear wheel 51); sprocket 13 is responsible for rotation of link 3 (bevel gear 52); while sprocket 14 transfers motion to bevel 54, which rotates link 4 around axis X-X. The reader can follow the kinematic chain and figure out the transfer of movement to links 5, 6, and 7 in the same manner. At this point the reader must say "What a mess!"—which is right! This is why *usually*:

- The number of degrees of freedom does not exceed 6 and often is only 2 or 3;
- The drives are located in the joints—simplicity is "purchased" at the expense of "bad" dynamics;
- Compensation for the mutual influence of the links' displacements is made by software in the control system;
- Cylindrical Cartesian manipulators with pneumatic or hydraulic drives are used.

Figure 9.19 illustrates the wide possibilities permitted by the use of pneumatic drives. (The figure is based on an example produced by PHD, Inc., P.O. Box 9070, Fort Wayne, Indiana 46899.) The main idea is to construct the desired manipulator from standard modules. The figure presents nine examples of this type (the number of possibilities is theoretically infinite), which are various combinations of three types of mechanisms, forming manipulators of two, three, and four degrees of freedom. Three of the combinations are simple duplications of the same module (combinations AA, BB, and CC). These manipulators are both Cartesian and cylindrical. The use of pneu-'

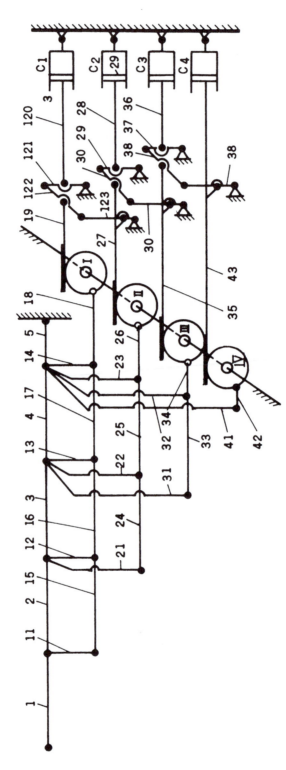

FIGURE 9.17 Layout of a spherical manipulator combining the advantages of both approaches, with the drives located on the base.

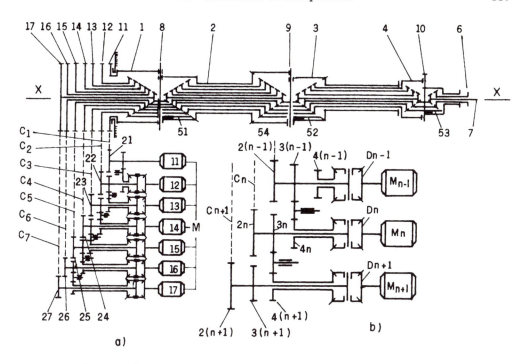

FIGURE 9.18   Design of a manipulator embodying the features of Figure 9.17. a) General layout of the seven-degree-of-freedom manipulator; b) Detail of the layout showing the connections between the drives number n-1, n, and n+1.

matic drives improves the dynamic properties of the manipulators (see Figure 9.16) for drives located on the links.

Another example is a design for a manipulator with two degrees of freedom, driven by two electric motors (preferably stepping motors) placed on the base of the device. The device is shown in Figure 9.20. On base 1 two geared bushings 3 and 4 are mounted on a pair of ball bearings 2. The inner surfaces of these bushings are threaded. One of the bushings, say 3, has a right-handed thread while the other (4) has a left-handed thread of the same pitch. These bushings, in turn, fit the threads on rod 5. (Rod 5 is threaded in two directions for its whole length.) On the end of this rod, arm 6 is fastened and serves for fastening a gripper or some other tool. This device operates in the following way: the gears of bushings 3 and 4 are engaged by transmissions with motors (not shown in Figure 9.20). A combination of the speeds and directions of rotation of the bushings forces rod 5 to perform a combined movement along and around the axis of rotation. For instance, driving bushings 3 and 4 in one direction with the same speed causes pure rotation of rod 5 at the same speed (remember, the bushings possess opposite threads). When the speed values of bushings 3 and 4 are equal but the directions of their rotation are opposite, pure axial movement of rod 5 will occur. (The bushings work in concert, pushing the rod the same distance for each revolution.) Every other response of the rod is mixed rotational and translational movement. This design has good dynamic properties because the motors' masses do not take part in the motion, and the rotating masses are concentrated along the axis of rotation, thus causing low moments of inertia.

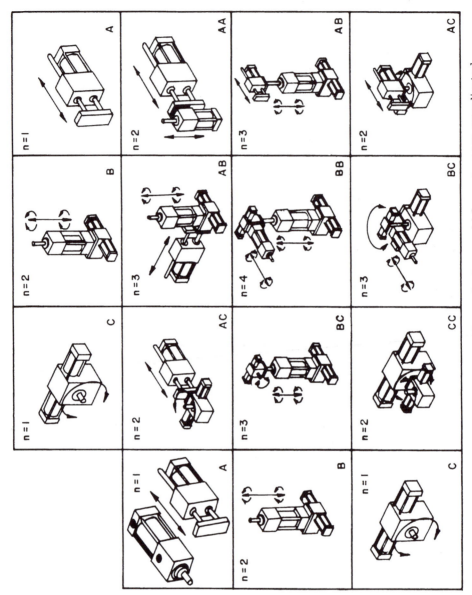

FIGURE 9.19 Combinations of simple pneumatic manipulators which create more complicated manipulators. Here n = the number of degrees of freedom.

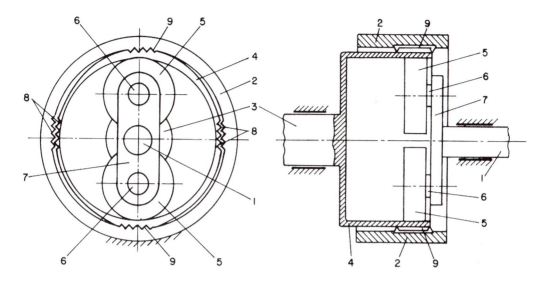

FIGURE 9.19a) Layout of a harmonic drive.

As was already mentioned, one of the most powerful means of providing high torque in small volumes and low weights is the use of the so-called harmonic drives. Such a drive is schematically shown in Figure 9.19a). It consists of driving shaft 1 on which transverse 7 is fastened. The latter holds two axes 6 on which rollers 5 are freely rotating. These rollers roll inside an elastic ring 4 fixed on the driven shaft 3 (located in the plane behind this figure). The outer perimeter of ring 4 is provided by teeth 8. Another rigid ring 2, fastened to the base, also has teeth on its inner surface.

The teeth of both rings are engaged at the points where rollers 5 deform the elastic ring 4 (see location 9). The number of teeth on ring 2 is close to that on ring 4—for instance, 252 teeth on the resting ring and 250 on the elastic one. This fact obviously results in rotation of shaft 3 for an angle corresponding to the difference in the number of teeth (in our case, two) during one revolution of driving shaft 1. Thus, the ratio achieved in our example is about 1:125.

A high-speed electric motor in concert with this kind of drive presents a compact, effective drive which is applicable to manipulators and which may be directly incorporated in its joints.

We finish this section with a concept for a manipulator drive which, to some extent, resembles biological muscles. An experimental device of this sort, built in the Mechanical Engineering Department of Ben-Gurion University of the Negev, is shown in Figure 9.21, and schematically in Figure 9.22. The "muscle" consists of elastic tube l sealed at the ends with corks 2. Ring 3 divides the tube into two (or more, with more rings) parts. Tube l is reinforced by longitudinal filaments. Thus, when inflated the tube will deform transversally, changing its diameter as shown in Figure 9.22b). The filaments provide the following relation between the section $l \cong L/2$ of the tube and its radius R when maximally inflated:

$$\pi R \cong l. \qquad [9.45]$$

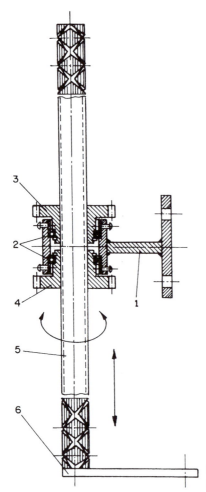

FIGURE 9.20 Low-inertia manipulator with two degrees of freedom.

FIGURE 9.20a) General view of the manipulator having two degrees of freedom, which is shown schematically in Figure 9.20. Built in the Mechanical Engineering Department of Ben-Gurion University. This device was patented by RoBomatex Company in Israel.

FIGURE 9.21   Photograph of the artificial muscle.

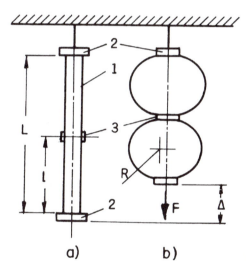

a)   b)

FIGURE 9.22   Artificial "muscle."

Thus, the maximum deformation Δ of the "muscle" is:

$$\Delta \cong L - 4R = L - \frac{4l}{\pi} \cong 2l\left(1 - \frac{2}{\pi}\right) = 0.726l. \tag{9.46}$$

In our experimental device the length L of the tube is about 170 mm; thus, the value of $\Delta \cong 60$ mm. In reality we do not reach this maximum deformation. Our experiments at a pressure of about two atmospheres gave a deformation of about 30 mm and a lifting force T of about 35 N. Figure 9.21 shows a photograph of the experimental set of "muscles." One of the muscles is inflated while the other is relaxed. Figure 9.23 shows the results of experimental measurements where the elongations L/L_0 versus the weight lifted by the muscle for different inflation pressures were determined. These characteristics are nearly linear. (Here, L_0 = initial length of the muscle under zero load at the indicated pressures, and L = length of the muscle at the indicated loads.)

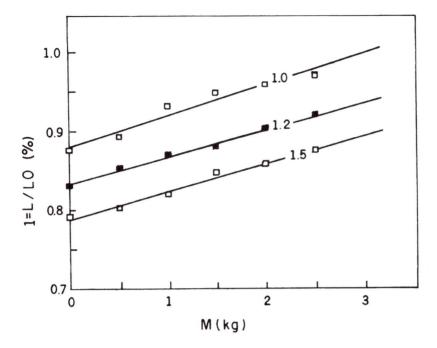

FIGURE 9.23 Elongation of the muscle measured at different inflation pressures (P = 1, 1.2, and 1.5 atm) while lifting weights of 0 to 3 kg.

This sort of drive (artificial muscle) is an interesting low-inertia device which is not widespread. However, it is worthwhile to develop and study it. Some companies and people involved in this work are "ROMAC" McDonald Detwiller & Associates of Richmond, British Columbia, Canada; SMA, Yoshiyuki Nakano, Masakatsu Fujie, and Yuji Hosada at the Mechanical Engineering Research Laboratory, Hitachi, Ltd., in Japan; and Brightston, Ltd., Great Britain.

Free vibrations of an arm of a manipulator (robot)

The limited stiffness of a real manipulator causes the appearance of harmful and undesired vibrations of the device. These vibrations reduce the accuracy of performance and productivity of the machine. In addition, due to these vibrations we observe a considerable increase of the dynamic loads acting on the system.

After the positioning of the arm is completed (in the absence of external forces) free vibrations of the mechanism occur. We now show how to estimate the parameters of these vibrations. These parameters are the natural frequency and the main shapes of the vibrations. Our explanation is based on an example given in Figure 9.23a). Here a manipulator consists of a base 1, two levers 2 and 3, and the end effector 4. The motors we consider stopped near the stable position of the manipulator.

We denote following concepts:

l = length of the levers;
m_1, m_2 = masses of the levers correspondingly;

FIGURE 9.23a)   Layout of a manipulator.

m = mass of the manipulated body;
φ_1 and φ_2 = deflection angles (generalized coordinates) of the levers from the ideal, programmed positions q_1 and q_2 correspondingly.

We consider the levers 1 and 2 to be thin, homogeneous rods. Obviously, we deal with a linear dynamic model of a two-mass system in which we must express the inertial and elastic coefficients. For the dynamic investigation purpose we use here the Lagrange equation in the usual form.

First step: writing the expressions for the kinetic energy:

$$T = T_1 + T_2 + T_m . \tag{9.46a}$$

Here

$$T_1 = \frac{1}{2}\frac{m_1 l^2}{3}\dot{\varphi}_1^2 ;$$

$$T_2 = \frac{1}{2}m_2\left[l^2\dot{\varphi}_1^2 + \frac{l^2}{4}\left(\dot{\varphi}_1 + \dot{\varphi}_2\right)^2 + 2l\dot{\varphi}_1\frac{l}{2}\left(\dot{\varphi}_1 + \dot{\varphi}_2\right)\cos q_2 \right]$$

$$+ \frac{1}{2}\frac{m_2 l^2}{12}\left(\dot{\varphi}_1 + \dot{\varphi}_2\right)^2 ;$$

$$T_m = \frac{1}{2}m\left[l^2\dot{\varphi}_1^2 + l^2\left(\dot{\varphi}_1 + \dot{\varphi}_2\right)^2 + 2l\dot{\varphi}_1 l\left(\dot{\varphi}_1 + \dot{\varphi}_2\right)\cos q_2 \right].$$

We neglect the members of third order of infinitesimality in these expressions and rewrite (9.46a) in the following form:

$$T \approx \frac{1}{2}\left(a_{11}\dot{\varphi}_1^2 + 2a_{12}\dot{\varphi}_1\dot{\varphi}_2 + a_{22}\dot{\varphi}_2^2 \right), \tag{9.46b}$$

and here are the inertial coefficients of the dynamic system, and they correspondingly are

$$a_{11} = l^2\left[\frac{m_1}{3} + m_2\left(\frac{4}{3} + \cos q_2 \right) + 2m\left(1 + \cos q_2 \right) \right];$$

$$a_{22} = l^2\left(\frac{m_2}{3} + m \right); \tag{9.46c}$$

$$a_{12} = l^2\left[m_2\left(\frac{1}{3} + \frac{1}{2}\cos q_2 \right) + m\left(1 + \cos q_2 \right) \right].$$

Stiffness can be introduced either as generalized forces

$$Q_r^c \approx -c_r \phi_r$$

or in the form of potential energy

$$V = \frac{c_r \varphi_r^2}{2},$$

where r is the number of the corresponding generalized coordinate. In this example $r = 1, 2$. Completing the procedure of writing the Lagrange equation we obtain the following dynamic equations system:

$$a_{11}\ddot{\varphi}_1 + a_{12}\ddot{\varphi}_2 + c_1\varphi_1 = 0,$$
$$a_{21}\ddot{\varphi}_1 + a_{22}\ddot{\varphi}_2 + c_2\varphi2 = 0. \tag{9.46d}$$

Further development is as usual; the solutions are sought in the form

$$\varphi_1 = B_1 \cos(kt + \mu), \qquad \varphi_2 = B_2 \cos(kt + \mu), \tag{9.46e}$$

where B_1, B_2, k, and μ are unknown parameters, depending upon the initial conditions. Substituting (9.46e) into Equations (9.46d) we obtain a system of algebraic equations

$$\left(c_1 - a_{11}k^2\right)B_1 - a_{12}k^2 B_2 = 0,$$
$$-a_{12}k^2 B_1 + \left(c_2 - a_{22}k^2\right)B_1 = 0. \tag{9.46f}$$

Nonzero solutions of these equations must be determined from the following equation:

$$\begin{vmatrix} c_1 - a_{11}k^2 & -a_{12}k^2 \\ -a_{12}k^2 & c_2 - a_{22}k^2 \end{vmatrix} = \left(a_{11}a_{22} - a_{12}^2\right)k^4 - \left(a_{11}c_2 - a_{22}c_1\right)k^2 + c_1 c_2 = 0. \tag{9.46g}$$

Solving this equation with respect to k we find two values for free vibration frequencies as follows:

$$k_{1,2}^2 = \frac{a_{11}c_2 + a_{22}c_1 \pm \sqrt{\left(a_{11}c_2 + a_{22}c_1\right)^2 - 4c_1 c_2 \left(a_{11}a_{22} - a_{12}^2\right)}}{2\left(a_{11}a_{22} - a_{12}^2\right)}. \tag{9.46h}$$

The next and the last step in this calculation is the determination of the vibration modes ξ_{rm} and the estimation of the order of the vibration amplitudes. Corresponding to the definition we have

$$\xi_{rm} = B_{rm} / B_{1m}$$

where r = number of the mode and m = number of the frequency k_m.

From the system of algebraic equations (9.46d), we respectively obtain

$$\xi_{21} = a_{12}k_1^2 \left(c_2 - a_{22}k_1^2\right)^{-1},$$
$$\xi_{22} = a_{12}k_2^2 \left(c_2 - a_{22}k_2^2\right)^{-1}. \tag{9.46i}$$

The general solution of the dynamic equations then is

$$\varphi_1 = B_{11} \cos(k_1 t + \mu_1) + B_{12} \cos(k_2 t + \mu_2),$$

$$\varphi_2 = B_{11}\xi_{21} \cos(k_1 t + \mu_1) + B_{12}\xi_{22} \cos(k_2 t + \mu_2). \tag{9.46j}$$

The parameters B_{11}, B_{12}, M_1, M_2 are determined through the initial conditions:

$$t = 0,$$
$$\varphi_1 = \varphi_{10},$$
$$\varphi_2 = \varphi_{20},$$
$$\dot{\varphi}_1 = \dot{\varphi}_{10}, \text{ and}$$
$$\dot{\varphi}_2 = \dot{\varphi}_{20}.$$

Thus, the free vibrations of the arm close to the average position consist of two oscillation processes with two different and usually not commensurable frequencies k_1 and k_2, which, in turn, means that in general these oscillations are not of a periodic nature. For further analyses it is convenient to use the so-called main or normal coordinates z_m which are defined in the following form:

$$\varphi_r = \sum_{m=1}^{2} \xi_{rm} z_m \qquad (r, m = 1, 2).$$

For any initial conditions these new variables change periodically, monoharmonically:

$$z_m = D_m \cos(k_m t + \psi_m) \quad (m = 1, 2),$$

where the constants D_m and ψ_m are determined by the initial conditions.

   An example follows. Supposing $m_1 = m_2 = m$ and $c_1 = c_2 = c$; we rewrite (9.46c) in the form

$$a_{11} = ml^2 \left[\frac{11}{3} + 3\cos q \right];$$

$$a_{22} = ml^2 \frac{4}{3}; \tag{9.46k}$$

$$a_{12} = ml^2 \left[\frac{4}{3} + \frac{3}{2}\cos q \right].$$

For $q = 0$ this becomes

$$a_{11} = ml^2 \frac{20}{3};$$

$$a_{22} = ml^2 \frac{4}{3}; \tag{9.46l}$$

$$a_{12} = ml^2 \frac{17}{6}.$$

For the natural frequencies in this case we obtain

$$k_1^2 = 0.1267 c/ml^2; \quad k_2^2 = 9.164 c/ml^2; \quad \text{and} \quad K_1 \cong 0.356 \frac{1}{l}\sqrt{\frac{c}{m}}; \quad K_2 = 3.027 \frac{1}{l}\sqrt{\frac{c}{m}}.$$

For $q = \pi/2$:

$$a_{11} = ml^2 \frac{11}{3};$$

$$a_{22} = ml^2 \frac{4}{3}; \qquad\qquad [9.46m]$$

$$a_{12} = ml^2 \frac{4}{3}.$$

For the natural frequencies in this case we obtain

$$k_1^2 = 0.2341\, c/ml^2, \qquad k_2^2 = 1.373\, c/ml^2.$$

The coefficients describing the shape of the oscillations are correspondingly

$$\xi_1 \begin{pmatrix} \xi_{11} = 1 \\ \xi_{21} = 0.4537 \end{pmatrix}; \quad \xi_2 \begin{pmatrix} \xi_{12} = 1 \\ \xi_{22} = -2.203 \end{pmatrix}.$$

It is possible to obtain an approximate estimation for the initial deformations appearing in the manipulator under discussion which determine the amplitudes of the free oscillations after the motors are stopped. For this purpose we write kinetostatical equations of acting torques Q_1 and Q_2.

$$Q_1 = m\varepsilon_0 \frac{l}{2}\frac{l}{2} + \frac{ml^2}{12}\varepsilon_0 + m\varepsilon_0 \frac{3l}{2}\frac{3l}{2} + \frac{ml^2}{12}\varepsilon_0 + m\varepsilon_0\, 2l\, 2l = \frac{20}{3}ml^2\varepsilon_0;$$

$$\qquad\qquad [9.46n]$$

$$Q_2 = m\varepsilon_0 \frac{3l}{2}\frac{l}{2} + \frac{ml^2}{12}\varepsilon_0 + m\varepsilon_0\, 2ll = \frac{17}{6}ml^2\varepsilon_0.$$

This is done with an assumption that the angle $q = 0$ and that angular acceleration before the stop was ε_0 = constant. From (9.46m) follows for the initial amplitudes correspondingly

$$\varphi_1 = \frac{Q_1}{c} = \frac{20}{3}\frac{ml^2}{c}\varepsilon_0; \qquad \varphi_2 = \frac{Q_2}{c} = \frac{17}{6}\frac{ml^2}{c}\varepsilon_0.$$

Some simple dependences for evaluating the behavior of a robot's arm can be helpful. We denote:

ϕ_{max} = maximal value of the oscillations amplitude;
J = moment of inertia with respect to the rotating shaft of the vibrating body;
q_{max} = maximal angle the arm travels;
t = travelling time;
k = frequency of the oscillations.

Then we can use the following approximations:

$$\varepsilon_0 \cong 4q_{max}/t^2,$$

$$\phi_{max} \cong J\varepsilon_0/c \cong 4Jq_{max}/t^2 c, \qquad\qquad [9.46o]$$

$$\phi_{max}/q_{max} \cong 10^{-4} \geq 4/t^2 k^2.$$

An illustration of the process of vibrations calculated for Case (9.46l) is given in Figure 9.23a).

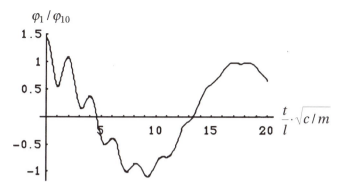

$$\varphi_1 / \varphi_{10}$$

$$\frac{t}{l} \cdot \sqrt{c/m}$$

FIGURE 9.23b)   Oscillations of the two-freedom degree manipulator.

fi1 = Plot[Cos[.356 t]+.454 Cos[3.027 t] Exp[-.15 t],{t,0,20}]

Robots with parallel connections of links

We consider here some aspects of the principles used in the device called the Stewart platform (SP) in regard to manipulators. We begin with a description of the main properties of this device.

The definition of SP is: *two rigid basic bodies connected by six rods (noncomplanar and noncolinear) with variable lengths.*

Such a structure is shown in Figure 9.23c) and possesses the following main properties:

1. This system has six degrees of freedom;
2. One-to-one unambiguous connection between the lengths of the rods and the mutual positions of the two bodies;
3. Two main problems may be formulated in this regard:
   a. The length of the rods is given—what is the relative position of the bodies? (FKT—Forward Kinematic Transform.)
   b. The relative position of the bodies is given—what is the length of the rods needed for this situation? (IKT—Inverse Kinematic Transform.)

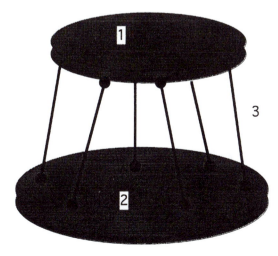

FIGURE 9.23c)   Layout of the Stewart platform: 1) upper body; 2) lower body; 3) variable rods.

4. The length of the rods is changed in parallel, i.e., each rod is controlled independently relative to the base, which makes this kinematic solution preferable because of the lack of cumbersome kinematic chains.

The vectors appearing in Figure 9.23d) are defined as follows:

Q = vector between a certain point in the basic coordinate system and another certain point on the platform. This vector represents one of the six rods connecting the two bodies comprising the mechanism;

R_0 = vector defining the begining of the coordinates of the platform relative to the basic coordinate system;

R = vector defining the position of the point belonging to the platform (the point the rod is fastened to the platform) in the basic coordinate system;

p = vector of the same point in the coordinate system of the platform;

b = vector of the point in the basic coordinate system (the lower fastening point of the rod); and

K = cosine matrix.

Then the connection between all these vectors looks as follows:

$$Q = K \cdot p + R_0 - b. \tag{9.46p}$$

Now we define K:

$$K \equiv \begin{vmatrix} C\psi \cdot C\theta & -S\psi \cdot C\phi + C\psi \cdot S\theta \cdot S\phi & S\psi \cdot S\phi + C\psi \cdot S\theta \cdot C\phi \\ S\psi \cdot C\theta & C\psi \cdot C\phi + S\psi \cdot S\theta \cdot S\phi & -C\psi \cdot C\phi + S\psi \cdot S\theta \cdot C\phi \\ -S\theta & C\theta \cdot S\phi & C\theta \cdot C\phi \end{vmatrix} \equiv \begin{vmatrix} r_{11} r_{12} r_{13} \\ r_{21} r_{22} r_{23} \\ r_{31} r_{32} r_{33} \end{vmatrix}. \tag{9.46q}$$

Here:

S = sin and C = cos of a corresponding angle;

ϕ = rotation angle relative to the axis x (roll);

θ = rotation angle relative to the axis y (pitch);

ψ = rotation angle relative to the axis z (azimuth).

Now it is possible to solve the IKT which is the real case in most applications of this kind of mechanism-as-a-manipulator (robot) for manipulating parts for processing or other purposes. We then need to calculate the length of the rods l_i ($i = l, \ldots, 6$) knowing

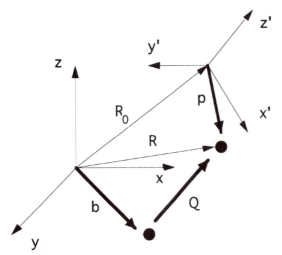

FIGURE 9.23d) Coordinates describing the position of the platform relative to the base.

the desired position of the platform. This is done, for instance, in the following way using the above-shown Expressions (9.44q):

$$l_i = |Q_i| = \sqrt{q_{ix}^2 + q_{iy}^2 + q_{iz}^2}$$
$$Q_i = \{q_{ix}, q_{iy}, q_{iz}\}$$
$$q_{ix} = r_{11} \cdot p_{ix} + r_{12} \cdot p_{iy} + r_{13} \cdot p_{iz} + x - b_{ix}$$
$$q_{iy} = r_{21} \cdot p_{ix} + r_{22} \cdot p_{iy} + r_{23} \cdot p_{iz} + y - b_{iy}$$
$$q_{iz} = r_{31} \cdot p_{ix} + r_{32} \cdot p_{iy} + r_{33} \cdot p_{iz} + z - b_{iz}.$$

[9.46r]

In the photograph shown in Figure 9.23e) there is an embodiment of a Stewart platform where the rods are pneumatic cylinders. Controlling the positions of the piston rods correspondingly to the calculated values l_i, we obtain the desired location and orientation of the platform.

(By the way, the FKT is more unpleasant. The expressions we get in this case are nonlinear and the equations have a number of formal solutions, which makes the procedure of finding the practical one complicated enough.)

The structure of the SP has a wide potential of creative possibilities in theoretical, design, and application domains. For instance, the micro domain of applications opens interesting theoretical and design alternatives. For small displacements of parts, especially when we deal with dimensions of the order of 10^{-4}, 10^{-7} m, the description of the movement can be simplified. The cosine matrix in this case is

$$K \equiv \begin{bmatrix} 1 & -\psi + \theta \cdot \phi & \psi \cdot \phi + \theta \\ \psi & 1 + \psi \cdot \theta \cdot \phi & -1 + \psi \cdot \theta \\ -\theta & \phi & 1 \end{bmatrix}.$$

[9.46s]

Another field for new SP applications occurs when combinations of these devices are investigated. One such idea given in Figure 9.23f), interesting for the robotics field, is to create a "trunk"-like structure by using a series of SP for robotics applications.

FIGURE 9.23e)   Embodiment of a Stewart platform built in the Mechanical English Department of Ben-Gurion University (Israel).

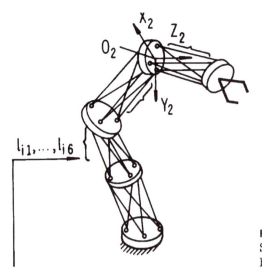

$l_{i1}, \ldots, l_{i6}$

FIGURE 9.23f) Idea of a "trunk" made of Stewart-platform–like elements. This idea belongs to Dr. A. Sh. Kiliskor.

9.4 Grippers

In previous sections we have discussed the kinematics and dynamics of manipulators. Now let us consider the tool that manipulators mainly use—the gripper. To manipulate, one needs to grip and hold the object being manipulated. Grippers of various natures exist. For instance, ferromagnetic parts can be held by electromagnetic grippers. This gripping device has no moving parts (no degrees of freedom and no drives). It is easily controlled by switching the current in the coil of the electromagnet on or off. However, its use is limited to the parts' magnetic properties, and magnetic forces are sometimes not strong enough. When relatively large sheets are handled, vacuum suction cups are used; for instance, for feeding aluminum, brass, steel, etc., sheets into stamps for producing car body parts. Glass sheets are also handled in this way, and some printing presses use suction cups for gripping paper sheets and introducing them into the press. Obviously, the surface of the sheet must be smooth enough to provide reliability of gripping (to seal the suction cup and prevent leakage of air and loss of vacuum). Here, also, no degrees of freedom are needed for gripping. The vacuum is switched on or off by an automatically controlled valve. (We illustrated the use of such suction cups in the example shown in Figure 2.10.)

Grippers essentially replace the human hand. If the gripping abilities of a mechanical five-finger "hand" are denoted as 100%, then a four-finger hand has 99% of its ability, a three-finger hand about 90%, and a two-finger hand 40%.

We consider here some designs of two-fingered grippers. In the gripper shown in Figure 9.24, piston rod 1 moves two symmetrically attached connecting links 2 which in turn move gripping levers 3, which have jaws 4. (Cylinder 5 can obviously be replaced by any other drive: electromagnet, cable wound on a drum driven by a motor, etc.) The jaws shown here are suitable for gripping cylindrical bodies having a certain range of diameters. Attempts to handle other shapes or sizes of parts may lead to asymmetrical gripping by this device, because the angular displacements of jaws may not be parallel. To avoid skewing in the jaws, solutions like those shown in Figure 9.25a) or b) are used. In Case a) a simple cylinder 1 with piston 2 and jaws 3 ensures parallel

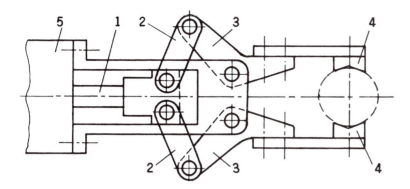

FIGURE 9.24 Design of a simple mechanical gripper.

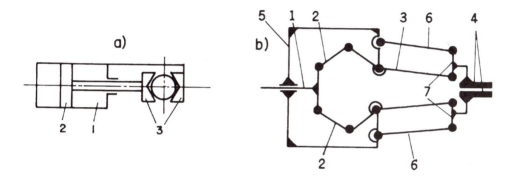

FIGURE 9.25 Grippers with translational jaw motion.

displacement of the latter. In case b) a linkage as in Figure 9.24, but with the addition of connecting rods 6 and links 7 with attached jaws 4, provides the movement needed. These additional elements create parallelograms which provide the transitional movement of the jaws.

Various other mechanical designs of grippers are possible. For instance, Figure 9.26 shows possible solutions a) and b) with angular movement of jaws 1, while cases c) and d) provide parallel displacement of jaws 1. In all cases the gripper is driven by rod 2. All the cases presented in Figure 9.26 possess rectilinear kinematic pairs 3. Introduction of higher-degree kinematic pairs are shown in Figure 9.27. In case a) cam 1 fastened on rod 2 moves levers 3 to which jaws 4 are attached. Spring 5 ensures the contact between the levers and the cam. In case b) the situation is reversed: cams 1 are fastened onto levers 3 and rod 2 actuates the cams, thus moving jaws 4. Spring 5 closes the kinematic chain. In case c), which is analogous to case b), springs 5 also play the role of joints. In case d) the higher-degree kinematic pair is a gear set. Rack 1 (moved by rod 2) is engaged with gear sector 3 with jaws 4 attached to them. Cases a) to d) have dealt with angular displacement of jaws. In case e) we see how the addition of parallelograms 5 (as in the example in Figure 9.25b)) to the mechanism shown in Figure 9.27d) makes the motion of the jaws translational. The last two cases do not need springs, since the chain is closed kinematically.

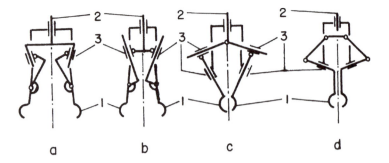

FIGURE 9.26   Designs of grippers using low-degree kinematic pairs.

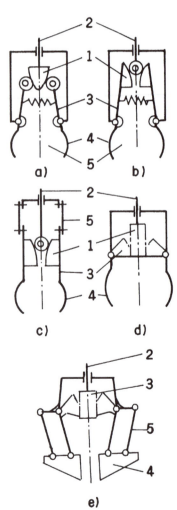

FIGURE 9.27   Designs of grippers using high-degree kinematic pairs.

To describe these mechanisms quantitatively we use the relationships between:

1. Forces F_G which the jaws develop, and the force F_d which the driving rod applies; and

2. The displacements S_d of the driving rod and the jaws of the gripper S_G.

Figure 9.28 illustrates these parameters and graphically shows the functions $S_G(S_d)$ and $F_G/F_d = f(S_d)$ for a gripper.

This discussion of grippers has been influenced by the paper by J. Volmer, "Technische Hochschule Karl-Marx-Stadt, DDR, Mechanism fur Greifer von Handhabergeraten," Proceedings of the Fifth World Congress on Theory of Machines and Mechanisms, 1979, ASME. We should note that the examples of mechanical grippers discussed above permit a certain degree of flexibility in the dimensions of parts the gripper can deal with. This property allows using these grippers for measuring. For instance, by remembering the values of S_d by which the driving rod moves to grip the parts, the system can compare the dimensions of the gripped parts.

When the manipulated parts are relatively small and must be positioned accurately, miniaturization of the gripper is required. A solution of the type shown in Figure 9.29 can be recommended, for example, in assembly of electronic circuits. Here, the gripper

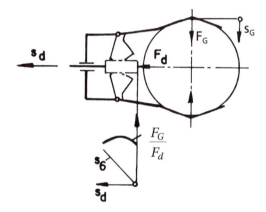

FIGURE 9.28 Characteristics of a mechanical gripper.

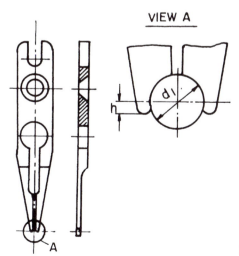

VIEW A

FIGURE 9.29 Miniature elastic gripper.

is a one-piece tool made of elastic material that can bend and surround the gripped part, of diameter *d,* to create frictional force to hold the part, and then to release it when it is fastened on the circuit board. The overlap $h \cong 0.2d$ serves this purpose.

Three-fingered grippers are also available (or can be designed for special purposes). Figure 9.30 shows a concept of a three-fingered gripper. Part a) presents a general view and part b) shows a side view. Here, 1 is the base of the gripper and 2 the driving rod, which is connected by joints and links to fingers 3. When rod 2 moves right, the fingers open, and when it moves left, they close. This gripper (as well as some considered earlier) can grip a body from both the outside and the inside. (Such grippers are produced by Mecanotron Corporation, South Plainfield, New Jersey, U.S.A.)

One of the most serious problems that appears in manipulators equipped with different sorts of grippers is control of the grasping force the gripper develops. Obviously, there must be some difference between grasping a metal blank, a wine glass, or an egg, even when all these objects are the same size. This difference is expressed in the different amounts of force needed to hold the objects and (what is more important) the limited pressure allowed to be applied to some objects. Figure 9.31 shows a possible solution for handling tender, delicate objects. Here, hand 1 is provided with two elastic

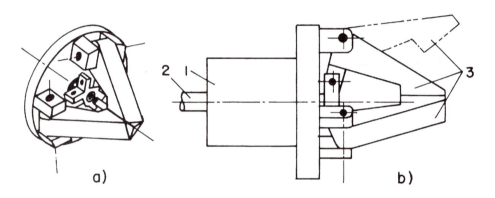

FIGURE 9.30   Three-fingered gripper.

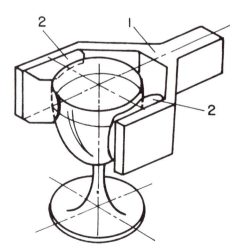

FIGURE 9.31   A soft gripper for grasping delicate objects.

pillows 2. When inflated by a controlled pressure, they develop enough force to hold the glass, while keeping the pressure on it small enough to prevent damage. (The small pressure creates considerable holding force due to the relatively large contact area between the glass and the pillows.) It is a satisfying solution when modest accuracy of positioning is sufficient.

A more sophisticated approach to the problem of handling delicate objects is the Utah-MIT dextrous hand which is described in the *Journal of Machine Design* of June 26, 1986. This is a four-fingered hand consisting of three fingers with four degrees of freedom and one "thumb" with four degrees of freedom. The "wrist" has three degrees of freedom. The thumb acts against the three fingers. Thus, the hand consists of 16 movable links driven by a system of pneumatically operated "tendons" and 184 low-friction pulleys. The joints connecting the links include precision bearings. The problem of air compressibility is overcome by use of special control valves. Figure 9.32a) shows a general design of one finger. Here links A, B, and C can rotate around their joints. The space inside the links is hollow and contains the pulleys and the tendons, which go around the pulleys and are fastened to the appropriate links. Figure 9.32b) shows the drive of link C. Tendons I and Ia run around pulleys 7, 8, and 9 and are fastened to the center of pulley 6. Thus, pulling tendons I and Ia causes bending and straightening of link C. Figure 9.32c) shows the control of link B by tendons II and IIa, and Figure 9.32d) shows the control of link A by tendons III and IIIa. A pair of tendons IV and IVa are used for turning the whole finger around the X–X axis, as shown in Figure 9.32e). The Utah-MIT hand has 16 position sensors and 32 tendon-tension sensors. Thus its grasping force can be controlled, and the object handled by the gripper with a light or heavy touch.

For simpler grippers (as in Figures 9.24, 9.28, and 9.30), force-sensitive jaws can be made as shown in Figure 9.33. Here, part 1 is grasped by jaws 2 which develop grasping force F_G. The force is measured by sensor 3 located in base 4 which connects the gripper with drive rod 5. The latter moves rack 6 and the kinematics of the gripper. Force F_d, which is developed by rod 5, determines grasping force F_G. Sensor 3 enables the desired ratio F_G/F_d to be achieved. The sensor can be made so as to measure more than one force, say, three projections of forces and torques relative to a coordinate axis.

These devices help to control the grasping force; however, its value must be predetermined (before using the gripper) and the system tuned appropriately. Serious efforts are being devoted to simulating the behavior of a human hand, which "knows" how to learn the required grasping force during the grasping process itself. This ability of a live hand is due to its tactile sensitivity. Next, we consider some concepts of artificial tactile sensors installed inside the gripper's fingers or jaws. Figure 9.34 illustrates a design for a one-dimensional tactile sensor. Jaws 1 develop grasping force F_G which must cause enough frictional force F_u (vertically directed) to prevent object 2 from falling due to gravitational force P. The sensor consists of roller 3 mounted on shaft 4 by means of bearings. Shaft 4 is mounted on jaw 1 by flat spring 5, which presses roller 3 against object 2 through a window in the jaw. When $F_u < P$, slippage occurs between the gripper and object, and the object moves downward for a distance X, thus rotating roller 3 (see the arrow x in the figure). This rotation is translated into electric signals (say, pulses, due to an encoder located between shaft 4 and the inner surface of hollow roller 3), which cause the control system to issue a command to increase force F_G until the slippage stops (but no more than that, to prevent any damage to the object). In

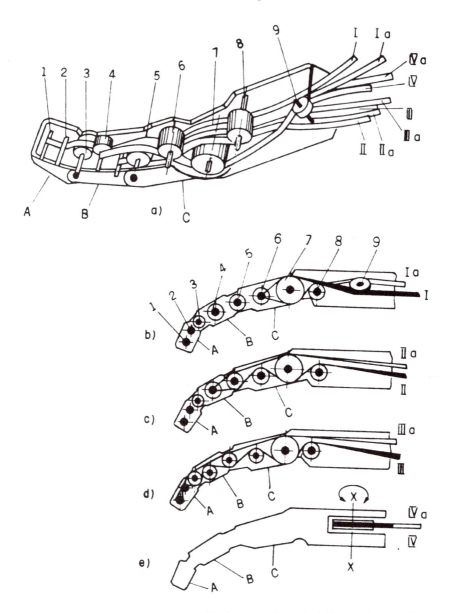

FIGURE 9.32 Design of the Utah-MIT dextrous hand: a) General view of one finger; b) Drive of link C; c) Drive of link B; d) drive of link A; e) Turning around the X-X axis.

addition, the control system also gives a command to lift the gripper for a distance Y to compensate for the displacement X due to the slippage.

For two-dimensional compensation, the concept shown in Figure 9.35 can be proposed. Here conducting sphere 1 (instead of a roller) is used. The surface of this sphere is covered with an insulating coating in a checkered design. Three (at least) contacts 2, 3, and 4 touch the sphere and create a circuit in which a constant voltage V energizes the system. When slippage occurs between object 5 and the gripper, the sphere

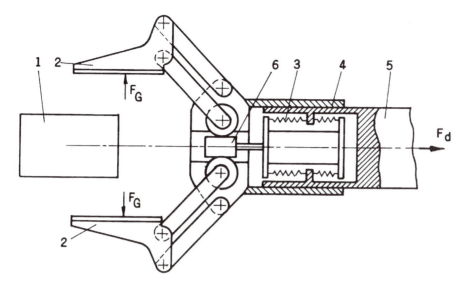

FIGURE 9.33 Design of a grasp-force-sensitive gripper.

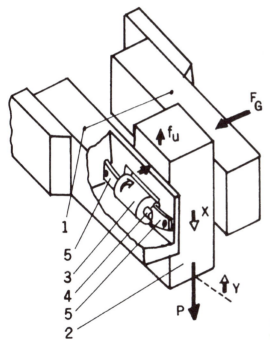

FIGURE 9.34 One-dimensional tactile sensor.

rotates and voltage pulses V_1 and V_2 correspond to the direction of the slippage vector S relative to the X- and Y-coordinates. A layer of soft material 6 is used to protect the sphere from mechanical damage.

The jaws or fingers discussed in this section can be provided with special inserts and straps to better fit the specific items the grippers must deal with. For handling tools like drills, cutters, probes, etc., the straps must go round their shaft and provide

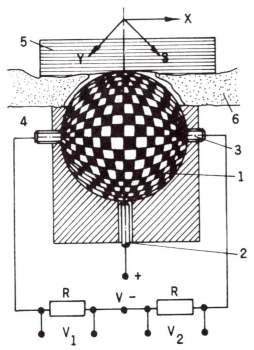

FIGURE 9.35 Two-dimensional tactile sensor.

accuracy and reliability of grasping. The same idea is used for making the jaws correspond to other specific shapes, dimensions, and materials of items being processed. Special devices can be considered for holding exchangeable grippers, say, to replace a two-finger gripper with a three-finger one during the processing cycle, which may be effective in some cases.

9.5 Guides

The problem of designing guides is mainly specific for X–Y tables which, according to our classification, belong to Cartesian manipulators with two degrees of freedom. However, the concept of guides can be generalized and applied more broadly (except for translational movement) also to polar or rotating elements as well as to spiral guides (screws). Guides must provide:

- Stable, accurate, relative disposition of elements;
- Accurate performance of relative displacements, whether translational or angular;
- Low frictional losses during motion;
- Wear resistance for a reasonable working lifetime;
- Low sensitivity to thermal expansion (and compression) to maintain the required level of accuracy.

These properties must be achieved within the limits of reasonable expense and technical practicality. The designer faces contradictory conditions in trying to meet these requirements. In certain cases the weight of the structure must be minimized, e.g., for moving links such as manipulator links. For accuracy, the guides must be rigid to prevent deflections. For heavier loads, the area of contact between the guide and the moving part must be larger. To prevent excess wear, the guides must apply low pressure to the moving part, which also entails a certain width of the guide and length of the support (to create the required contact area). It is important to mention that, above all, wear of the guides depends on the maintenance and operating conditions. Wear varies from 0.02 mm per year for good conditions to 0.2 mm per year for careless operation. We discuss here some ideas and concepts for overcoming some of these technical obstacles.

Figure 9.36 shows a typical example of a Cartesian guide system for a lathe and the scheme of forces acting in the mechanism. Guides 1 along axis X–X (main guides of the bed shown in projection b)) and guides 2 along axis Y–Y in dovetail form (its cross section is shown in projection a)) direct the support 4 of cutter 3. The cutter develops force P at the cutting point. Decomposition of this force yields its three components P_x, P_y, and P_z. Together with the weight G of the moving part, these forces cause the guides to react with forces A, B, and C in the Z–Y plane and frictional forces f_A, f_B, and f_C along the X-axis (when movement occurs). Statics equations permit finding the reactive forces A, B, C, and Q:

$$\Sigma X = 0; \quad \Sigma Y = 0; \quad \Sigma Z = 0;$$
$$\Sigma T_x = 0; \quad \Sigma T_y = 0; \quad \Sigma T_z = 0. \tag{9.47}$$

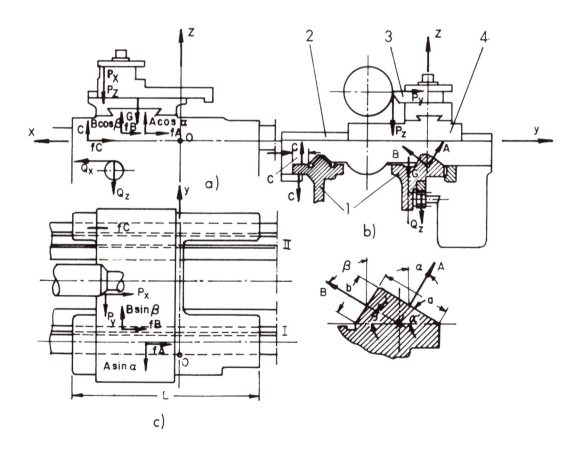

FIGURE 9.36 Two-dimensional Cartesian guide system and forces acting in it.

Here, X, Y, and Z are components of acting forces, and T_X, T_Y, T_Z are components of acting torques. (Two other equations and one additional condition permit figuring out the coordinates X_A, X_B, and X_C where the forces are applied, but we do not consider this calculation here.) When A, B, and C are defined, the corresponding pressures can be calculated:

$$P_A = \frac{A}{aL}; \quad P_B = \frac{B}{bL}; \quad P_C = \frac{C}{cL}. \tag{9.48}$$

Here a, b, c, and L are geometrical dimensions of the guides and are shown in Figure 9.36.

The obtained pressure values are average values, and the real local pressure might not be uniformly distributed along the guides. The allowed maximum pressures depend on the materials the guides are made of and their surfaces, and are about 300 N/cm² for slow-moving systems to 5 N/cm² for fast-running sliders. Obviously, the lower the pressure, the less the wear and the thicker the lubricant layer and, as a result, smaller frictional forces f_A, f_B, and f_C appear in the mechanism.

Figure 9.37 shows some common shapes of heavy-duty translational guides. The prismatic guides in cases a) and b) are symmetrically shaped and those in cases c) and d) are asymmetrical. Cases b) and d) are better for holding lubricant; however, these shapes collect dirt of various kinds, which causes increased wear. In contrast, cases a) and c) have less ability to hold lubricant, but do not suffer from trapped dirt. Cases a) and f) are dovetail-type guides. This type of guide can be used not only for guiding horizontal movement (like cases a), b), c), and d)) but also for vertical or even upside-down orientation of the slider. The difference between cases e) and f) is the pairs of mating surfaces: lower and side surfaces in case e) and upper and side surfaces in case f). Case e) is more expensive to produce but easier to lubricate, while case f) is easier to produce but worse at holding lubricant. Rectangular guides—cases g) and h)—are cheaper and simpler to produce and also provide better precision. However, this shape is worse for lubrication and is sensitive to dirt, especially when the dirt is metal chips which scratch the surface, causing wear and increasing friction. The cylindrical guides in cases i) and j) have the same properties as the prismatic guides but are simpler to produce.

To provide the required level of precision and smoothness of action, special devices are used to decrease play. Figure 9.38 illustrates some common means of backlash adjustment. Cases a), b), and c) show rectangular guides. In case a) vertical and hori-

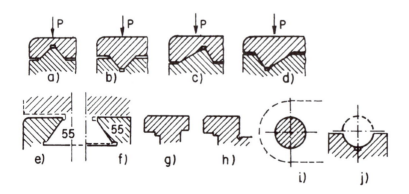

FIGURE 9.37 Cross sections of translational guides.

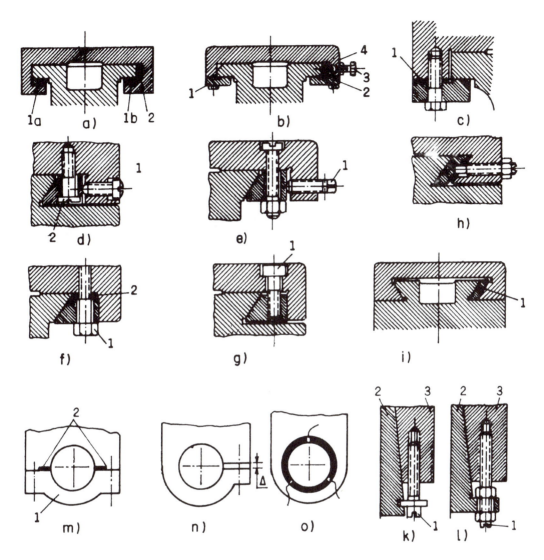

FIGURE 9.38 How to decrease play and adjust backlash in translational guides to the required values: a), b), c) Flat rectangular guides; d), e), f), g), h), i) Dovetail guides; m), n), o) Cylindrical guides; k), l) Wedges for adjustment.

zontal backlash is eliminated by wedges 1a, 1b, and 2, respectively. Purely horizontal movement, case b), can be controlled with only one wedge 4, while vertical play is taken up by straps 1 and 2. Sometimes spacers 1 (case c)) are used for more precise limitation of backlash. The wedges are usually mounted with special bolts or screws (3 in Figure 9.38b or as shown in Figures 9.38k) and l)). Screwing (or unscrewing) bolts 1 moves wedge 2 in the desired direction relative to housing 3, closing or opening the gap. In Figures 9.38d), e), f), g), and h) are shown various ways to adjust the wedges via bolts 1 and spacers 2. For dovetail guides (case i)), only one wedge 1 is needed to solve the play problem. To control play in cylindrical guides (case m)), strap 1 with spacers 2 can be used, or an elastic design with a bolt closing gap Δ (as in case n)), or a split conical bushing 1 (case o)).

A serious problem arises when these frictional guides are used, which is associated with frictional forces and leads to not only driving power losses but also (and often this is more important) limited accuracy. It is worthwhile to analyze this problem in greater depth. Frictional force F_F appearing in a slide pair depends on the speed of relative motion x, as shown in Figure 9.39. This means that, when the speed is close to 0, the frictional force F_{ST} is higher than it is at faster speeds. Thus,

$$\dot{x} = \dot{x}_{ST} \cong 0; \quad F < F_{ST};$$
$$\dot{x} > \dot{x}_{ST}, \quad F_{ST} \geqslant F > F_{din}. \tag{9.49}$$

Here F is the driving force, and F_{din} is the frictional force at the final sliding speed.

This can be analyzed further with the help of Figure 9.40. Mass M of the slider is driven by force F through a rod with a certain stiffness c. (This can be and often is a lead screw, piston rod, rack, etc.) From the layout in Figure 9.40 it follows that the mass essentially does not move until F reaches F_{ST}. This entails deformation x_{ST} of the rod, which can be calculated as

$$x_{ST} = F_{ST} / c. \tag{9.50}$$

At the moment when movement begins ($\dot{x} > 0$), mass M is under the influence of a composite moving force:

$$c(x_{ST} - x) - F_{din}. \tag{9.51}$$

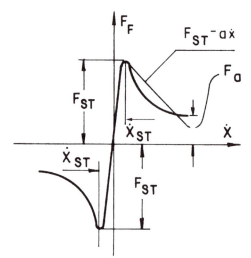

FIGURE 9.39   Frictional force versus speed.

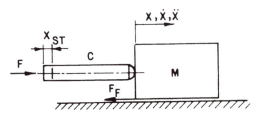

FIGURE 9.40   Calculation model for friction as in Figure 9.39.

It thus follows that, even if at that moment F is changed to 0, some displacement of the mass will take place. An equation approximately describing this movement and taking Expression (9.51) into account is

$$M\ddot{x} + cx = cx_{ST} - F_{din}.$$ [9.52]

Force F_{din} is a function of $\dot{x}$. Let us suppose that it is justified to express this function in the following way (see Figure 9.39):

$$F_{din} = F_{ST} - a\dot{x}.$$ [9.53]

Then we can rewrite and simplify Equation (9.52) as follows:

$$\ddot{x} - \frac{a}{M}\dot{x} + \frac{c}{M}x = 0.$$

(Here Expression 9.50 is substituted; therefore 0 appears on the right side.) The solution has the form

$$x \cong e^{\alpha t}[A\cos\omega t + B\sin\omega t].$$ [9.54]

By substituting this solution into Equation (9.53), we obtain the following expressions for α and ω:

$$\alpha = \frac{a}{2M}; \quad \omega \cong \sqrt{\frac{c}{M}}.$$ [9.55]

Under the initial conditions (when $t = 0$) the displacement $x = x_{ST}$, and speed $\dot{x} = 0$. So we obtain for the coefficients A and B

$$A = x_{ST}; \quad B = -\frac{ax_{ST}}{2M\omega}.$$ [9.56]

Thus, finally, the solution is

$$x = x_{ST}e^{\alpha t}\left[\cos\omega t - \frac{a}{2M\omega}\sin\omega t\right].$$ [9.57]

For instance, for $M = 100$ kg, c $= 10^4$ N/cm, $F_{ST} = 100$ N and $a = 1$ Nsec/m, we find from (9.50) that

$$x_{ST} = 10^{-4}\,\text{m}$$

and from (9.55) that

$$\alpha \cong 0.005^1/\sec, \quad \omega \cong 100^1/\sec.$$

The ratio $z = x/x_{ST}$ is shown in Figure 9.41 as a function of time. An analytical approximation expressing the dependence between the friction force F_F and the sliding speed

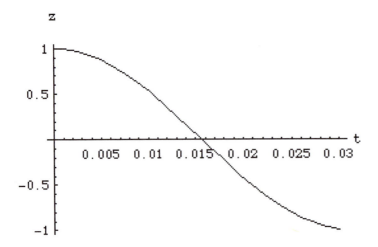

FIGURE 9.41   Motion-versus-time diagram from the calculation model shown in Figure 9.40.

V may be convenient in engineering applications. This approximation may have the following form:

$$F_F = \frac{1}{1+\exp[1/V]} - a - bV.$$

For $x(t)$ as displacement we have $V = \dot{x}(t)$.

When using computer means, for example, MATHEMATICA, we can simplify this computation by introducing this approximation for describing the friction versus speed behavior of the slider as follows:

q2=Plot[200 (((1 + E^v^(-1))^(-1)-.5)-.05 v),{v,-5,5}]

Figure 9.41a. shows the form of the "friction force versus speed" dependence which is close to the experimentally gained results.

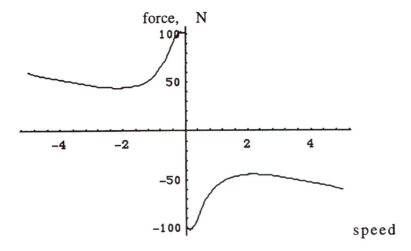

FIGURE 9.41a)   Friction force versus speed dependence using the above-given approximation; here a = 0.5 and b = 0.05.

In MATHEMATICA language we ask the numerical solution of the motion equation of the slider in a following form:

j10=NDSolve[{100 y"[t]+200 ((1 + E^y'[t]^(-1))^(-1)-.5-.05 y'[t])+
10^6 (y[t])==0,y[0]==0,y'[0]==0.0001},y,{t,0,.1}]

And for graphical representation of the displacement of the slider we have:

b10= Plot[Evaluate[y[t]/.j10],{t,0,.031},
AxesLabel->{"t","y[t]"},PlotRange->All]

For the speed of the slider we then obtain:

g10=Plot[Evaluate[y'[t]/.j10],{t,0,.031},AxesLabel->{"t","y'[t]"},PlotRange->All]

In Figures 9.41b) and c) the calculated results for displacement and speed of the driven mass are shown. The initial data are the same as in the manually calculated

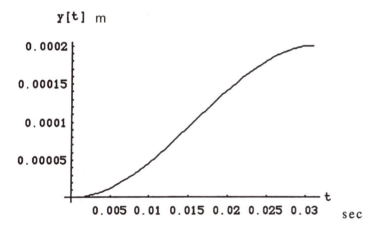

FIGURE 9.41b)   Displacement of the slider during one period of its motion.

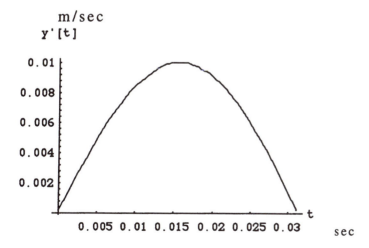

FIGURE 9.41c)   Development of the motion speed of the slider during the same period.

example. The graph shown in Figure 9.41. considers approximately half of the period of the displacement.

The model presented in Figure 9.40 also describes roughly the behavior of pneumo- or hydro-cylinders and lead screws. Here, the pistons and their rods behave according to the above explanation. This entails decreased accuracy of the whole system in which these drives are installed. All together (guides, cylinders, lead screws) cause limited reproducibility of manipulators. This is explained in Figure 9.42. Link 1 of the manipulator driven by cylinder 2 must repeatedly travel from point A to B. Mirror 3 is fastened to link 1 close to the joint. A light beam from laser source 4 hits this mirror and is reflected onto screen 5, thus amplifying any deviations of point B from its desired position x. A histogram of the x values is shown schematically. The desired value x indicates accurate positioning of link 1 at point B. The actual positions deviate from this desired value in a statistically random manner, as shown in the histogram. The conclusions we derive from this explanation and simplified example are:

- The described dynamic phenomenon means that the control system of the device cannot limit the movement of the driven mass within tolerances of less than about 0.01 mm;
- To increase accuracy and improve control sensitivity, frictional forces must be reduced. The smaller the value F_{ST}, the better is the performance of the mechanism.

The first means of reducing friction is to use rolling supports. Figure 9.43 presents a cross section of a rolling guide. This device guides the movement of slider 1 in the horizontal plane by means of two ball bearings 2 and 3, which are fastened onto shafts 4 and 5, respectively. Shaft 5 is made eccentric so that, by rotating pin 6, one can adjust the value of the play between the bearings and horizontal guide 7. In the vertical plane the rolling is carried out by balls 8 which are placed in a corresponding groove made in base 9 of the device (only one slot is shown in this figure).

Figure 9.44 shows a cylindrical rolling guide. Directed rod I is supported by balls 2 located in bushing 3. Spacer 4 is used to keep the balls apart. It is obvious that when the rod moves for a distance l, part 4 travels for $l/2$. This fact causes complications, especially where space must be conserved. Then, another concept can be proposed, as shown in Figure 9.45. Here moving body 1 (say, a rod) is supported by a row (or several rows) of balls 2, which run in closed-loop channels 3. Thus, no additional length

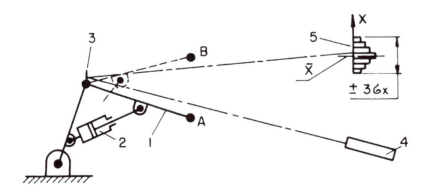

FIGURE 9.42   Reproducibility of manipulator link movement.

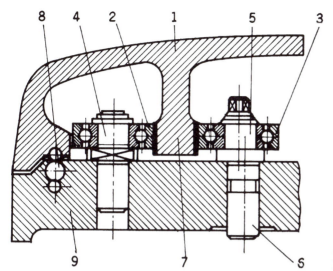

FIGURE 9.43 Design of a heavy-duty rolling support.

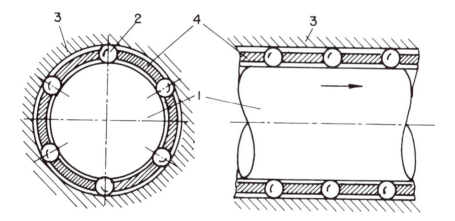

FIGURE 9.44 Cylindrical rolling support with a separator holding the balls.

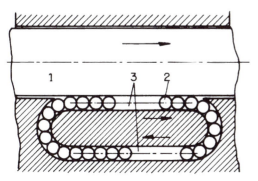

FIGURE 9.45 Rolling support with free-running balls and a channel for returning the balls to the supporting section.

is required (it does require additional width). This concept is useful for heavier-duty guides such as the dovetail table shown in Figure 9.46. Table 1 travels between bars 2 and 3 on rollers kept in separator 4. Play in the system is adjusted by screw 5. Shields 6 and 7 keep the guides clean.

Rolling guides have much lower friction than sliding guides, and therefore the F_{ST} values are much smaller. However, these guides employ more matching surfaces: between the housing and the rolling elements, and between the rolling elements and the moving part. In addition, deviations in the shapes and dimensions of the rolling elements affect the precision, and such guides have an accuracy ceiling of about 10^{-6} m. (Their load capacity is lower than that of sliding guides.)

The above discussion with regard to the effect of friction on accuracy can be extended also to lead screws. The model shown in Figure 9.40 is also suitable for the behavior of screw-nut pairs. Figure 9.47 presents a design for a lead screw and nuts, with rolling balls to minimize friction between the thread of the screw and that of the nut. When rolling along the thread, the balls enter the channels and are pushed back to the beginning of the thread in the nut. Figure 9.47 shows two such nuts. Obviously, the profile of the thread must match the running balls. By combining this kind of lead screw with, say, stepping motors, relatively high-precision performances can be achieved.

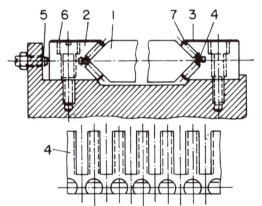

FIGURE 9.46 Dovetail rolling support: a) General view; b) Separator to keep rollers apart.

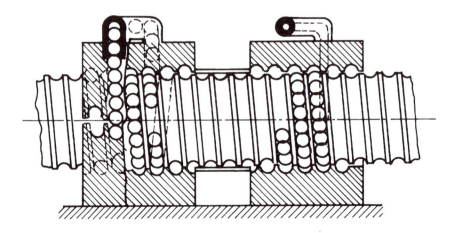

FIGURE 9.47 Rolling lead screw.

For almost complete elimination of friction, air-cushioned guides have recently been implemented. A top view of a schematic air-cushioned *X–Y* Cartesian manipulator is shown in Figure 9.48. Part 2 is supported on a granite table 1 by three air-cushion supports a, b, and c. Air-cushion supports d and e facilitate the motion of part 2 (together with part 3) along the *X*-axis. Part 3 also is supported by three air-cushion supports f, g, and h. Air-cushions i and k aid the motion along the *Y*-axis. This device is controlled by motors developing driving forces P_X and P_Y, while the feedback monitor that provides information about the real positions of parts 2 and 3 is usually an interferometer (see Figure 5.9).

The cushion of air is created by elements shown schematically in Figure 9.49. Each is about 3″ to 4″ in diameter. Compressed air (about 50 psi) is blown through channels 1 surrounding contact channel 2, where a vacuum is supplied. The ratio of the pressures in channels 1 and 2 is automatically controlled so as to provide an air layer with a stable and accurate thickness. The accuracy of this device is about 0.0001″.

The machines recently developed by ASET (American Semiconductor Equipment Technologies Company, 6110 Variel Ave., Woodland Hills, CA 91367) have achieved even higher accuracy which reaches 0.0001 mm or 0.00004″.

An attentive reader may ask at this point, "Well, air cushions a b, and c supporting part 2, and f, g, and h supporting part 3, act against gravitational force. Against which forces do air-cushions d, e, i, and k act?" What force pushes bodies 2 and 3 to the corresponding walls? A possible answer is a magnetic field that helps keep the bodies on track.

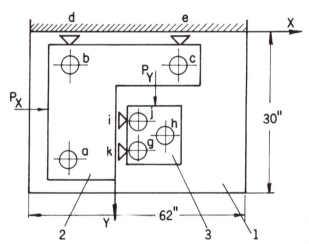

FIGURE 9.48 Air-cushion-supported *X-Y* Cartesian table.

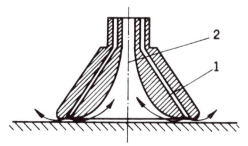

FIGURE 9.49 Air-cushion nozzle.

Another means to reduce frictional forces to practically zero is electrodynamic levitation. This is a phenomenon where a metallic (electroconductive or ferromagnetic) item is kept suspended by the interaction of magnetic fields. This kind of suspension has been applied to rapid trains that travel almost at aviation speeds. A design of such a suspension is shown in Figure 9.50. Above an electromagnet fed by alternating current with a frequency of 50–60 Hz is suspended aluminum disc 6, which is about 300 mm in diameter, and with its edges bent upwards. The gap Δ depends on the power consumed by the system. Magnet 1 consists of two cores 3 and 2 and two opposition coils 5 and 4 connected in phase. The left side of the figure shows the magnetic field without the disc while the right side shows it in the presence of disc 6.

Until now, we have discussed means to reduce or completely diminish friction (by putting the electromagnet levitation in vacuum, one achieves practically zero friction). Now we consider a design for guides where the frictional force is nearly linearly dependent on the speed (complete lubricational friction). Thus,

$$F_F = a\dot{x}. \tag{9.58}$$

The motion equation for the mass driven by force F takes the following form instead of (9.52):

$$M\ddot{x} + a\dot{x} = F. \tag{9.59}$$

(Here the deformation of the rod shown in Figure 9.40 is neglected.) For initial conditions

$$t = 0, \quad \dot{x} = 0, \tag{9.60}$$

the solution is (similar equations were solved in Chapter 3)

$$\dot{x} = \frac{FM}{a}\left[1 - e^{-(a/M)t}\right]. \tag{9.61}$$

This expression indicates the following facts:

- The smaller the acting force F, the smaller is the speed of mass M.
- The movement is smooth and begins from the very moment that the force is applied.

How to realize this condition of complete lubricational friction? One example is shown in Figure 9.51. Here, on plate 1, channels are drilled through which lubricant under high pressure is introduced so that part 2 is kept moving on a layer of liquid.

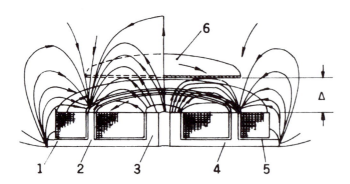

FIGURE 9.50 Design of an electromagnetic levitation device.

Another way to achieve this condition is shown in the plan in Figure 9.52. The guides here are two rapidly rotating shafts 1 on which slider 2 is located. Normal force N creates friction between the slider and the shafts, and force F causes its movement along the shafts. As a result, frictional forces act on slider 2 and are directed opposite to its relative movement on the shafts. The frictional force vectors F_V are directed opposite to the sliding speed vectors V. By analyzing the design shown in Figure 9.52, the following dependencies can be derived:

$$\tan\alpha = V_N / V_T. \qquad [9.62]$$

For very high speed V_T, when $V_T \gg V_N$, we can rewrite (9.62) as

$$\alpha \cong V_N / V_T. \qquad [9.63]$$

On the other hand,

$$F = 2F_V \tan\alpha \cong 2F_V\alpha = 2F_V V_N / V_T. \qquad [9.64]$$

For $V_T = $ const and $F_V = fN = $ const, Expression (9.64) can be rewritten

$$F = AV_N. \qquad [9.65]$$

Here $A = 2fN/V_T$, and $f = $ frictional coefficient.

This effect was mentioned in Chapter 5, where acceleration sensors were discussed (see Figure 5.29).

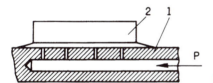

FIGURE 9.51 Full-lubrication guide.

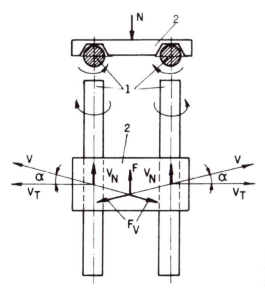

FIGURE 9.52 Full-lubrication friction conditions achieved by purely mechanical means.

9.6   Mobile and Walking Robots

A separate book could be dedicated to mobile and walking robots. However, a short review of mobile robots, including some walking problems, seems to be necessary to complete this book. First some ideas for wheeled mobile systems will be considered. The simplest concept is a three-wheeled bogie (truck, cart) such as in Figure 9.53a). Two wheels 1 rotate in one plane (parallel to the longitudinal axis symmetry of the bogie) and a third wheel 2 is placed in steering fork 3. The device is usually provided with a battery 4. Several alternatives for driving this bogie exist. For instance, Figure 9.53b) shows a plan where wheels 1 are driven by motors 5 so that, by controlling speeds V_1 and V_2, the direction of travel is determined. Thus, steering fork 3 automatically takes the correct direction and wheel 2 rolls by friction. Another alternative, shown in Figure 9.53c), uses wheel 2 as the driving one. Motor 6 is installed for this purpose. The direction of the bogie is determined by steering fork 3 which is driven, say, by special motor 7 controlled by the control unit. In this case wheels 1 roll freely.

A three-wheeled bogie has the advantage of theoretical stability. Three points determine a flat plane; thus, three wheels are stable on every surface. However, this bogie can be overturned by force F applied to corner A or B. Thus, for this design some load restrictions exist. A four-wheeled bogie, as shown in Figure 9.54, does not suffer from this disadvantage. This bogie consists of frame 1, four steering forks 2, energy source 3, and control unit 4. To make this device more maneuverable, all four wheels 5 can turn. In view a) the bogie is arranged for travelling straight ahead, while in view b) the behavior of the device depends on the direction of the wheels' rotation. When they all rotate in one direction and stay strictly parallel, the device moves sideways. When the pairs of wheels rotate in opposite directions, the device rotates in place around point 0. (The wheels in this case must be oriented tangentially to a circle with radius R.)

The advantages of a three-wheeled device, combined with greater maneuverability, are found in the Stanford Research Institute robot vehicle (Figure 9.55). The vehicle

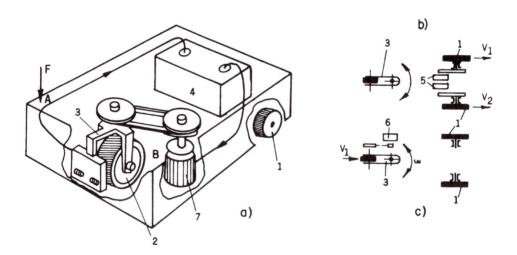

FIGURE 9.53   Three-wheeled bogie: a) General view; b) Two-wheel drive: c) Drive of the steering wheel.

FIGURE 9.53d)   General view of a three-wheeled cart that automatically follows a white stripe drawn on the floor. This device
corresponds to that shown schematically in Figure 9.53a). This
vehicle was designed and built in The Mechanical Engineering
Department of Ben-Gurion University and is used in the Robotics
teaching laboratory.

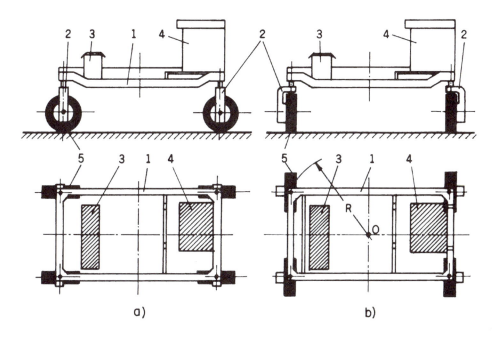

FIGURE 9.54   Four-wheeled bogie: a) Wheels in position for moving in longitudinal
direction; b) Wheels in position for travelling in transverse direction or turning in place.

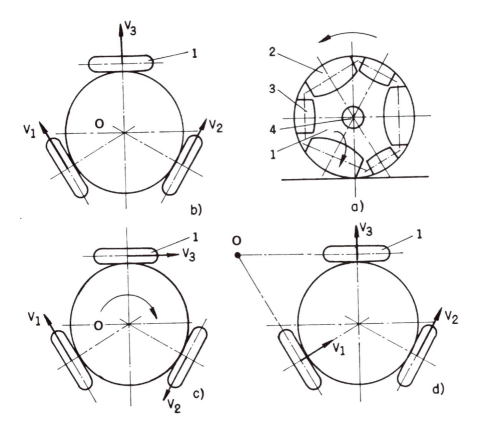

FIGURE 9.55 Stanford three-wheeled bogie: a) The wheel; b) Running along a straight line; c) Turning around center O; d) Running along a curved path.

has axial symmetry and is provided with three specially designed wheels 1. These wheels each consist of six barrel-like rollers 2 and 3 (Figure 9.55a)) rolling free in a frame fastened onto shaft 4. Obviously, rollers 2 and 3 rotate in the plane perpendicular to the rotational plane of the wheels. When two of the wheels are driven as shown in Figure 9.55b) with equal speeds V_1 and V_2 and the third wheel is immobile, the vehicle moves in the direction V_3. (The barrel-like rollers do not resist sideways movement of a wheel.) When all three wheels are driven as shown in Figure 9.55c) so that $V_1 = V_2 = V_3$, the vehicle turns around center O. When one wheel is driven with speed V_2 and the other wheels are braked, the vehicle travels around point O as illustrated in Figure 9.55d. Here, the stopped wheels roll in the directions perpendicular to their planes. In the intermediate cases, when the wheels are driven at different speeds, the motion of the vehicle will respond correspondingly.

All wheeled vehicles or bogies require specially prepared areas to function properly. Wheels are not usually adequate for moving across rugged terrain. The well-known solution for such purposes is the caterpillar-tracked vehicle. Such a vehicle is diagrammed in Figure 9.56. It consists of body 1 where the energy source, engines, and control unit are located, and two tracks 2 which are driven at speeds V_1 and V_2. Changing these speeds changes the travelling direction of the vehicle.

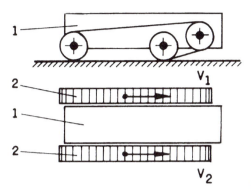

FIGURE 9.56   Caterpillar-driven vehicle.

How can the automatic navigation of mobile robots be arranged? The problem of flexible travelling was already mentioned in Chapter 1 (see Figure 1.28), where the problem of an automatic waiter was briefly discussed. Photosensors can be used for such purposes. Then, for instance, white strips must be painted on the floor between the tables and the sensors will control the vehicle so that it follows the strips. In addition, another sensor counts the number of intersections of strips, while a program in the system controls the steering wheel. A wheel-revolution counter stops the vehicle when a certain distance has been travelled. Thus, the vehicle reaches the required table. Instead of painted guides, metallic strips or wires can be installed under the floor or carpet. Then metal-sensitive sensors will work as navigation devices.

Next, we briefly consider a concept permitting independent navigation of a vehicle without interaction with the floor or anything else. Let us try to develop an algorithm that permits describing the position of the vehicle relative to some immovable coordinate system X, Y (with initial point 0) through the number of revolutions of the wheels and, say, the angle of the steering fork. (We assume that no slippage between the wheel and the floor occurs.) Thus, by continually calculating these two values. the control system knows the location of the vehicle at all moments. From Figure 9.57a) it follows that:

$$Y = V \int_0^t \sin \phi(t)\, dt,$$
$$X = V \int_0^t \cos \phi(t)\, dt,$$

[9.66]

where $V=$ constant speed of the vehicle, $\phi(t) =$ variable (in the general case) angle of the tangent to the vehicle's trajectory S, and $t =$ time. To complete the algorithm, angle ϕ must be expressed in terms of the steering angle ψ. From Figure 9.57b) it follows that

$$L/r = \tan \psi(t),$$

[9.67]

where L is a constant parameter for each vehicle.

In changing ψ, one changes the radius r of the trajectory S. Thus,

$$r(t) = \frac{L}{\tan \psi(t)}.$$

[9.68]

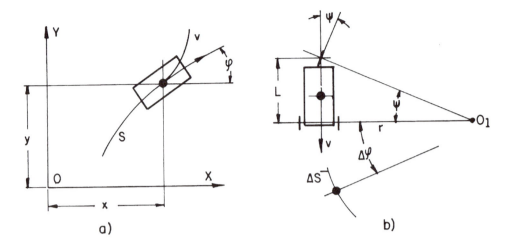

FIGURE 9.57 a) Layout of a three-wheeled vehicle in an absolute-coordinate system; b) The relation between the steering angle and the trajectory slope.

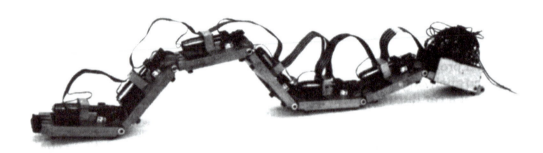

FIGURE 9.57c) General view of 7-link "caterpillar" device. This model was designed and built by a team of fourth-year students in the Mechanical Engineering Department of Ben-Gurion University under the author's supervision. It can move by using several bending modes, one of which appears in the photograph.

During time interval Δt, the vehicle travels a section ΔS of the trajectory. We can write

$$\Delta S = r\Delta\phi = V\Delta t \qquad [9.69]$$

or

$$\Delta\phi = \frac{V}{r}\Delta t. \qquad [9.70]$$

Substituting (9.68) into (9.70) we obtain:

$$\Delta\phi = \frac{V\Delta t}{L}\tan\psi(t). \qquad [9.71]$$

Thus,

$$\phi = \frac{V}{L} \int_0^t \tan\psi(t)\, dt. \tag{9.72}$$

By substituting the latter into (9.66), the position of the vehicle is determined. This algorithm can be used both when programming and controlling the trajectory of the vehicle and when analyzing the trajectory of the vehicle, for instance, for indicating the location of a car during its travel.

The problem of travelling across rugged terrain has prompted investigations in the field of walking vehicles. There is no real substitute for legs, which can step over local obstacles, find the optimum point on a surface for reliable support, change the length of the stride, and change the pace from walking to running and jumping or from an amble to a trot to a gallop.

It seems that investigators have covered all possibilities in seeking the optimal walking machine. In Figures 1.30 and 1.31 of Chapter 1 we briefly described the purely mechanical walking machine designed by the famous mathematician P. Chebyshev. This device can walk in an optimal way; however, it requires a flat surface and cannot make turns. Therefore, it does not fulfill the hopes a walking device has to realize. (By the way, this device clearly shows how cumbersome legs are in comparison with a wheel. The only reason the Creator (or evolution) used legs in his design of walking animals is the simplicity in providing reliable connection of blood vessels, nerves, and muscles. Otherwise we would have wheels, and shoes would be designed like tires.)

We begin our consideration with some six-legged devices, for which insects have served as a prototype. Figure 9.58a) shows an example. (This design was influenced by the article by A. P. Bessonov and N. V. Umnov, *Mechanism and Machine Theory,* Institute for the Study of Machines, Vol. 18, No. 4, pp. 261–265, Moscow, U.S.S.R, 1983.) The skeleton of the machine consists of rigid frame 1 to which links 3 are connected via joints 2. Another two pairs of joints 4 and rods 5 and 6 close the kinematic chains, creating two parallelograms. Hydrocylinders 7 and 8 control the position of these two parallelograms relative to frame 1. This arrangement serves for steering. Slide bushings 9 are mounted on rods 3 and frame 1. These bushings are driven along the rods by means of hydraulic cylinders 10. To each bushing 9 is fastened one of the six legs. Each leg includes a hydraulic cylinder 11 which is responsible for the height of the leg's foot. This is a design with 14 degrees of freedom. Figure 9.58b) explains the movement of a foot. By combining the movement of the cylinders in a certain time sequence, a step-over movement of the foot is carried out. Say, the foot at the beginning of the cycle is in its upper position I due to cylinder 11; then cylinder 10 moves it horizontally to point II. At this moment cylinder 11 lowers the foot to point III, from which cylinder 10 pulls the foot back (relative to the frame) to point IV, thus moving the "insect" one step leftward.

Other six-legged walker concepts using other leg kinematics are also possible. For instance, a leg with three degrees of freedom is illustrated in Figure 9.59a). Shaft 1 with joint 2 represents the hip, to which the thigh 3 with knee 4 is attached. The latter serves for connecting to the shin 5. These links can be driven by hydraulic cylinders. Cylinder 6 drives shaft 1 via rack-gear transmission 7. Cylinder 8 controls thigh 3 while cylinder 9 is responsible for the movement of shin 5. We can imagine three of this kind of leg placed on each side of a rectangular frame, simulating an insect. In this case the

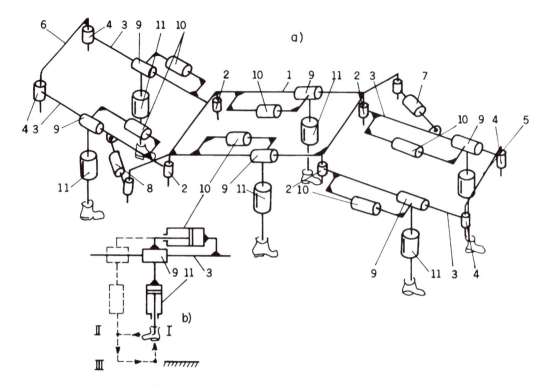

FIGURE 9.58   a) Six-legged walking machine with 12 degrees of freedom; b) Design of one leg.

system possesses 18 degrees of freedom. (This kind of machine was created by Ivan E. Sutherland (see *Scientific American*, January, 1983).) It can also be made in the form presented in Figure 9.59b), where the hip-shafts 1 are installed along radii 0-I, 0-II, 0-III, 0-IV, 0-V, and 0-VI. Thus, the axially symmetrical device is supported by at least three legs (black points in Figure 9.59c) at any moment. To move along vector V the sequence of lifting and lowering the legs is as shown in this figure when the circle moves rightward. Obviously, no preferred direction exists for this concept; it does not have a front or back. It can begin its travel in any direction and provides high maneu-

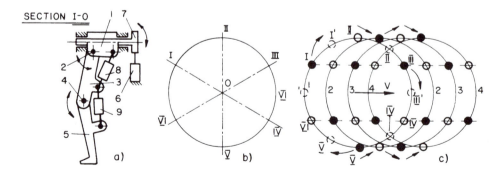

FIGURE 9.59   a) Leg with three degrees of freedom; b) Circular form of a vehicle; c) Walking tracks of this vehicle.

verability, including rotation in place. For this purpose foot I moves to point I', foot II moves to point II', and so on.

Using the leg with three degrees of freedom shown in Figure 9.59a, one can design four-legged devices, three-legged vehicles (A. Seireg, R. M. Peterson, University of Wisconsin, Madison, Wisconsin, U.S.A.) and, of course, try to create a two- or one-legged walker. Here there are problems with balance. Marc H. Ralbert and his colleagues at Carnegie-Mellon University have built and demonstrated a machine that hops on a single leg and runs like a kangaroo. This device consists of two main parts (Figure 9.60): body 1 and leg 2. The body carries the energy-distribution and control units. The leg can change its length, due to pressure cylinder 3, and pivots with respect to the body around the X- and Y-axes due to spherical hinge 4. A gyrosystem keeps the body horizontal while the leg bounces on the pneumatic spring-pneumatic cylinder 3. The pressure in the cylinder is controlled to dictate the bouncing dynamics. Another pneumosystem controls the pivoting of the leg during its rebound. A feedback system provides the pivoting required for balance when landing.

The balance problem is very important in walking devices. Solving it will permit four- and two-legged robots to be built. (The six-legged devices are always balanced on three supporting feet.) To solve the balancing problems a mathematical description of the device must be analyzed. Let us consider the model of the two-legged walker shown in Figure 9.61, which presumably reflects the main dynamic properties of natural two-legged creatures. (The following discussion is taken from the book *Two-legged Walking* by V. V. Beletzky, Moscow, Nauka, 1984 (in Russian).) Different levels of complexity can be considered for the model; however, it is always true that:

- The device possesses a certain mass, which is subject to gravitation;
- The device is controlled, which means that its links are driven and torques act in every joint; (Here we consider an additional degree of freedom between the shin and the foot, compared with the leg shown in Figure 9.59a).)
- Movement across a surface entails reactive forces against the feet.

All walking machines can act in two regimes:

- Walking—at any moment, at least one leg touches the ground;
- Running—there are moments when no contact between the ground and feet exists.

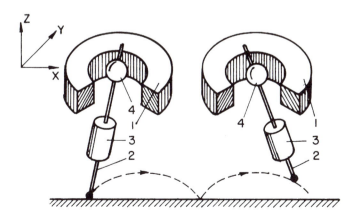

FIGURE 9.60   One-legged hopper.

The scheme shown in Figure 9.61 permits expressing the coordinates of the mass center C or any other point, through the parameters of the device. For instance, for the position in the figure we have for point O the coordinates are

$$z = z_v + 2(a\cos\alpha_v + b\cos\beta_v),$$
$$x = x_v + 2(a\sin\alpha_v + b\sin\beta_v),$$
[9.73]

where v = index of the supporting leg, and z_v and x_v are components of the r_v vector (position of the foot's contact point).

Expression (9.73) allows, for instance, answering the question of what the kinematic requirements are for providing constant height of point O above the ground (for better energy saving). This condition states

$$h = z - z_v = \text{const};$$

and from (9.73) it then follows that

$$\frac{[2\sin\alpha_v/2]^2}{[2(a+b)-h]/a} + \frac{[2\sin\beta_v/2]^2}{[2(a+b)-h]/b} = 1.$$
[9.74]

The opposite kinematic problem can arise: for a given location of point 0 what are the corresponding angles α and β? Assuming that $a = b = l$ (the links have equal lengths) and denoting

$$(x - x_v)^2 + (z - z_v)^2 = r_0^2,$$

we obtain

$$\alpha_v = \arctan\frac{x - x_v}{z - z_v} - \arctan\sqrt{\frac{4l^2}{r_0^2} - 1},$$
$$\beta_v = \arctan\frac{x - x_v}{z - z_v} + \arctan\sqrt{\frac{4l^2}{r_0^2} - 1}.$$
[9.75]

Obviously, the speeds and accelerations of the links can be calculated if the functions $z(t)$ and $x(t)$ are known.

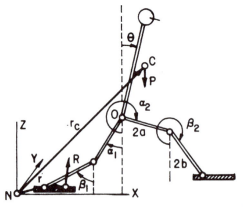

FIGURE 9.61   Mathematical model of a two-legged anthropoid walking machine.

The dynamics of a two-legged robot present a complex problem for the general case. For the designations in Figure 9.61, the equations can be written in vector form as follows:

$$M\frac{d^2r_c}{dt^2} = R + P; \qquad [9.76]$$

$$\frac{dK}{dt} = (r_T - r_c) \times R. \qquad [9.77]$$

Here,

> M = the mass of the system;
> r_c = radius vector of the mass center C from the coordinates' initial point N;
> P = gravity force;
> R = reaction force;
> r_T = radius vector of the point where the force R crosses the supporting surface;
> K = angular momentum of the system (moment of momentum).

Let the reader try to solve Equations (9.76) and (9.77). Here the simplified model shown in Figure 9.62—a two-legged "spider"—will be considered. This spider consists of a massive particle and two three-link legs. Then vector R is the reactive force acting at the initial point. Gravity force P acts at point O where mass center C is also located. For this case Equations (9.76) and (9.77) take the following scalar form:

$$M\ddot{x} = R\frac{x}{r} \quad \text{and} \quad M\ddot{z} = R\frac{z}{r} - Mg. \qquad [9.78]$$

Here x and z are coordinates of particle C, and M is the mass of the particle.

It is supposed that at every moment the spider is supported by only one leg. The exchange of legs happens instantly and mass center C does not change its height. Thus, $z = h$ = const and therefore $\dot{z} = 0$. For periodical walking with step length L and duration $2T$, the following limit conditions are valid:

$$\text{For } t = (2n-1)T; \quad x = -\frac{L}{2}; \quad \dot{x} = V_0; \quad z = h; \quad \dot{z} = 0;$$

$$[9.79]$$

$$\text{For } t = (2n+1)T; \quad x = \frac{L}{2}; \quad \dot{x} = V_0; \quad z = h; \quad \dot{z} = 0.$$

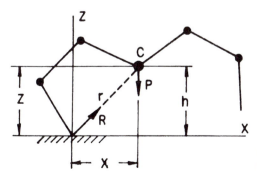

FIGURE 9.62 Simplified model of a two-legged walker or "spider."

For these assumptions it follows from the second Equation (9.78) that

$$Mg = R\frac{h}{r};$$ [9.80]

and the first equation can be rewritten in the form (taking 9.80 into account)

$$\ddot{x} - \frac{g}{h}x = 0.$$ [9.81]

The solution for (9.81) for the conditions (9.79) has the following form:

$$x = \frac{L}{2sh\sqrt{g/h}\ T}\ sh\sqrt{g/h}(t - 2nT),$$

$$\dot{x} = \frac{L}{2sh\sqrt{g/h}T}\sqrt{g/h}\ ch\sqrt{g/h}(t - 2nT), \quad n = 0,1,2,\cdots,$$ [9.82]

if only

$$V_0 = \frac{L}{2}\sqrt{g/h}\ cth\sqrt{g/h}\ T.$$ [9.83]

Thus, V_0, L, and T are related by Expression (9.83). The average marching speed V can be calculated from

$$V = \frac{1}{2T}\int_{-T}^{T}\dot{x}\,dt = L/2T.$$ [9.84]

The work A that the reaction force's horizontal component produces is determined with the help of the following formula:

$$A = \int_{-T}^{T}|\frac{R}{r}x\cdot\dot{x}|\,dt.$$ [9.85]

From (9.85) we obtain

$$A = \frac{PL^2}{4h}, \quad P = mg.$$ [9.86]

Obviously, the power required for walking can be obtained from (9.86) and expressed as follows:

$$N = \frac{A}{2T} = \frac{PL}{4h}V.$$ [9.87]

Now let us consider running for this spider model. We assume momentary foot contact with the support surface. Between these contact instants, $R \cong 0$ (there is no reaction force), and the spider, in essence, flies. The trajectory of the particle in this case consists of parabolic sections, as shown in Figure 9.63. In this figure

V_0 = initial speed of the particle,
$\dot{x}_0$ = horizontal component of speed V_0,

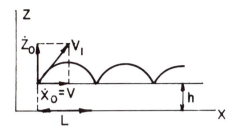

FIGURE 9.63   Running of a two-legged "spider."
The trajectory of the particle.

$\dot{z}_0$ = vertical component of speed V_0,
T = flying time, and
L = flying distance.

These concepts are related by known formulas

$$T = \frac{2\dot{z}_0}{g} \quad \text{and} \quad L = VT, \tag{9.88}$$

where V = average speed of translational movement.

At the end of the flying period, the vertical speed component $\dot{z}_1 = -\dot{z}_0$. To accelerate the particle again to speed $\dot{z}_0$, an amount of energy A_1 must be expended by the foot:

$$A_1 = \frac{1}{2} M\dot{z}_0^2.$$

In stopping at the end of the flying period, the amount of energy A_2 expended by the other foot is

$$A_2 = \frac{1}{2} M\dot{z}_1^2.$$

Thus, the energy A expended for one step is

$$A = A_1 + A_2 = M\dot{z}_0^2. \tag{9.89}$$

Substituting Expression (9.88) into (9.89), we obtain

$$A = \frac{PgL^2}{4V^2}. \tag{9.90}$$

Obviously, as follows from (9.90), the power spent in running can be calculated from the expression

$$N = \frac{PgL}{8V}. \tag{9.91}$$

Comparing the latter with Expression (9.87), we discover that when

$V > \sqrt{gh}$, it is worthwhile to run instead of walk;

$V < \sqrt{gh}$, it is worthwhile to walk instead of run.

Above, we considered the energy consumption of the walking or running body. However, to be more accurate, one must also take into account the power spent for moving the feet. This power can be estimated with the formula

$$N_F = \frac{4\mu P}{gL} V^3.$$ [9.92]

Here μ = the relation of foot mass m to particle mass M.

Thus, together with (9.87), we have an expression for the total power expended for walking:

$$N_\Sigma = \frac{PL}{4h} V + \frac{4\mu P}{gL} V^3;$$ [9.93a]

and for running:

$$N_\Sigma = \frac{PgL}{8V} + \frac{4\mu P}{gL} V^3.$$ [9.93b]

The latter formula (9.93a and b) enables the value of the optimum length L_0 of a walking step to be derived:

$$L_0 = 4V \sqrt{\frac{\mu h}{g}},$$ [9.94a]

or, for running:

$$L_0 = \frac{V^2}{g} 4\sqrt{2\mu}.$$ [9.94b]

The computation model presented here can give a rough estimation of power consumption in multi-legged vehicles by simply multiplying the results and distributing the mass of the moving body among all the pairs of legs.

Using the derived formulas we can recommend that the reader walk with an optimum step which is, for an average person ($h = 1$ m, $\mu = 0.2$, $V = 1.25$ m/sec), $L_0 = 0.7$ m. Then he or she will expend about 150 watts (0.036 kcal/sec) of power. We also recommend changing from walking to running when a speed of 11.3 km/hr is reached. However, if the reader is overweight, let him or her continue to walk with higher speed (more energy will be expended). The speed record for walking is about 15.5 km/hr.

On this optimistic tone we finish this chapter, the final one in the book.

Solutions to the Exercises

1   Solution to Exercise 3E-1

The first step is to reduce the given mechanism to a single-mass system. The resistance torque T_r on drum 1, obviously, varies in inverse proportion to the ratio $i = 1:3$. Thus,

$$T = T_r / 3 = 5\,\text{Nm} / 3 \cong 1.66\,\text{Nm}.$$

The procedure of reducing inertia I_2 of drum 2 to the axes of drum 1 requires calculation of the common kinetic energy of the mechanism, which is

$$\left(I_1 \omega_1^2 + I_2 \omega_2^2 \right) / 2,$$

where ω_1 and ω_2 are the angular velocities of drums 1 and 2, respectively. (The inertia of the gears and the shafts is neglected.) The kinetic energy of the reduced system with moment of inertia I is:

$$I \omega_1^2 / 2.$$

Thus,

$$I \omega_1^2 = I_1 \omega_1^2 + I_2 \omega_2^2$$

and

$$I = \left(I_1 \omega_1^2 + I_2 \omega_2^2 \right) / \omega_1^2 = \left(I_1 \omega_1^2 + I_2 (\omega_1 / 3)^2 \right) / \omega_1^2 = I_1 + I_2 / 9 \cong$$

$$\cong 0.01 + 0.045 / 9 \cong 0.015 \ \text{kg\,m}^2.$$

The motion equation may then be written as follows:

$$Ix'' / R + cRx = -T \quad \text{or} \quad x'' + cR^2 x / I = -T \frac{R}{I}, \qquad \text{[a]}$$

where x is the displacement of point K on the rope (see Figure 3E-1.1). Substituting the numerical data into (a) we obtain

$$\ddot{x} + \frac{500 \cdot 0.05^2}{0.015} x = -\frac{1.66 \cdot 0.05}{0.015}$$

or

$$\ddot{x} + 83.33x = -5.553 \, \text{m/sec}^2 .$$

The solution x is made up of two components: $x = x_1 + x_2$.
   The homogeneous component is sought in the form:

$$x_1 = A\cos kt + B\sin kt,$$

where k is the natural frequency of the system.
   Here, obviously,

$$k^2 = cR^2 / I$$

and

$$k = \sqrt{83.33} \cong 9.13 \, 1/\text{sec.}$$

The partial solution x_2, as follows from (a), is sought in the form of a constant X:

$$x_2 = X = -T/cR = -\frac{1.66}{500 \cdot 0.05} \cong -0.0664 \, \text{m.}$$

Thus

$$x = A\cos 9.13t + B\sin 9.13t - 0.0664. \hspace{2cm} \text{[b]}$$

   From the initial conditions given in the formulation of the problem, it follows that for time $t = 0$, the spring is stretched for $x_0 = 2\pi R = 2 \cdot \pi \cdot 0.05 \cong 0.314 \, \text{m}$, while the speed $\dot{x}_0 = 0$. Thus, from (b) we derive

$$0.314 = A - 0.0664 \quad \text{or} \quad A = 0.3804 \, \text{m.}$$

Differentiating (b) in terms of speed, we obtain

$$\dot{x} = -9.13 A \sin 9.13t + 9.13 B \cos 9.13t. \hspace{2cm} \text{[c]}$$

Substituting the initial conditions, we obtain $9.13 B = 0$ or $B = 0$. Finally, from (b) we obtain the following expression for the solution:

$$x \cong 0.38 \cos 9.13t - 0.0664 \, \text{m.} \hspace{2cm} \text{[d]}$$

   To answer the question formulated in the problem, we find t from (d), substituting the value X' (location of the point K after the rope had been rewound half a perimeter around drum 1). Obviously,

$$x' = \pi R \cong 3.14 \cdot 0.05 \cong 0.157 \, \text{m.}$$

And from (d), it follows that

$$t = \frac{1}{9.13} \arccos\left[\frac{x' + 0.0664}{0.38}\right] = \frac{1}{9.13} \arccos\left[\frac{0.157 + 0.0664}{0.38}\right] \cong 0.103 \, \text{sec.}$$

Now, we illustrate the same solution in MATHEMATICA language.

f1=x"[t]+83.3 x[t]+5.553
y1=DSolve[{f1==0,x[0]==.314,x'[0]==0},{x},{t}]

The solution corresponds to (d).

j1=Plot[Evaluate[x[t]/.y1],{t,0,.2},AxesLabel->{"t","x"}]
j2=Plot[x=-.0664,{t,0,.2}]
Show[j1,j2]

The curve in the graphic representation begins at the point 0.314 m. The horizontal lines $x^* = -0.0664$ m and $x^{**} = 0.157 - 0.0664 = 0.0906$ m, in turn, indicate: x^* is the zone where the point K does not reach because of the resistance torque T_r (from 0 to -0.0664 m); it is the new zero point relative to which the value x^{**} (the position of the point K after the rope is rewound for half a revolution) is defined and which is achieved at $t = 0.103$ second.

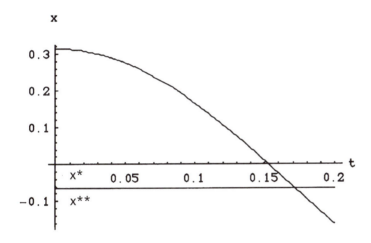

FIGURE 3E-1.1   Displacement (of the point K) x versus time.

2   Solution to Exercise 3E-2

The solution is divided into two stages: stage a) the blade's movement from the initial point until it comes into contact with the wire (here we neglect the frictional resistance), and stage b) the cutting of the wire.

Stage a)

We use Equations (3.27) to (3.81). Thus,

$$m\ddot{x} + cx = mg,$$ [a]

which means that the weight mg of the blade aids its downward motion. From (a) we obtain

$$\ddot{x} + \omega^2 x = g,$$

where

$$\omega^2 = c/m = \frac{5000\,\text{N/m}}{1\,\text{kg}} = 5000\,1/\text{sec}^2 \quad \text{or} \quad \omega \cong 70.7\ 1/\text{sec},$$

where ω is the natural frequency of the mechanism.

The solution x is made up of two components:

$$x = x_1 + x_2.$$

The homogeneous component is expressed as

$$x_1 = A\cos\omega t + B\sin\omega t.$$

The partial solution is given by

$$x_2 = X = \text{const.}$$

Substituting X into Equation (a) we find

$$X = g/\omega^2 \cong \frac{9.8\,\text{m/sec}^2}{5000\,1/\text{sec}^2} \cong 0.002\,\text{m.}$$

Thus,

$$x = A\cos 70.7t + B\sin 70.7t + 0.002. \qquad [\text{b}]$$

Substituting the initial conditions into (b), we obtain the coefficients A and B.

When $t = 0$ and $x = L_0 = 0.2$ m, then $0.2 = A + 0.002$, and finally we obtain

$$A = 0.198\,\text{m.}$$

Differentiating (b), we obtain

$$\dot{x} = -70.7A\sin 70.7t + 70.7B\cos 70.7t, \qquad [\text{c}]$$

where $t = 0$, $\dot{x} = 0$, and we thus obtain

$$B = 0.$$

Finally, we obtain

$$x = 0.198\cos 70.7t + 0.002\,\text{m.} \qquad [\text{d}]$$

We now calculate the time t_1 needed by the blade to reach the point at which it comes into contact with the wire, i.e., $x = 0.1\,\text{m}$. From (d) we obtain

$$0.1 = 0.198\cos 70.7t_1 + 0.002$$

or

$$t_1 = \frac{1}{70.7}\arccos\left[\frac{0.1 - 0.002}{0.198}\right] \cong 0.0149\,\text{sec.}$$

The speed $\dot{x}_1$ developed by the blade at this moment in time is calculated from (c):

$$\dot{x}_1 = -0.198 \cdot 70.7 \sin 70.7 \cdot 0.0149 \cong -12.17\,\text{m/sec}.$$

In MATHEMATICA language, we solve the above-derived equation as follows:

```
f2=x1"[t] +5000 x1[t] -10
y2=DSolve[{f2==0,x1[0]==.2,x1'[0]==0},{x1},{t}]
j21=Plot[Evaluate[x1[t]/.y2],{t,0,.025},
    AxesLabel->{"t","x1"}]
g1=Plot[{x1=.1},{t,0,.02}]
Show[g1,j21]
```

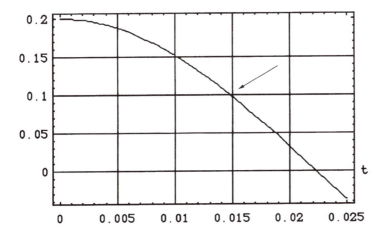

FIGURE 3E-2.1 Movement of the blade $x_1[t]$ until it makes contact with the wire ($x = 0.1\,\text{m}$).

Stage b)

We consider two ways to solve this stage.

I. We begin with a simple physical estimation of the time needed for cutting the wire. The whole energy E (kinetic plus potential components) of the blade at the moment in time when it comes into contact with the wire is

$$E = 0.5(m\dot{x}_1^2 + cx^2) \cong 0.5(1 \cdot 12.17^2 + 5000 \cdot 0.1^2) \cong 99.05\,\text{Nm}.$$

The work A that must be expended for cutting the wire is expressed as

$$A = Ph \cong 800\,\text{N} \cdot 0.004\,\text{m} \cong 3.2\,\text{Nm}.$$

The saved energy E^* after the cutting is accomplished is given by

$$E^* = E - A \cong 99.05 - 3.2 \cong 95.85\,\text{Nm},$$

and this energy (a remaining sum of kinetic and potential components after the cutting is accomplished) is given by

$$E^* = 0.5\left(m\dot{x}_1^{*2} + c[x - 0.004]^2\right),$$ [e]

where $\dot{x}^*$ is the speed of the blade after cutting the wire.

From (e) we derive:

$$\dot{x}_1^* = \left\{\left(2E^* - c[x - 0.004]^2\right)/m\right\}^{0.5} \cong$$

$$\cong \left\{\left(2 \; 95.85 - 5000\,[0.1 - 0.004]^2\right)/1\right\}^{0.5} \cong 12.07 \, \text{m/sec}.$$

We now express the loss of the momentum M as

$$M = m(\dot{x}_1 - \dot{x}_1^*) \cong 1(12.17 - 12.07) \cong 0.10 \, \text{N/sec},$$

and, finally, the impulse of force

$$I = P^* t = M.$$

Here,

$$P^* \cong P - 0.5c(0.1 + [0.1 - 0.004]),$$

and, therefore,

$$P^* \cong 800 - 0.5 \; 5000\,(0.1 + [0.1 - 0.004]) \cong 310 \, \text{N}.$$

Thus,

$$t = M/P^* \cong 0.10/310 \cong 0.000322 \, \text{sec}.$$

II.   Now let us solve this part of the problem describing the process of the blade's motion by a differential equation. This latter is

$$m\ddot{x} + cx = mg - P$$

or

$$\ddot{x} + \omega^2 x = g - P/m$$

and

$$\ddot{x} + 5000 \, x = 9.8 \, \text{m/sec}^2 - \frac{800 \, \text{N}}{1 \, \text{kg}} = -790.2 \, \text{m/sec}^2.$$ [f]

The solution x is made up of two components:

$$x = x_1 + x_2.$$

The homogeneous component is given by

$$x_1 = A\cos(\omega t + \gamma).$$

The partial solution is expressed as

$$x_2 = X = \text{const.}$$

Substituting X into Equation (f), we find:

$$X = (g - P/m)/\omega^2 \cong \frac{9.8 - 800/1}{5000} \cong -0.158 \text{ m.}$$

Thus,

$$x = A\cos(70.7t + \gamma) - 0.158. \tag{g}$$

Differentiating (g), we obtain

$$\dot{x} = -70.7A\sin(70.7t + \gamma). \tag{h}$$

Substituting the initial conditions into (g), we obtain the coefficients γ and A. When $t = 0$ and $x = L_1 = 0.1$ m, then the speed is $\dot{x} = -12.17$ m/sec, and we obtain from (g) and (h), respectively,

$$0.258 = A\cos(\gamma) \quad \text{and} \quad -12.17 = -A70.7\sin(\gamma);$$

thus,

$$tn(\gamma) \cong 12.17/0.258 \; 70.7 \cong 0.667 \quad \text{and} \quad \gamma \cong 0.588.$$

Now from either Equation (g) or Equation (h), we derive A as follows:

$$0.258 \cong A\cos(0.588) \quad \text{and} \quad A \cong 0.31.$$

Finally, we have

$$x \cong 0.31\cos(70.7t + 0.588) - 0.158. \tag{i}$$

We now calculate the time t_1 needed by the blade to cut the wire, which takes place when $x = 0.096$ m. Therefore, we may write

$$0.096 \cong 0.31\cos(70.7t_1 + 0.588) - 0.158$$

or

$$t_1 \cong \frac{\arccos\left(\dfrac{0.096 + 0.158}{0.31}\right) - 0.588}{70.7} \cong 0.0003 \text{ sec.}$$

In the MATHEMATICA language, we solve the above-derived equation as follows:

```
f1=x"[t]+5000 x[t]+790.2
y1=DSolve[{f1==0,x[0]==0,x'[0]==-12.7},{x[t]},{t}]

j1=Plot[Evaluate[x[t]/.y1],{t,0,.0005},AxesLabel->{"t","x"}
```

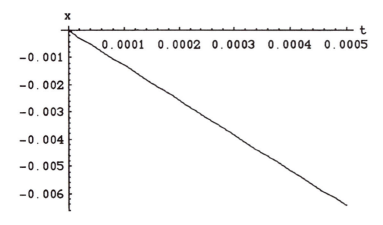

FIGURE 3E-2.2 Movement of the blade $x_2[t]$ during the cutting process (from $x_1 = 0.1$ m to $x_1 = 0.096$ m).

3 Solution to Exercise 3E-3

In this case, the total moment of inertia I of the masses driven by the electromotor is calculated from

$$I = I_0 + mr^2 = 0.005 + 10 \ 0.035^2 \cong 0.01725 \, \text{k gm}^2.$$

The differential equation, according to equation (3.41), takes the following form:

$$\dot{\omega} + \frac{T_1}{I}\omega = \frac{T_0}{I} - \frac{mgr}{I}.$$

Substituting the numerical data into this equation, we rewrite it as

$$\dot{\omega} + \frac{0.05}{0.01725}\omega = \frac{5}{0.01725} - \frac{10 \cdot 9.8 \cdot 0.035}{0.01725} \qquad [a]$$

or

$$\dot{\omega} + 2.9\omega = 91.$$

The solution is sought as a sum $\omega = \omega_1 + \omega_2$, where ω_1 is the homogeneous solution in the form $\omega_1 = Ae^{\alpha t}$.

Substituting this expression into Equation (a), we obtain

$$Ae^{\alpha t} + 2.9 \ Ae^{\alpha t} = 0 \quad \text{or} \quad \alpha = -2.9 \ 1/\text{sec}.$$

The partial solution is then sought as a constant $\omega_2 = \Omega = $ const.
Substitution of this solution into Equation (a) yields

$$\Omega = \frac{91}{2.9} \cong 31 \ 1/\text{sec}.$$

Thus,

$$\omega = \left\{ A\exp[-2.9t] + 31 \right\} \ 1/\sec. \tag{b}$$

From the initial conditions, we find the coefficient A. For time $t = 0$, the speed $\omega = 0$. Therefore,

$$0 = A\exp[0] + 31 \quad \text{or} \quad A = -31 \ 1/\sec,$$

and finally

$$\omega = 31\left\{ 1 - \exp[-2.9t] \right\}. \tag{c}$$

From here,

$$t = \frac{1}{2.9}\ln\frac{1}{1 - \omega/31}. \tag{d}$$

Substituting the desired speed $\omega = 10 \ 1/\sec$ into (d), we may rewrite this expression as

$$t \cong \frac{1}{2.9}\ln\frac{1}{1 - 10/31} \cong 0.134 \sec.$$

Integrating Expression (b), we find the rotational angle $\phi(t)$ of the motor:

$$\omega = \frac{d\phi}{dt} \quad \text{or} \quad \phi = \int_0^t \omega\, dt.$$

Thus,

$$\phi = 31\int_0^t \left\{ 1 - \exp[-2.9t] \right\} dt = $$
$$= 31\left\{ 1/2.9\exp[-2.9t] + t - 1/2.9 \right\}. \tag{e}$$

Substituting $t = 0.134 \sec$ into (e), we obtain

$$\phi = 31\left\{ 1/2.9\exp\left[-2.9\ 0.134\right] + 0.134 - 1/2.9 \right\} \cong 0.776 \text{ rad} \cong 0.123 \, rev.$$

Taking into account the radius r of the drum, we obtain the height h that the mass m has travelled:

$$h = \phi r \cong 0.776\ 0.035 \cong 0.027 \text{ m}.$$

An illustration of the solution in MATHEMATICA language follows.

```
f1=w'[t]+2.9 w[t]-91
y1=DSolve[{f1==0,w[0]==0},{w},{t}]
j1=Plot[Evaluate[w[t]/.y1],{t,0,2},AxesLabel->{"t","w"}]
```

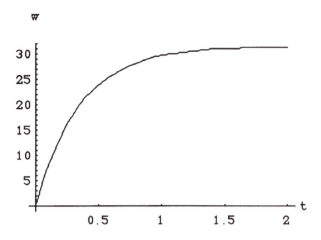

FIGURE 3E-3.1   Rotational speed of the motor versus time.

4   Solution to Exercise 3E-3a)

The first step is to reduce the given mechanism to a single-mass system. The resistance torque T_r on the axes of the electromotor, obviously, varies in inverse proportion to the ratio $i = 1{:}4$. Thus,

$$T = T_r / 4 \cong m_2\, g R / 4 \cong 5\ 9.8\ 0.02/4 \cong 0.245\,\text{Nm}.$$

The procedure of reducing the inertia of all the moving parts with respect to the axes of the motor requires calculation of the common kinetic energy E of the mechanism, which is

$$E = \left(I_0\omega^2 + I_1[\omega/4]^2 + [m_1 + m_2][0.02\omega/4]^2 \right)/2,$$

where ω is the angular speed of the shaft of the motor. The kinetic energy of the reduced system with moment of inertia I with respect to the axle of the motor is

$$E = I\omega^2 / 2$$

or

$$I\omega^2 / 2 = \left(I_0\omega^2 + I_1[\omega/4]^2 + [m_1 + m_2][0.02\omega/4]^2 \right)/2$$

or

$$I = I_0 + I_1[1/4]^2 + [m_1 + m_2][0.02/4]^2 \cong$$

$$\cong 0.001 + 0.01\ 0.0625 + [5+5]0.000025 \cong 0.001875\,\text{kg m}^2.$$

The differential equation according to Expression (3.41) takes the following form:

$$\dot{\omega} + \frac{T_1}{I}\omega = \frac{T_0}{I} - \frac{T}{I}.$$

where the characteristic of the motor gives $T = T_0 + T_1\omega = 4 - 0.1\,\omega$.

Substituting the numerical data into this equation, we may rewrite it as

$$\dot{\omega} + \frac{0.1}{0.001875}\omega = \frac{4}{0.001875} - \frac{0.245}{0.001875}$$

or

$$\dot{\omega} + 53.3\omega = 2003. \qquad \text{[a]}$$

The solution is sought as a sum $\omega = \omega_1 + \omega_2$, where ω_1 is the homogeneous solution in the form: $\omega_1 = Ae^{\alpha t}$.

Substituting this expression into Equation (a), we obtain

$$Ae^{\alpha t} + 53.3 \ Ae^{\alpha t} = 0 \quad \text{or} \quad \alpha = -53.3 \ 1/\text{sec}.$$

The partial solution is sought as a constant $\omega_2 = \Omega = \text{const.}$
Substituting this constant into Equation (a) yields

$$\Omega = \frac{2003}{53.3} \cong 37.6 \ 1/\text{sec}.$$

Thus,

$$\omega = \left\{ A\exp[-53.3t] + 37.6 \right\} \ 1/\text{sec}. \qquad \text{[b]}$$

From the initial conditions, we find the coefficient A. For the moment $t = 0$, the speed $\omega = 0$. Therefore,

$$0 = A\exp[0] + 37.6 \quad \text{or} \quad A = -37.6 \ 1/\text{sec},$$

and finally,

$$\omega = 37.6\left\{ 1 - \exp[-53.3t] \right\}. \qquad \text{[c]}$$

To answer the question formulated in the problem, we substitute $t = 0.1$ sec into Expression (c):

$$\omega = 37.6\left\{ 1 - \exp[-53.3 \ 0.1] \right\} \cong 37.4 \ 1/\text{sec}.$$

Integrating Expression (c), we find the angle of rotation $\phi(t)$ of the motor:

$$\omega \frac{d\phi}{dt} \quad \text{or} \quad \phi = \int_0^t \omega \, dt.$$

Thus,

$$\phi = 37.6\int_0^t \left\{ 1 - \exp[-53.3 \ t] \right\} dt = \qquad \text{[d]}$$
$$= 37.6\left\{ 1/53.3\exp[-53.3t] + t - 1/53.3 \right\}.$$

Substituting $t = 0.1$ sec into (d), we obtain

$$\phi = 37.6\left\{ 1/53.3\exp[-53.3 \ 0.1] + 0.1 - 1/53.3 \right\} \cong 3.6 \ \text{rad}.$$

The distance mass m_1 travels then is $3.6 \cdot 0.02/4 = 0.018 \text{m}$.

Now let us consider the case with an AC electromotor as the driving source. In this case the driving torque T is

$$T = 2 \cdot T_m \frac{s_m \cdot s}{s_m^2 + s^2} = 2 \cdot T_m \frac{s_m \cdot (\omega_0 - \omega)/\omega_0}{s_m^2 + \{(\omega_0 - \omega)/\omega_0\}^2} \cong$$

$$\cong 2 \cdot 1 \frac{0.1 \cdot (157 - \omega)/157}{0.01 + \{(157 - \omega)/157\}^2}.$$

The equation taking into the account the given characteristic, (as it is shown in the Expression (3.48)) and considering the definition of the concept s is as follows:

$$\dot{\omega} + \frac{2 \cdot 4}{24900 - 314 \ \omega + \omega^2} \ \omega = \frac{2 \cdot 4 \cdot 1160}{24900 - 314 \ \omega + \omega^2}.$$

We solve it using the MATHEMATICA package.

f1=0.001875 w'[t]- 2 15.7 (157- w[t])/(24900-314 w[t]+w[t]^2)
j1=NDSolve[{f1==0,w[0]==0},{w[t]},{t,0,1}]
y1=Plot[Evaluate[w[t]/.j1],{t,0,1},AxesLabel>{"t","w"}]
z1=Plot[2 15.7 (157- w)/(24900-314 w+w^2),{w,0,157}]

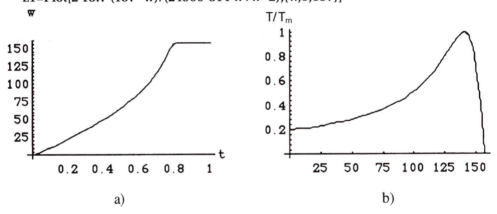

a)   b)

FIGURE 3E-3a).1   a) Speed versus time for AC motor drive; b) Characteristic of the electromotor.

5   Solution to Exercise 3E-3b)

The first step is to reduce the given mechanism to a single-mass system. The resistance torque T on the axes of the electromotor, obviously, varies in inverse proportion to the total ratio of the gears ($i = 1:3$) and of the screw-nut transmission. The latter transforms the resistance force Q to the needed torque on the screw T_s which, respectively, is

$$T_s = Qh/2\pi \cong 200 \ 0.01 \ 9.8/2\pi \cong 3.12 \, \text{Nm}.$$

Thus,

$$T = T_s / 3 = 3.12 \, \text{Nm} / 3 \cong 1.04 \, \text{Nm}.$$

The procedure of reducing the inertia of all moving parts with respect to the axes of the motor requires calculation of the common kinetic energy E of the mechanism, which is

$$E = \left(I_0 \omega^2 + I_1 [\omega / 3]^2 + m[h\omega / 6\pi]^2 \right) / 2,$$

where ω is the angular speed of the shaft of the motor.

The kinetic energy of the reduced system with a moment of inertia I respective to the axes of the motor is

$$E = I\omega^2 / 2$$

or

$$I\omega^2 / 2 = \left(I_0 \omega^2 + I_1 [\omega / 3]^2 + m[0.01\omega / 6\pi]^2 \right) / 2$$

or

$$I = I_0 + I_1 [1/3]^2 + m[0.01/6\pi]^2 \cong$$

$$\cong 0.001 + 0.01 \; 0.111 + 200 \; 0.28 \; 10^{-6} \cong 0.002166 \, \text{kg m}^2.$$

The differential equation, according to Expression (3.41), takes the following form:

$$\dot{\omega} + \frac{T_1}{I} \omega = \frac{T_0}{I} - \frac{T}{I},$$

where the characteristic of the motor gives $T = T_1 + T_0 = 4 + 0.1 \, \omega$.

Substituting the numerical data into this equation, we may rewrite it as

$$\dot{\omega} + \frac{0.1}{0.002166} \omega = \frac{4}{0.002166} - \frac{1.04}{0.002166}$$

or

$$\dot{\omega} + 46.17\omega = 1374. \tag{a}$$

The solution is sought as a sum $\omega = \omega_1 + \omega_2$, where ω_1 is the homogeneous solution in the form

$$\omega_1 = Ae^{\alpha t}.$$

Substituting this expression into Equation (a), we obtain

$$Ae^{\alpha t} + 46.17 Ae^{\alpha t} = 0 \quad \text{or} \quad \alpha \cong -46.17 \; 1/\text{sec}.$$

The partial solution is sought as a constant $\omega_2 = \Omega = \text{const}$.

Substituting this constant into Equation (a) yields

$$\Omega = \frac{1374}{46.17} \cong 29.76 \; 1/\text{sec}.$$

Thus,

$$\omega = \left\{ A\exp[-46.17t] + 29.76 \right\} \; 1/\sec. \tag{b}$$

From the initial conditions we find the coefficient A. For time $t = 0$, the speed $\omega = 0$. Therefore,

$$0 = A\exp[0] + 29.76 \quad \text{or} \quad A = -29.76 \; 1/\sec,$$

and finally,

$$\omega = 29.76 \left\{ 1 - \exp[-46.17t] \right\}. \tag{c}$$

Integrating Expression (c), we find the angle of rotation $\phi(t)$ of the motor:

$$\omega = \frac{d\phi}{dt} \quad \text{or} \quad \phi = \int_0^t \omega \, dt.$$

Thus,

$$\phi = 29.76 \int_0^t \left\{ 1 - \exp[-46.17\,t\,] \right\} dt = \tag{d}$$
$$= 29.76 \left\{ 1/46.17 \exp[-46.17t] + t - 1/46.17 \right\}.$$

Substituting $t = 0.5$ sec into (d), we obtain

$$\phi = 29.76 \left\{ 1/46.17 \exp[-46.17 \; 0.5] + 0.5 - 1/46.17 \right\} \cong 75 \, \text{rad},$$

and the travelling distance s is derived from the explanation given above as

$$s = \phi h/3 \; 2\pi \cong 75 \; 0.01/3 \; 2\pi \cong 0.04 \, \text{m}.$$

6   Solution to Exercise 3E-4

To solve this problem, we use Expression (3.103), which is derived from Equations and Formulas (3.99), (3.100), (3.101), and (3.102). Substituting the given parameters into the above-mentioned expression, we obtain

$$m = \frac{2\psi}{M} \cong \frac{2 \cdot 150}{200} \cong 1.5 \; 1/m,$$

$$A = \frac{pF - Q}{M} \cong \frac{500 \cdot 50 - 5000}{200} \cong 100 \; m/\sec^2,$$

$$t = \frac{1}{\sqrt{2mA}} \ln \frac{1 + V\sqrt{m/2A}}{1 - V\sqrt{m/2A}} \cong$$

$$\cong \frac{1}{\sqrt{2 \cdot 1.5 \cdot 100}} \ln \frac{1 + 5\sqrt{1.5/2 \cdot 100}}{1 - 5\sqrt{1.5/2 \cdot 100}} \cong 0.05 \, \text{sec}.$$

Now let us take Expression (3.104) and rewrite it in the following form:

$$ds = \sqrt{2A/m}\int_0^t \frac{e^{\beta t}-1}{e^{\beta t}+1}\,dt = \sqrt{2A/m}\int_0^t th\,\frac{\beta t}{2}\,dt$$

$$s = \sqrt{2A/m}\ \frac{2}{\beta}\ln ch\,\frac{\beta t}{2} = \frac{2}{m}\ln ch\,\frac{\beta t}{2} \qquad [a]$$

$$\beta = \sqrt{2Am} = \sqrt{2\cdot100\cdot1.5} = 17.32\,1/\sec.$$

Obviously, from (a), the distance s the piston travels during time t can be derived, but, the opposite function, i.e., the time t needed to accomplish a certain travelling distance s is much easier to obtain by computerized means. Using MATHEMATICA language, we build a graphic representation for the function $s(t)$ that answers both questions.

```
s=1.333 Log[Cosh[8.66 t]]
j=Plot[s,{t,0,.1},AxesLabel->{"t","s"}]
```

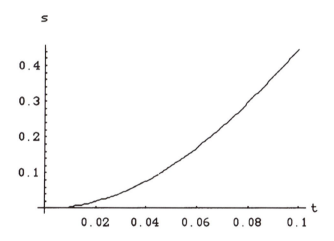

FIGURE 3E-4.1   Displacement of the piston rod versus time.

7   Solution to Exercise 3E-4a)

The first step is to reduce the given mechanism to a single-mass system. The force of resistance Q on the piston-rod, obviously, varies in inverse proportion to the ratio $R/r = 1{:}2.5$. Thus,

$$Q = m_2 g\ R/r \cong 50\ \text{kg}\ \ 9.8\ \text{m}/\sec^2\ \ 2.5 \cong 1225\ \text{N}.$$

The procedure of reducing inertia I of the wheels with respect to the piston-rod requires calculation of the common kinetic energy E of the mechanism, which is

$$E = \left(I[V/r]^2 + m_1 V^2 + m_2 [V R/r]^2 \right)/2,$$

where V is the speed of the piston-rod. The mass M of the reduced system is

$$MV^2 = \left(I[V/r]^2 + m_1 V^2 + m_2 [V R/r]^2 \right)$$

or

$$M = I[l/r]^2 + m_1 + m_2 [R/r]^2 \cong$$

$$\cong 0.2[1/0.04]^2 + 100 + 50[2.5]^2 \cong 537.5 \text{ kg}.$$

The active area F of the piston when lifting the mass m_2 is, obviously,

$$F = \pi \left(D_0^2 - D^2 \right)/4 \cong \pi \left(0.08^2 - 0.02^2 \right)/4 \cong 47.12 \text{ cm}^2.$$

To solve this problem, we use Expression (3.103), which is derived from Equations and Formulas (3.99), 3.100), 3.101), and (3.102). Substituting the given parameters into the above-mentioned expression, we obtain

$$m = \frac{2 \cdot \psi}{M} = \frac{2 \cdot 120}{537.5} \cong 0.44 \, 1/m,$$

$$A = \frac{pF - Q}{M} = \frac{200 \cdot 47.12 - 1225}{537.5} \cong 15.25 \, m/\sec^2,$$

$$t = \frac{1}{\sqrt{2mA}} \ln \frac{1 + V\sqrt{m/2A}}{1 - V\sqrt{m/2A}} \cong$$

$$\cong \frac{1}{\sqrt{2 \cdot 0.44 \cdot 15.25}} \ln \frac{1 + 2\sqrt{0.44/2 \cdot 15.25}}{1 - 2\sqrt{0.44/2 \cdot 15.25}} \cong 0.134 \text{ sec}.$$

Now the travel distance can be re-estimated by using the following formula:

$$\beta = \sqrt{2mA} = \sqrt{2 \cdot 0.44 \cdot 15.25} \cong 3.66 \ 1/\sec,$$

$$s = \sqrt{2A/m} \, \frac{2}{\beta} \ln ch \frac{\beta t}{2} = \frac{2}{m} \ln ch \frac{\beta t}{2}.$$

Substituting into the latter formula the values for time $t = 0.134$ sec and $\beta = 3.66 \ 1/\sec$, we obtain the distance s travelled by the piston during this time:

$$s = 0.135 \text{ m}.$$

Using MATHEMATICA language, we build a graphic representation for the function $s(t)$:

```
s=4.54 Log[Cosh[1.83 t]]
j=Plot[s,{t,0,.1},AxesLabel->{"t","s"}]
```

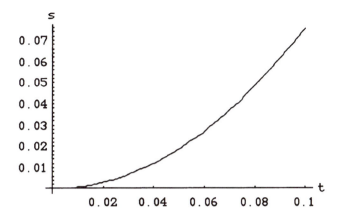

FIGURE 3E-4a).1 Displacement of the piston-rod versus time.

8   Solution to Exercise 3E-5

The time sought is a sum of three or four components.

Component 1

The first component t_0 is the time needed by the pressure wave to travel from the valve to the inlet of the cylinder:

$$t_0 = L / V_s \cong 10\,\text{m} / 340\,\text{m} / \text{sec} \cong 0.0294\,\text{sec},$$

where V_s is the speed of sound.

Component 2

The second component t_1 is the time needed to reach a pressure P_c in the cylinder volume that develops a force equal to the resisting force acting on the piston-rod (in our case this is the weight mg of the lifted mass m). Obviously,

$$P_c = 4\,mg / \pi D^2,$$

and for the problem under consideration this is either

$$P_{c400} \cong 4\frac{400 \cdot 9.8}{\pi 0.15^2} \cong 2.2 \cdot 10^5 \; N/m^2$$

or

$$P_{c500} \cong 4\frac{500 \cdot 9.8}{\pi 0.15^2} \cong 3 \cdot 10^5 \; N/m^2.$$

Corresponding to these two situations, we obtain for the coefficients β either

$$\beta_{400} \cong P_{c400} / P_r \cong 2.2/5 \cong 0.44 \le 0.528$$

or

$$\beta_{500} \cong P_{c500} / P_r \cong 3/5 \cong 0.61 \ge 0.528. \qquad [a]$$

A comparison of these results shows that the first belongs to a supercritical air flow regime while the second belongs to a subcritical regime.

A. We begin with the supercritical case. Let us calculate the second time component t_1. To do so, we use the Formula (3.120):

$$t_1 \cong \frac{V_c(P_{c400} - P_0)}{0.67\, T\alpha F_p P_r} \cong \frac{0.00177(2.2-0)\cdot 10^5}{0.67\cdot 293\cdot 0.5\cdot 11.3\cdot 10^{-5}\cdot 5\cdot 10^5} \cong 0.070\,\text{sec},$$

where V_c is the volume of the cylinder at the beginning of the action,

$$V_c = \frac{\pi D^2}{4}\, l \cong \frac{\pi\, 0.15^2}{4}\, 0.1 \cong 0.00177\, m^3,$$

and F_p is the cross-sectional area of the air pipe:

$$F_p = \frac{\pi d^2}{4} \cong \frac{\pi\, 0.012^2}{4} \cong 11.3\cdot 10^{-5} m^2.$$

Component 3

To find the third component of the time, t_2, we must solve Equation (3.133) describing the dynamics of a piston under supercritical air flow conditions. The solution will depict the displacement $s(t)$ of the piston in our case. This computation follows in MATHEMATICA terms:

```
f1=550 s"[t]-(.023 .5 28.7 293 .00011 600000 t +.05 100000 .01)/s[t]+ 5500
j1=NDSolve[{f1==0,s[0]==.05,s'[0]==0},{s[t]},{t,0,2}]
b1=Plot[Evaluate[s[t]/.j1],{t,0,1.5},AxesLabel->{"t","s"},
Plot Range->All]
```

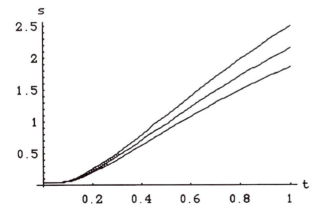

FIGURE 3E-5.1 Displacement of the piston versus time. The upper curve corresponds to $T = 400°\,K$, the middle is $T = 340°\,K$, and the lower is $T = 293°\,K$.

As follows from Figure 3E-5.1, the displacement for 1.5 m of the piston takes about $t_2 = 0.8$ sec. Thus, the total time t_{400} for this supercritical case is

$$t_{400} = t_0 + t_1 + t_2 \cong 0.0294 + 0.070 + 0.8 \cong 0.9 \text{ sec.}$$

B. Now we consider a combination of the supercritical and subcritical cases. Here, the duration t_1 consists of two components; thus, $t_1 = t_1' + t_1''$. The component t_1' is the time needed to reach critical pressure P_{cr} in the cylinder's volume (supercritical regime), while the component t_1'' is the time needed subsequently to reach the pressure $P_{c\,500}$ (subcritical regime).

1. The value

$$P_{cr} = \beta_{cr} \, P_r = 0.528 \cdot 5 \cdot 10^5 \cong 2.64 \; 10^5 \text{ N/m}^2.$$

Then, from Expression (3.120), we obtain for t_1'

$$t_1' \cong \frac{V_c \,(P_{cr} - P_0)}{0.67 \cdot T \alpha F_p P_r} \cong \frac{0.00177 \cdot (2.64 - 0) \cdot 10^5}{0.67 \cdot 293 \cdot 0.5 \cdot 11.3 \cdot 10^{-5} \cdot 5 \cdot 10^5} \cong 0.0846 \text{ sec.}$$

The component t_1'' is calculated from Expression (3.127), where $\beta_{500} \cong 0.61$ (see Expression (a)):

$$t_1'' \cong \frac{0.154 V_c}{\alpha T F_p} \sqrt{T_r} \left(0.41 - \sqrt{1 - \beta_{500}{}^{0.29}}\right) \cong$$

$$\cong \frac{0.154 \cdot 0.00177}{0.5 \cdot 11.3 \cdot 10^{-5} \cdot 293} \sqrt{293} \left(0.41 - \sqrt{1 - 0.61^{0.29}}\right) \cong 0.0125 \text{ sec.}$$

2. We estimate the lifting time t_2 by supposing that all the pressure P_r acts during the entire lifting process. Of course, this will be the lower estimate for the time the lifting will take. Thus, the acting force F_a is then:

$$F_a = P_r \pi D^2 / 4 - mg \cong 5 \; 10^5 \pi \, 0.15^2 / 4 - 500 \cdot 9.8 \cong 3931 \text{ N,}$$

and the acceleration a is then

$$a = F_a / m \cong 3931 \, N / 500 \text{ kg} \cong 7.86 \text{ m/sec}^2,$$

and the lifting time t_2 is

$$t_2 \cong \sqrt{\frac{2s}{a}} \cong \sqrt{\frac{2 \cdot 1.5}{7.86}} \cong 0.6177 \text{ sec.}$$

The total time t_{500} is

$$t_{500} \cong t_0 + t_1' + t_1'' + t_2 \cong 0.0294 + 0.0846 + 0.0125 + 0.6177 \cong 0.744 \text{ sec.}$$

9   Solution to Exercise 3E-5a)

The time sought here is a sum of three components.

Component 1

The first component t_0 is the time needed by the pressure wave to travel from the valve to the inlet of the volume:

$$t_0 = L/V_s \cong 20\,\text{m}/340\,\text{m}/\text{sec} \cong 0.0588\,\text{sec},$$

where V_s is the speed of sound.

Component 2

The second component t_1 is the time needed to reach a value of pressure P_r in the given volume. Here, we consider the combination of supercritical and subcritical air flow regimes. The duration t_1 consists, therefore, of two components; thus, $t_1 = t_1' + t_1''$. The component t_1' is the time needed to reach critical pressure P_{cr} in the given volume (supercritical regime), while the component t_1'' is the time needed to subsequently reach the pressure $P_{c\,500}$ (subcritical regime).

The value

$$P_{cr} = \beta_{cr}\, P_r = 0.528 \cdot 5 \cdot 10^5 \cong 2.64\ 10^5\,\text{N}/\text{m}^2.$$

Then, from Expression (3.120), we obtain for t_1'

$$t_1' \cong \frac{V_c\,(P_{cr} - P_0)}{0.67 \cdot T\alpha F_p P_r} \cong \frac{0.2 \cdot (2.64 - 0) \cdot 10^5}{0.67 \cdot 293 \cdot 0.5 \cdot 11.3 \cdot 10^{-5} \cdot 5 \cdot 10^5} \cong 9.5\,\text{sec}.$$

Here, as in solution 3E-5, $F_p = 11.3\ 10^{-5}\,\text{m}^2$.

The component t_1'' is calculated from Expression (3.127). We now need to determine the value of the pressure P_c: where $\beta \cong 1$ (same pressures in the receiver and the filled volume),

$$t_1'' \cong \frac{0.154 V_c}{\alpha T F_p} \sqrt{T_r}\,(0.41 - \sqrt{1 - \beta^{0.29}}) \cong$$

$$\cong \frac{0.154 \cdot 0.2}{0.5 \cdot 11.3 \cdot 10^{-5} \cdot 293} \sqrt{293}\,(0.41 - \sqrt{1 - 1^{0.29}}) \cong 13.0\,\text{sec}.$$

The total time t needed to fulfill the task is

$$t = t_0 + t_1' + t_1'' \cong 0.06 + 9.5 + 13.0 \cong 23\,\text{sec}.$$

10   Solution to Exercise 3E-5b)

The time sought here is comprised of three components.

Component 1

The first component t_0 is the time needed by the pressure wave to travel from the valve to the inlet of the cylinder of the jig:

$$t_0 = L/V_s \cong 20 \, \text{m} / 340 \, \text{m} / \sec \cong 0.0588 \, \text{sec},$$

where V_s is the speed of sound.

Component 2

The second component t_1 is the time needed to reach a value of the pressure P_c in the cylinder that develops the needed force $Q = 5000$ N. We consider here the combination of supercritical and subcritical air flow regimes. The duration t_1 consists of two components: $t_1 = t_1' + t_1''$. The component t_1' is the time needed to reach critical pressure P_{cr} in the given volume V_c of the cylinder (supercritical regime), while the component t_1'' is the time needed subsequently to reach the pressure P_{c500} (subcritical regime).

The value

$$P_{cr} = \beta_{cr} P_r = 0.528 \cdot 6 \cdot 10^5 \cong 3.168 \ 10^5 \, \text{N} / \text{m}^2.$$

Then, from Expression (3.120) we obtain for t_1'

$$t_1' \cong \frac{V_c \, (P_{cr} - P_0)}{0.67 \cdot T \alpha F_p P_r} \cong \frac{0.002 \cdot (3.168 - 0) \cdot 10^5}{0.67 \cdot 293 \cdot 0.5 \cdot 11.3 \cdot 10^{-5} \cdot 6 \cdot 10^5} \cong 0.095 \, \text{sec}.$$

Here, as in solution 3E-5, $F_p = 11.3 \ 10^{-5} \, \text{m}^2$.

The component t_1'' is calculated from Expression (3.127). We now need to determine the value of the pressure P_c:

$$P_c = 4Q / \pi D^2 \cong 4 \ 5000 / \pi \ 0.125^2 \cong 4.08 \ 10^5 \, \text{N} / \text{m}^2$$

and

$$\beta \cong P_c / P_r \cong 4.08/6 \cong 0.679 \quad \text{(see definition for } \beta\text{):}$$

$$t_1'' \cong \frac{0.154 V_c}{\alpha T F_p} \sqrt{T_r} \, (0.41 - \sqrt{1 - \beta^{0.29}}) \cong$$

$$\cong \frac{0.154 \cdot 0.002}{0.5 \cdot 11.3 \cdot 10^{-5} \cdot 293} \sqrt{293} \, (0.41 - \sqrt{1 - 0.679^{0.29}}) \cong 0.026 \, \text{sec}.$$

The total time t needed to fulfill the task is:

$$t = t_0 + t_1' + t_1'' \cong 0.06 + 0.095 + 0.026 \cong 0.18 \, \text{sec}.$$

11   Solution to Exercise 3E-5c)

The first step is to reduce the given mechanism to a single-mass system. The rotating speed ω_b of the brake drum, obviously, varies in inverse proportion to the ratio $i = 1 : 3.16$:

$$\omega_b = 3.16\,\omega.$$

Thus, the initial speed ω_0 of the brake drum is

$$\omega_0 = 3.16\,\omega \cong 3.16 \ 1500\,\text{rpm} \ 2\pi / 60 \cong 522.5 \ 1/\text{sec}.$$

The procedure of reducing the inertia of all moving parts with respect to the shaft of the brake requires calculation of the common kinetic energy E of the mechanism, which is

$$E = \left(I_0 \omega_b^2 + I_1 [\omega / 3.16]^2\right) / 2,$$

where ω_b is the angular speed of the shaft of the brake drum. The kinetic energy of the reduced system with a lumped moment of inertia I respective to the shaft of the brake drum is

$$E = I\omega_b^2 / 2$$

or

$$I\omega_b^2 / 2 = \left(I_0 \ \omega_b^2 + I_1 [\omega_b / 3.16]^2\right) / 2$$

or

$$I = I_0 + I_1 [1 / 3.16]^2 \cong$$
$$\cong 0.01 + 0.1 \ 1 / 3.16^2 \cong 0.02 \ \text{kg m}^2.$$

The differential equation for the first case to be solved takes the following form:

$$I\,\ddot{\phi} = -T_0 - T_1\phi = -5 - 4\,\phi$$

or

$$\ddot{\phi} + \frac{4\phi}{I} = -\frac{5}{I} \quad \text{and} \quad \ddot{\phi} + 200\phi = -250.$$

The solution then is made up of

$$\phi = \phi_1 + \phi_2,$$

where the solution of the homogeneous equation is

$$\phi_1 = A\cos(kt + \gamma)$$

and the partial solution is

$$\phi_2 = -\frac{T_0 I}{T_1 I} = -\frac{250}{200} = -1.25.$$

Obviously, here

$$k^2 = 200 \, 1/\sec^2 \quad \rightarrow \quad k \cong 14.14 \, 1/\sec.$$

Thus, the solution is

$$\phi = A\cos(14.14\,t + \gamma) - 1.25. \tag{a}$$

We seek the constants A and γ through the initial conditions, which are

when $t = 0$, then $\phi = 0$ and $\dot{\phi} = 522 \, 1/\sec$,

$$\dot{\phi} = -14.14 \, A\sin(14.14\,t + \gamma). \tag{b}$$

From (a) and (b), we obtain for $t = 0$ the value of the phase angle γ:

$$\tan\gamma = \frac{\sin\gamma}{\cos\gamma} \cong -\frac{522}{14.14 \cdot 1.25} \cong -29.5 \quad \rightarrow \quad \gamma \cong -1.54.$$

Now, we can find the value of A from either (a) or (b):

$$0 = A\cos(\gamma) - 1.25 \quad \rightarrow \quad A \cong 1.25/\cos 1.53 \cong 41 \, \text{rad}.$$

To reach a situation at which $\omega = 0$, let us substitute this condition into Expression (b) and find the value t

$$0 = -14.14 \cdot 41 \cdot \sin(14.14\,t - 1.54)$$

and

$$(14.14\,t - 1.53) = 0 \quad \rightarrow \quad t \cong 1.53/14.14 \cong 0.11 \, \sec.$$

In MATHEMATICA language the solution is expressed as:

```
b1=p[t]-f'[t]
b2=p'[t]+200 f[t]+250
y3=DSolve[{b1==0,b2==0,f[0]==0,p[0]==522},{f,p},{t}]
y1=Plot[Evaluate[f[t]/.y3],{t,0,.15},
AxesLabel->{"t","f"}]
```

The second case is solved as follows. We begin by writing the motion equation, which is

$$I\ddot{\phi} = -T_0 - T_1\dot{\phi} \quad \rightarrow \quad I\dot{\omega} = -T_0 - T_1\omega,$$

or by substituting numerical data and rearranging we obtain

$$\dot{\omega} + \frac{T_1}{I}\omega = -\frac{T_0}{I} \quad \rightarrow \quad \dot{\omega} + \frac{4}{0.02}\omega = -\frac{5}{0.02}.$$

Finally,

$$\dot{\omega} + 200\,\omega = -250. \tag{c}$$

The solution is again composed of two components

$$\omega = \omega_1 + \omega_2.$$

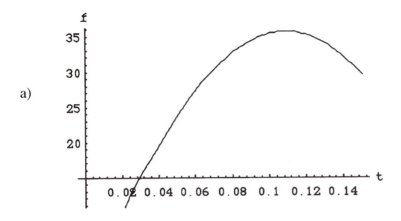

a)

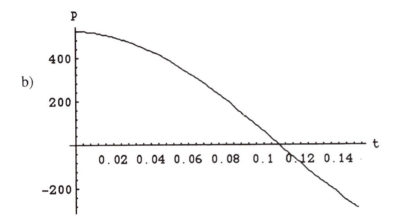

b)

FIGURE 3E-5c).1   a) Rotation angle f; b) Speed p versus time when braking.

The solution for the homogeneous equation ω_1 is sought in the form

$$\omega_1 = A\exp[\alpha t],$$

where, as follows from (c), $\alpha \cong -200$ 1/sec.

The component ω_2 is sought, obviously as a constant, and from (c) it follows that

$$\omega_2 = -\frac{T_0 I}{T_1 I} = -\frac{250}{200} = -1.25.$$

Thus, the complete solution is

$$\omega = A\exp[-200t] - 1.25.$$

From the initial condition, we find the value of A. For $t = 0$, the speed $\omega_0 = 522$ 1/sec. Therefore,

$$522 = A - 1.25 \quad \rightarrow \quad A = 523.25 \text{ rad/sec},$$

and, finally, the solution is

$$\omega = 523.5\exp[-200t]-1.25. \qquad [d]$$

Now to obtain the braking time, from (d) we have

$$0 \cong 523.5\exp[-200\,t]-1.25,$$

$$\ln(1.25/523.5) \cong -200t \quad \rightarrow \quad t \cong 0.030\,\text{sec}.$$

In MATHEMATICA language the solution is:

b1=w'[t] +200 w[t]+250
y1=DSolve[{b1==0,w[0]==522},{w},{t}]
j1=Plot[Evaluate[w[t]/.y1],{t,0,.1}]

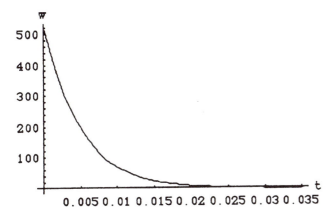

FIGURE 3E-5c).2   Rotation speed versus time when braking (case b).

12   Solution to Exercise 3E-6

The question relates the rotation angle ϕ of the column to the torques acting on the mechanism. The driving torque T must be equalized by inertia torques, in keeping with the expression (3.165):

$$T = d[I\dot{\phi}]/dt,$$

where I is the common moment of inertia of the column I_0 and the lever I_r. Obviously:

$$I = I_0 + I_r = I_0 + m[r\cos(\alpha t)]^2.$$

Therefore,

$$T = d\{[I_0 + m(r\cos\alpha t)^2]\dot{\phi}\}/dt$$

or

$$T = (I_0 + m[r\cos\alpha t]^2)\ddot{\phi} - m\alpha r^2 \sin 2\alpha t \cdot \dot{\phi}. \qquad [a]$$

For a DC motor with a characteristic $T = T_1 - T_0\dot{\phi}$, Expression (a) becomes

$$T_0 = (I_0 + m[r\cos\alpha t]^2)\ddot{\phi} - m\alpha r^2 \sin 2\alpha t \cdot \dot{\phi} + T_1\dot{\phi}.$$

A solution using MATHEMATICA language follows. The numerical data given in the problem are substituted by symbols (here, for reasons of convenience, we denote $\phi = v$ and $\dot{\phi} = w$).

```
f1=.01 v"[t]+.5 (Cos[.5 t])^2 v"[t]-
.125 (Sin[t]) v'[t] -.1
j1=NDSolve[{f1==0,v[0]==0,v'[0]==0},{v[t]},{t,0,2}]
b1=Plot[Evaluate[v[t]/.j1],{t,0,2},AxesLabel->{"t","v"}]
```

```
f2=.01 w'[t]+.5 (Cos[.5 t])^2 w'[t]-
.125 (Sin[t]) w[t]-.1
j2=NDSolve[{f2==0,w[0]==0},{w[t]},{t,0,2}]
b2=Plot[Evaluate[w[t]/.j2],{t,0,2},AxesLabel->{"t","w"}]
```

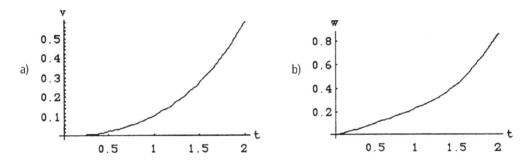

FIGURE 3E-6.1   a) Rotation angle v; b) Speed w of the column versus time.

13   Solution to Exercise 3E-6a)

The rotation angle ϕ of the disc depends upon the torques acting on the mechanism. The driving torque T must be equalized by inertia torques, in keeping with expression (3.165):

$$T = d[I\dot{\phi}]/dt,$$

where I is the common moment of inertia of the disc I_0 and the moving mass I_r. Obviously:

$$I = I_0 + I_r = I_0 + m[R_0\cos(\alpha t)]^2.$$

Therefore,

$$T = d\left\{\left[I_0 + m(R_0 \cos \alpha t)^2\right]\dot\phi\right\}/dt. \qquad [a]$$

For a DC motor with a characteristic $T = T_1 - T_0\dot\phi$, the Expression (a) becomes

$$T_0 = \left(I_0 + m[R_0 \cos\alpha t]^2\right)\ddot\phi - m\alpha R_0^{\,2}\sin 2\alpha t \cdot \dot\phi + T_1\dot\phi.$$

For the same data as in solution 3E-6, and where $T_1 = 0.1$ Nm and $T_0 = 0.025$ Nm/sec and $R_0 = 0.5$ m, the following solution using MATHEMATICA is given. The equation is

$$0.01\ddot\phi + 0.5\,(\cos 0.5\,t)^2\,\ddot\phi - 0.125 \sin t\,\dot\phi - 0.1 + 0.025\,\dot\phi = 0.$$

```
f01=.01 v"[t]+.5 (Cos[.5 t])^2 v"[t]-.125 (Sin[t]) v'[t] -.1 +0.025 v'[t]
j01=NDSolve[{f01==0,v[0]==0,v'[0]==0},{v[t]},{t,0,2}]
b01=Plot[Evaluate[v[t]/.j01],{t,0,2},AxesLabel->{"t","v"}]

f02=.01 w'[t]+.5 (Cos[.5 t])^2 w'[t]-.125 (Sin[t]) w[t] -.1 +0.025 w[t]
j02=NDSolve[{f02==0,w[0]==0,w'[0]==0},{w[t]},{t,0,2}]
b02=Plot[Evaluate[w[t]/.j02],{t,0,2},AxesLabel->{"t","w"}]
```

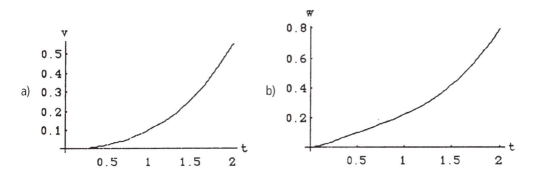

FIGURE 3E-6a).1   a) Rotation angle v; b) Speed w of the column versus time.

14   Solution to Exercise 3E-7

To answer the questions we use Formula (3.39). Thus, from this formula, it follows that when the number of winds $W_s = 2$ is doubled, we have the following expressions for the response time:

$$t_s = t_2/t_1;$$

$$W_s = t_s u_s/\sqrt{m_s\delta_s S_s};$$

$$t_s = W_s\sqrt{m_s\delta_s S_s}/u_s \Rightarrow 2\sqrt{m_s\delta_s S_s}/u_s = t_s = 2.$$

From these formulas, it follows that when the voltage is doubled, $u_s = 2$, we have the following expressions for the response time:

$$t_s = t_2 / t_1;$$

$$W_s = t_s u_s / \sqrt{m_s \delta_s S_s};$$

$$t_s = W_s \sqrt{m_s \delta_s S_s} / u_s \Rightarrow W_s \sqrt{m_s \delta_s S_s} / 2 = t_s = 0.5.$$

From these formulas it follows that when the mass of the armature is doubled, $m_s = 2$, we have the following expressions for the response time:

$$t_s = t_2 / t_1;$$

$$W_s = t_s u_s / \sqrt{m_s \delta_s S_s};$$

$$t_s = W_s \sqrt{m_s \delta_s S_s} / u_s \Rightarrow W_s \sqrt{2 \delta_s S_s} / u_s = t_s \cong 1.41.$$

15   Solution to Exercise 4E-1

Case a)

From geometrical considerations, the motion function $\Pi(x)$ becomes

$$y = \sqrt{L^2 - x^2}.$$ [a]

Differentiating (a), we obtain

$$\dot{y} = \frac{-x}{\sqrt{L^2 - x^2}} \dot{x}.$$ [b]

Thus,

$$\Pi'(x) = -\frac{x}{\sqrt{L^2 - x^2}}.$$ [c]

Substituting the given data into (c), we obtain for $\dot{y}$

$$\dot{y} = \frac{-0.05}{\sqrt{0.15^2 - 0.05^2}} 0.1 \cong -0.0353 \, \text{m} / \text{sec}.$$

By differentiating (b), we obtain the following dependence from Expression (4.3) [the case where $\ddot{x} = 0$]:

$$\ddot{y} = \Pi''(x) \dot{x}^2 = -\frac{L^2}{(L^2 - x^2)^{3/2}} \dot{x}^2 \cong$$

$$\cong -\frac{0.15^2}{(0.15^2 - 0.05^2)^{3/2}} 0.1^2 \cong -0.07955 \, \text{m} / \text{sec}^2.$$

From (c) and the Relationship (4.4) we obtain

$$F_{input} / F_{output} = \Pi'(x) = 0.353. \qquad [d]$$

Substituting the numerical data into (c) and (d), we obtain

$$F_{output} = \frac{-5 \cdot 0.05}{\sqrt{0.15^2 - 0.05^2}} \cong -14.16 \, \text{N}.$$

Case b)

From the geometry of the given mechanism, we have $AD = CE$. Then, the motion function $\Pi(x)$ is defined as follows:

$$y = \Pi(\phi) = \overline{AO} \sin\phi = 0.2 \sin 30° = 0.1 \, \text{m}.$$

Thus,

$$\dot{y} = \Pi'(\phi)\dot{\phi} = 0.2 \cos 30° \, 5 \cong 0.866 \, \text{m/sec},$$

and

$$\ddot{y} = \Pi''(\phi)\dot{\phi}^2 = -0.2 \sin 30° \, 5^2 \cong -2.5 \, \text{m/sec}^2.$$

16 Solution to Exercise 4E-2

From the Formula (4.24) and its derivatives, we have:

$$\tan\alpha = \frac{\Pi'(\phi)}{r_0 + \Pi(\phi)}. \qquad [a]$$

Here, it follows from the description of the problem that

$$y = \Pi(\phi) = \frac{h}{2}\left[1 - \cos(4 \cdot \phi)\right],$$

and therefore

$$\Pi'(\phi) = 2h \sin(4 \cdot \phi).$$

Thus, from (a) we obtain

$$\tan\alpha = \frac{2h \sin(4 \cdot \phi)}{0.08 + \dfrac{h}{2}\left[1 - \cos(4 \cdot \phi)\right]}. \qquad [b]$$

To find the angle ϕ corresponding to the maximum pressure angle α_{max}, we differentiate (b):

$$\frac{d(\tan\alpha)}{d\phi} = \frac{8h\cos4\phi\left[0.08+\frac{h}{2}(1-\cos4\phi)\right]-4h^2\sin^2 4\phi}{\left[0.08+\frac{h}{2}(1-\cos4\phi)\right]^2} = 0. \qquad [c]$$

From (c), it follows that

$$2\cdot0.08\cos(4\cdot\phi)+h\cos(4\cdot\phi)-h[\cos^2(4\cdot\phi)+\sin^2(4\cdot\phi)]=0$$

or

$$h=\frac{0.16\cos(4\cdot\phi)}{1-\cos(4\cdot\phi)}. \qquad [d]$$

On the other hand, from (b), we have

$$2h\sin(4\cdot\phi)-tg\,\alpha\left[0.08+\frac{h}{2}(1-\cos4\cdot\phi)\right]=0. \qquad [e]$$

Substituting $\tan\alpha = \tan 20° \cong 0.324$ into (e), and from (d), we obtain

$$\frac{0.16\sin 8\phi}{1-\cos4\phi}-0.0291-\frac{0.0291\cos4\phi}{1-\cos4\phi}-0.08\frac{\cos^2 4\phi}{1-\cos4\phi}=0. \qquad [f]$$

Solving Equation (f) by any method (for instance, graphically, by the method of Newton, or by computer) we obtain

$$\phi\cong0.37\text{ rad,}$$

which from (d) gives for h

$$h=\frac{0.16\cos 4\cdot0.37}{1-\cos 4\cdot0.37}\cong0.016\,m.$$

The solution in MATHEMATICA language is

a1=(2 Sin[8 f]-.364 (Cos[4 f]+(Cos[4 f])^2))/(1-Cos[4 f])-.364
b1=FindRoot[a1==0,{f,.5}]
{f -> 0.369625}

17   Solution to Exercise 6E-1

Condition (6.17) states that horizontal component A_h of the acceleration takes the form

$$A_h > \mu(g-A_v), \qquad [a]$$

where

$$A_h = A\cos\alpha \quad \text{and} \quad A_v = A\sin\alpha.$$

A is the vibrational amplitude. Assuming that the vibrations S have the form:

$$S = a\sin\omega t.$$

Then the accelerations are

$$A = -a\omega^2 \cos\omega t.$$

Thus,

$$A_h = a\omega^2 \cos\alpha \quad \text{and} \quad A_v = a\omega^2 \sin\alpha.$$

Case a)

Condition (a) can then be rewritten in a form that takes into account that $\omega = 2\pi f = 2\pi 50 \cong 314 \ 1/\sec$:

$$A_h = a314^2 \cos 10^0 \cong 97098 \, a\,\mathrm{m}/\sec^2,$$
$$A_v = a314^2 \sin 10^0 \cong 17121 \, a\,\mathrm{m}/\sec^2.$$

[b]

From (a) follows

$$97098\,a > 0.2(9.8 - 17121\,a)$$

or

$$a > 0.2 \ 9.8/(97098 + 0.2 \ 17121) \cong 0.00002 \,\mathrm{m}.$$

Case b)

From the condition (a) and taking into account (b) of the previous case, it follows also for this case that

$$A_h = 0.00001\omega^2 \cos 10^0 \cong 0.97098 \ 10^{-5}\omega^2 \,\mathrm{m}/\sec^2,$$
$$A_v = 0.00001\omega^2 \sin 10^0 \cong 0.17121 \ 10^{-5}\omega^2 \,\mathrm{m}/\sec^2.$$

Now we obtain

$$0.97098 \ 10^{-5}\omega^2 = 0.2(9.8 - 0.17121 \ 10^{-5} \ \omega^2)$$

or

$$\omega^2 \cong 0.2 \ 9.8/(0.97098 + 0.2 \ 0.17121)10^5 \cong 195000 \ 1/\sec^2$$

and

$$\omega \cong 441 \ 1/\sec \quad \text{or} \quad f \cong 70\,\mathrm{Hz}.$$

18   Solution to Exercise 7E-1

Here we use Equation System (7.1). Since the two levers press the strip from both sides (upper and lower), the mechanism must develop a friction force $P = F/2$ at every contact point. Thus, the equations for forces and torques with respect to point O become

$$P = \mu N = F/2, \tag{a}$$

$$Ql - NL + P\left(\frac{H-h}{2}\right) = 0, \tag{b}$$

and

$$P = R_x; \quad N = R_y. \tag{c}$$

Here, R_x and R_y are the reaction forces in hinge 0; N is the normal force at the contact point between the strip and the lever. From (a), we express the normal force N as

$$N = \frac{F}{2\mu} = \frac{100}{2 \cdot 0.15} \cong 333 \text{ N.}$$

From the Equation (b), we express the force Q developed by the spring as

$$Q = \frac{NL - \dfrac{F}{2}\left(\dfrac{H-h}{2}\right)}{l} \cong$$

$$\cong \frac{333 \cdot 0.1 - \dfrac{100}{2}\left(\dfrac{0.06 - 0.004}{2}\right)}{0.05} \geq 638 \text{ N.}$$

Reactions R_x and R_y are, respectively,

$$R_x = \frac{F}{2} \cong 50 \text{N}, \qquad R_y = N \cong 333 \text{N.}$$

19   Solution to Exercise 7E-1a)

Here we use equation system (7.6). Since the two rollers press the strip from both sides (upper and lower), the mechanism must develop a friction force $F_b = Q/2$ at every point of contact with the strip. Thus, the equations for forces and torques with respect to point O become

$$N_b - N_a \cos 15° - F_a \sin 15° = 0, \tag{a}$$

$$-F_b - F_a \cos 15° - N_c = 0, \tag{b}$$

and

$$\mu N_b \ge F_b, \quad \mu N_a \ge F_a. \qquad [c]$$

From the Equation (a), it follows that

$$\frac{N_b}{N_a} = \cos 15 + \mu \sin 15° \cong 1.044.$$

From the Equation (b), it follows that

$$N_c = -\mu \left(N_b + N_a \cos 15° \right) = -\mu N_b \left(1 + \frac{N_a}{N_b} \cos 15° \right)$$

or

$$N_b = \frac{Q}{\mu \cdot \left(1 + \dfrac{N_a}{N_b} \cos 15° \right)} \cong \frac{20}{0.3 \cdot \left(1 + \dfrac{1}{1.044} \cos 15° \right)} \cong 34.65 \text{ N}.$$

From the Equation (c) and the given mechanism it follows that

$$Q = 2 F_b = 2 \mu N_b = 2 \cdot 0.3 \cdot N_b = 0.6 N_b$$

and finally

$$Q \cong 0.6 \cdot 34.65 \text{ N} = 20.79 \text{ N}.$$

20   Solution to Exercise 7E-1b

We continue to use Equation System (7.1). Since the two levers press the strip from both sides (right and left), the mechanism must develop a friction force $F = Q/2$ at every contact point. Thus, the equations for forces and torques with respect to point O become

$$F = \mu N = \frac{Q}{2} = \frac{40}{2} = 20 N \quad \rightarrow \quad N = \frac{F}{\mu} = \frac{20}{0.4} = 50 N,$$

$$P\frac{l}{2} + Fl - NA = 0,$$

and

$$A = \frac{l}{2 \cdot 2N}(P+Q) = \frac{H-h}{2 \cdot 2N}(P+Q) =$$

$$\cong \frac{80-20}{2 \cdot 2 \cdot 50}(0.8+40) \cong 12.25 \text{ mm}.$$

21   Solution to Exercise 7E-2

The angular frequency ω of the oscillations of the bowl is

$$\omega = 2\pi f = 2\pi 50 \cong 314 \ 1/\text{sec}.$$

The motion S of the bowl is: $S = 0.0001 \sin 314 \, t$ m. The acceleration $\ddot{S}$ of the bowl obviously is

$$\ddot{S} = -a\omega^2 \sin \omega t \cong -0.0001 \cdot 314^2 \sin 314 t \ \text{m}/\text{sec}^2.$$

The maximal value of the acceleration $\ddot{S}_{\text{max}}$ is

$$\ddot{S}_{\text{max}} = a\omega^2 = 0.0001 \cdot 314^2 \cong 10 \ \text{m}/\text{sec}^2.$$

The angle $\beta = \gamma - \alpha = 30° - 2° = 28°$. From Expressions (7.33–7.34), we calculate the values of critical accelerations for the half-periods of both positive and negative oscillations. Thus,

$$\ddot{S}_{cr}' \geq g\,\frac{\mu \cos \alpha + \sin \alpha}{\mu \sin \beta + \cos \beta} = 9.8\,\frac{\sin 2° + 0.6 \cos 2°}{0.6 \sin 28° + \cos 28°} \approx 5.34 \ \text{m}/\text{sec}^2,$$

and

$$\ddot{S}_{cr} \geq g\,\frac{\sin \alpha - \mu \cos \alpha}{\mu \sin \beta - \cos \beta} = 9.8\,\frac{\sin 2° - 0.6 \cos 2°}{0.6 \sin 28° - \cos 28°} \approx 9.4 \ \text{m}/\text{sec}^2.$$

The latter expression means that during the second half-period of oscillations slide conditions practically do not occur for the body on the tray . By applying Expression (7.35), we check whether rebound conditions exist on the tray, a situation that occurs when the acceleration exceeds the value S_r. Thus,

$$\ddot{S}_r = g\,\frac{\cos \alpha}{\sin \beta} \cong 9.8\,\frac{\cos 2°}{\sin 28°} \cong 20.85 \ \text{m}/\text{sec}^2.$$

At any point of movement, no point of the bowl reaches this acceleration value. Therefore, there is no rebound in the discussed case.

We can now proceed to calculate the displacement of the items. From the curves in Figure 7.25 it follows that the time t_1, at which the slide begins (section EM) and the groove lags behind the item, is defined as

$$t_1 = \frac{1}{\omega} \arcsin \frac{\ddot{S}_{cr}}{\ddot{S}_{\text{max}}}. \qquad\qquad [a]$$

At this moment in time, the speed V_0 of the item (and the bowl) is defined as

$$V_0 = \dot{S}(t_1) = a\omega \cos\left[\arcsin\frac{\ddot{S}_{cr}}{\ddot{S}_{max}}\right] \cong 0.0001 \cdot 314\cos\left[\arcsin\frac{5.34}{10}\right] \cong 0.027\,\text{m/sec.} \quad \text{[b]}$$

The slide begins with this speed and is under the influence of the friction force $F \cong -\mu m\,(g+\ddot{y})$ acting backwards. For our engineering purposes, we simplify this definition to the form $F \cong \mu mg$. This force causes deceleration:

$$W_L = -\mu \cong -0.6 \cdot 9.8 \cong 5.88\,\text{m/sec}^2.$$

This assumption gives a lower estimation of the displacement. The following gives the upper estimation:

$$W_u = -\mu(g - \ddot{S}_{cr}\sin\beta) \cong -0.6(9.8 - 5.34\sin 28°) \cong 4.37\text{m/sec}^2.$$

This condition exists during time t_2, which is defined as

$$t_2 = \frac{1}{\omega}\left[\pi - 2\arcsin\frac{\ddot{S}_{cr}}{\ddot{S}_{max}}\right] \cong \frac{1}{314}\left[\pi - 2\arcsin\frac{5.34}{10}\right] \cong 0.0066\,\text{sec.} \quad \text{[c]}$$

The displacement δ_1 is then

$$V_0 t_2 - \frac{W_L t_2^2}{2} \le \delta_1 \le V_0 t_2 - \frac{W_U t_2^2}{2} \quad \text{[d]}$$

or

$$0.027 \cdot 0.0066 - \frac{5.88 \cdot 0.0066^2}{2} \le \delta_1 \le 0.027 \cdot 0.0066 - \frac{4.37 \cdot 0.0066^2}{2}$$

or

$$0.000053\,m \le \delta_1 \le 0.000083\,m.$$

It is interesting to observe the influence of the friction coefficient μ on the values of the critical accelerations for both oscillation directions. We show here the computation in MATHEMATICA language. Results are given in Figure 7E-2. (For convenience in MATHEMATICA we use m for the friction coefficient.)

```
g1=Plot[9.8 (Sin[2 Degree]+m Cos[2 Degree])/
(m Sin[30 Degree]+Cos[30 Degree]),
{m,.2,1},AxesLabel->{"m","s""}]

g2=Plot[9.8 (Sin[2 Degree]-m Cos[2 Degree])/
(m Sin[30 Degree]-Cos[30 Degree]),
{m,.2,1},AxesLabel->{"m","s""}]

Show[g1,g2]
```

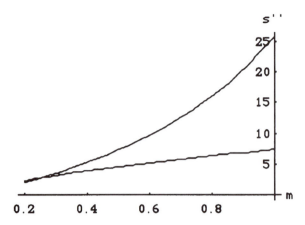

FIGURE 7E-2.1 Dependence "critical acceleration s" versus friction coefficient "m" for the specific design of the vibrofeeder as in this and next exercises.

This displacement takes place 50 times every second. Therefore, the total displacement H during one second is

$$0.00265m \le H_1 \le 0.0041\,m. \qquad [e]$$

22   Solution to Exercise 7E-3

When increasing the vibrations amplitude "a" to 0.00015 m for the same vibrofeeder as in the previous Exercise 7E-2, obviously, we change the dynamics of the device. However, its characteristics remain the same. Therefore, we have the same values of the critical accelerations $\ddot{S}_{cr}$ and $\ddot{S}'_{cr}$ as before. The changes in the dynamic behavior of the device take place because of the fact that the maximal value of the acceleration of the bowl $\ddot{S}_{max}$ becomes higher, due to the increased oscillations amplitude "a." In this case, we have

$$\ddot{S}_{max} = a \cdot \omega^2 \cong 0.00015 \cdot 314^2 \cong 15\,\text{m}/\text{sec}.$$

We follow the same procedure as in the previous problem, and for the value of the speed at the moment in time that the slide begins, we obtain

$$V_0 = \dot{S}(t_1) = a\omega\cos\left[\arcsin\frac{\ddot{S}_{cr}}{\ddot{S}_{max}}\right] \cong 0.00015 \cdot 314\cos\left[\arcsin\frac{5.34}{15}\right] \cong 0.044\,\text{m}/\text{sec},$$

$$t_2 = \frac{1}{\omega}\left[\pi - 2\arcsin\frac{\ddot{S}_{cr}}{\ddot{S}_{max}}\right] \cong \frac{1}{314}\left[\pi - 2\arcsin\frac{5.34}{15}\right] \cong 0.0077\,\text{sec},$$

and

$$0.044 \cdot 0.0077 - \frac{5.88 \cdot 0.0077^2}{2} \le \delta_1 \le 0.044 \cdot 0.0077 - \frac{4.37 \cdot 0.0077^2}{2},$$

$$0.000165m \le \delta_1 \le 0.00021\,m.$$

The conditions of this exercise result in the appearance of a backslide in the domain EK. To calculate the back displacement δ_2, we apply the ideas of the forward displacement in formulas (b), (c), (d), and (e):

$$V_0' = -a\omega \cos\left[\arcsin\frac{\ddot{S}_{cr}'}{\ddot{S}_{max}}\right] \cong -0.00015\cdot 314 \cos[\arcsin\frac{9.4}{15}] \cong -0.032\,\text{m/sec}$$

$$t_4 = \frac{1}{\omega}\left[\pi - 2\arcsin\frac{\ddot{S}_{cr}'}{\ddot{S}_{max}}\right] \cong \frac{1}{314}\left[\pi - 2arc\sin\frac{9.4}{15}\right] \cong 0.0057\,\text{sec}$$

$$W = \mu(g + \ddot{S}_{cr}'' \sin\beta) \cong 0.6(9.8 + 9.4\sin 28^{o}) \cong 8.53\,\text{m/sec}^2$$

$$-0.032\cdot 0.0057 + \frac{5.88\cdot 0.0057^2}{2} \geq \delta_2 \geq -0.032\cdot 0.0057 + \frac{8.53\cdot 0.0057^2}{2}$$

or

$$-0.000087\,m \geq \delta_1 \geq -0.000044\,m.$$

The total displacement δ_3 during one period of the bowl's vibrations is obviously

$$\delta_3 = \delta_1 + \delta_2 = 0.00021 + 0.000044 \cong 0.000166\,m$$

and

$$\delta_3{}' = \delta_1{}' + \delta_2{}' = 0.000165 - 0.000087 \cong 0.000078\,m.$$

Finally,

$$0.0039\,m \leq H_1 \leq 0.0083\,m.$$

And, for the last time, we put forward an illustration of the displacement of a body on the tray of the feeder calculated in keeping with Equation (7.34) by means of the MATHEMATICA language. The result is shown in Figure 7E-3.

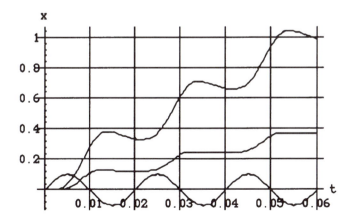

FIGURE 7E-3.1 Body movement on the tray of the vibrofeeder.

23   Solution to Exercise 7E-4

To answer the question we use Figure 7.28. We begin with the simplest case—a cube with a hole drilled symmetrically in the middle of it ($A = B = C$ and $H = 2h$). This case is analogous to case 4 in the figure (the hole makes a difference to one of the dimensions). Therefore, it has three different positions on the tray. When a right parallelepiped with a symmetrically located hole ($H = 2h$)—for both cases: $A \neq B = C$ and $A \neq B \neq C$—is considered, we have a body possessing three planes of symmetry—line 3 in the figure. This gives six different positions of the body on the tray. Finally, the most common case, when the hole is located so that $H \neq 2h$, fits line 2 in the figure for both cases. The body possesses two planes of symmetry and, therefore, 12 different positions on the tray are possible. This results are presented in the following table.

	$A = B = C$	$A \neq B = C$	$A \neq B \neq C$
$H = 2h$	3	6	12
$H \neq 2h$	12	12	12

24   Solution to Exercise 7E-5

To answer the question we use Figure 7.28. This is the case that corresponds to line 2 in the figure. The body possesses two planes of symmetry and, therefore, 12 different positions on the tray are possible. Because of its internal asymmetry, this body requires special means for its orientation. These means are, for instance, a) utilization of the location of the asymmetrical mass center, and b) means of electrodynamic or magnetic orientation.

Recommended Readings

Fu, K. S., R. C. Gonzales, and C. S. G. Lee, *Robotics: Control, Sensing, Vision and Intelligence,* McGraw-Hill Book Company, New York, 1987.

Pessen, D. W., *Industrial Automation,* John Wiley & Sons, New York, 1989.

Ogata, Katsuhiko, *System Dynamics: Second Edition,* Prentice-Hall, Englewood Cliffs, New Jersey, 1992.

Dieter, George, *Engineering Design: A Materials and Processing Approach: Second Edition,* McGraw-Hill, Inc., New York, 1991.

Schey, John A., *Introduction to Manufacturing Processes: Second Edition,* McGraw-Hill International Editions, New York, 1987.

Powers Jr., John H., *Computer-Automated Manufacturing,* McGraw-Hill International Editions, New York, 1987.

Critchlow, Arthur J., *Introduction to Robotics,* Macmillan Publishing Company, New York, Collier Macmillan Publishers, London, 1985.

Bradley, D. A., D. Dawson, N. C. Burd, and A. J. Loader, *Mechatronics: Electronics in Products and Processes,* Chapman & Hall, London, 1996.

Slocum, Alexander H., *Precision Machine Design,* Prentice-Hall, Englewood Cliffs, New Jersey, 1992.

Erdman, Arthur G., George N. Sandor, *Mechanism Design: Analyses and Syntheses,* Prentice-Hall International, Inc., Simon & Schuster/ A Viacom Company, Upper Saddle River, New Jersey, 1997.

Rampersad, Hubert K., *Integrated and Simultaneous Design for Robotic Assembly,* John Wiley & Sons, Chichester, New York, 1993.

Groover, Mikell P., *Fundamentals of Modern Manufacturing: Materials, Processes and Systems,* Prentice-Hall International, Inc., Simon & Schuster, Upper Saddle River, New Jersey, 1996.

Lindberg, Roy A., *Processes and Materials of Manufacture: Fourth Edition,* Allyn and Bacon, Boston, 1990.

Krar, S. F., J. W. Oswald, and J. E. St. Amand, *Technology of Machine Tools: Third Edition,* McGraw-Hill International Editions, New York, 1984.

Miu, Denny K., *Mechatronics: Electromechanics and Contromechanics,* Springer-Verlag, New York, Berlin, 1992.

Groover, Mikell P., *Automation, Production Systems, and Computer Integrated Manufacturing,* Prentice-Hall International, Inc., Simon & Schuster, Englewood Cliffs, New Jersey, 1987.

Brown, James, *Modern Manufacturing Processes,* Industrial Press Inc., New York, 1991.

Fawcett, J. N., J. S. Burdess, *Basic Mechanics with Engineering Applications,* Edward Arnold, A division of Hodder and Stoughton, London, New York, 1988.

Birmingham, R., G. Cleland, R. Driver, and D. Maffin, *Understanding Engineering Design: Context, Theory and Practice,* Prentice-Hall, London, New York, 1997.

Mabie, Hamilton H., Charles F. Reinhololtz, *Mechanisms and Dynamics of Machinery,* John Wiley & Sons, New York, 1987.

Meriam, J. L., and L. G. Kraige, *Engineering Mechanics: Dynamics, SI Version, vol. 2,* John Wiley & Sons, Inc., New York, 1993.

Askeland, Donald R., *The Science and Engineering of Materials: Third Edition,* PWS Publishing Company, Boston, 1994.

List of Main Symbols

Chapter 1

L	distance, length
P	productivity of a machine
T	pure processing time
t_1	duration of a movement
t_2	duration of a pause
V	speed of the product in the machine
τ	time losses

Chapter 2

Chapter 3

A, B, C, D	integration constants
a_1, a_2, b_1, b_2, q	constants
a	linear acceleration
c	stiffness of a spring or elastic link
F, P	force
f	dry friction coefficient
g	acceleration of gravitation
I	moment of inertia of a link

L	distance, length
M, m	mass
p	pressure in a hydraulic or pneumatic system
$r, r(t)$	radius or variable distance of a rotating mass
s	slip in a synchronous electromotor
s, x	linear displacement, deflection
T	torque
t	time
α	angular acceleration
α_1, α_2	constants
δ	clearance or gap
ϕ	inclination angle, angular displacement
ω	frequency of oscillations, angular speed

Chapter 4

a, D, e, h, l, r	geometrical dimensions
b, c	damping coefficients
c, k	stiffness of springs or elastic links
E	Young modulus
F	force
f	dry friction coefficient
I	moment of inertia of a cross-section of a link
J	moment of inertia of a massive body
m	mass
q	deflection of an elastic link
s, x, y	linear displacement
T	torque
t	time
V	speed
z_1, z_2, z_3, z_4	number of teeth
α, β, γ	geometrical angular dimensions
Π	symbol of the position function
ϕ, ψ	angular displacements
ω	angular speed, frequency of oscillations

Chapter 5

a, d, l	geometrical linear dimensions
c	capacitance
E	electromotive force
f	area of a pipe's cross section

H, h	height, pressure
L	inductance
R	electrical resistance
ΔR	increment of electrical resistance
s, x	linear displacements or distance
T	temperature
t	time
V	speed
v	voltage
Δv	increment of voltage
W	energy
w	number of winds
x_l	inductive resistance
z	complete electrical resistance
α	angular displacement
α_1, α_2	coefficient of flow rates in pipe sections 1 and 2, respectively
ε	dielectric permitivity
μ	magnetic permeability
Φ	magnetic flow
p	density of flowing liquid
ω	frequency of oscillations

Chapter 6

A	linear acceleration
g	gravitation acceleration
L	geometrical dimensions
m	mass
P	force
T	time, period
V	speed
x	displacement
z	number of slots or teeth
μ	friction coefficient
ϕ	angular displacement
Ω, ω	angular speed

Chapter 7

a	linear acceleration
a, b, c, d, D, h, l, L	geometrical dimensions

g	acceleration of gravity
F, N, P, Q	forces
m	mass
n	revolutions per minute
s, x, y	displacements
T	torque
t	time
Δt	value of a clearance
V	speed
W	energy
α, β, γ	angles of inclination
Δ	value of a geometrical gap
μ	friction coefficient
ρ	friction angle
ω	frequency of oscillations

Chapter 8

Chapter 9

a	linear acceleration
A, B, C	components of opposed forces, constants of integration
A_1, A_2, W	energy
a, b, c, h, l, L	geometrical dimensions of links
C	number of combinations
c	stiffness of elastic elements
F	force
f_A, f_B, f_C	friction forces corresponding to the opposing forces
J	moment of inertia
K	constant
M, m	moving mass
N	power
P	weight of an element
R	radius of an element
$r(t)$	variable distance of a rotating mass from the rotation axis
s	displacement of a gripper's jaw
T	torque
t, τ	time

u_1, u_2	control functions
V	velocity of a moving body
x	displacement of a body
x_1, x_2, x_3, x_4	auxiliary variables
X, Y, Z	components of forces
$z(t)$	variable heights
α	angular acceleration
α_ν, β_ν	binding angles of the links of the ν-th leg
$\phi(t)$	variable azimuth
$\Delta\phi, \Delta\psi$	increments of angles
$\psi(t)$	variable angle of a link
ω	frequency of oscillations

Index